The Record of the Past

The Record of the Past

An Introduction to Physical Anthropology and Archaeology

CHRISTOPHER R. DeCORSE

Syracuse University

PRENTICE HALL
Upper Saddle River, New Jersey 07458

Library of Congress Cataloging-in-Publication Data

DeCorse, Christopher R.
    The record of the past: an introduction to physical
anthropology and archaeology/Christopher R. DeCorse.
        p. cm.
    Includes bibliographical references and index.
    ISBN 0-13-490335-8 (alk. paper)
    1. Physical anthropology. 2. Archaeology. 3. Antiquities,
Prehistoric. I. Title.
    GN60 .D43 1999
    599.9—dc21

                                                    99-34374
                                                    CIP

Editorial Director: Charlyce Jones Owen
Editor-in-Chief: Nancy Roberts
Managing Editor: Sharon Chambliss
AVP, Director of Production and Manufacturing: Barbara Kittle
Production Editor: Barbara Reilly
Copy Editor: Mary Louise Byrd
Proofreader: Karen Bosch
Production Assistant: Kathleen Sleys
Manufacturing Manager: Nick Sklitsis
Prepress and Manufacturing Buyer: Ben Smith
Creative Design Director: Leslie Osher

Art Director: Nancy Wells
Cover Design: Ximena Tamvakopoulos
Cover Art: José de P. Machado/Photonica
Director, Image Resource Center: Melinda Lee Reo
Image Specialist: Beth Boyd
Manager, Rights and Permissions: Kay Dellosa
Photo Researcher: Tobi Zausner
Line Art Coordinator: Guy Ruggiero
Director of Marketing: Gina Sluss
Marketing Manager: Christopher DeJohn
Marketing Assistant: Judie Lamb

This book was set in 10/12 ITC Garamond Light
by RoRo Productions and Laurel Road Publishing
Services, and was printed and bound by Press of
Ohio. The cover was printed by Press of Ohio.

© 2000 by Prentice-Hall Inc.
Upper Saddle River, New Jersey 07458

Printed in the United States of America
10 9 8 7 6 5 4 3 2 1

ISBN 0-13-490335-8

Prentice-Hall International (UK) Limited, *London*
Prentice-Hall of Australia Pty. Limited, *Sydney*
Prentice-Hall Canada Inc., *Toronto*
Prentice-Hall Hispanoamericana, S.A., *Mexico*
Prentice-Hall of India Private Limited, *New Delhi*
Prentice-Hall of Japan, Inc., *Tokyo*
Pearson Education Asia Pte. Ltd., *Singapore*
Editora Prentice-Hall do Brasil, Ltda., *Rio de Janeiro*

Brief Contents

Contents

PART II
PHYSICAL ANTHROPOLOGY

CHAPTER 4
The Primates   65

CHAPTER 5
Hominid Evolution   92

Boxes

ANTHROPOLOGISTS AT WORK

CRITICAL PERSPECTIVES

Preface

The world is getting smaller. Global communications, trade among far-flung nations, geopolitical events affecting countries hemispheres apart, and the ease of international travel are bringing people and cultures into more intimate contact than ever before, forcing us to become ever more knowledgeable about societies other than our own. With that in mind, this book is grounded in the belief that an enhanced global awareness is essential for people preparing to take their place in the fast-paced, increasingly interconnected world of the twenty-first century. The anthropological perspective, which stresses critical thinking, the evaluation of competing hypotheses, and the skills to generalize from specific data and assumptions, contributes significantly to a well-rounded education. This text engages readers in the varied intellectual activities underlying an anthropological approach by delving into the major themes in physical anthropology and archaeological study.

A holistic interdisciplinary outlook resonates throughout this book. Contemporary anthropologists draw on the findings of biologists, paleontologists, geologists, economists, historians, psychologists, sociologists, political scientists, religious studies specialists, philosophers, and researchers in other fields whose work sheds light on anthropological inquiry. In considering different topics, this book often refers to research conducted in these other fields. In addition to enlarging the scope and reach of the text, an exploration of interactions between anthropology and other fields sparks the critical imagination that brings the learning process to life.

The comparative approach, another traditional cornerstone of the anthropological perspective, is spotlighted in this text as well. When anthropologists assess fossil evidence, artifacts, or cultural beliefs, they weigh comparative evidence while acknowledging the unique elements of each fossil find, archaeological site, society, or culture. The book casts an inquiring eye on materials from numerous geographic regions and historical eras. Such examination allows us to evaluate interpretations and findings, and also to perceive change through time.

ORGANIZATION OF THE BOOK

The arrangement and treatment of topics in this text are intended to introduce the reader to anthropological inquiry and to examine the major issues and themes in physical anthropological and archaeological investigation. The book has four overarching sections. Basic anthropological concepts are introduced in Part I. Chapter 1 introduces the field of anthropology and explains how it relates to the sciences and humanities. This lead-in chapter also delves into how anthropologists use the scientific method. Chapter 2 examines how paleoanthropologists and archaeologists locate and interpret fossils and the record of past human behavior. This chapter is intended to provide background information on topics such as dating techniques and excavation methods, which will be mentioned in later chapters. Chapter 3 presents basic evolutionary concepts, focusing on evolutionary processes and the origins of life on earth. Principles of heredity and molecular genetics are also briefly introduced.

Parts II and III underscore research topics in physical anthropology (Part II) and archaeology (Part III). Part II begins with Chapter 4 on the primates, discussing taxonomy and the fossil finds that allow researchers to trace primate ancestry. It also includes a discussion of living nonhuman primates and the primate features found in humans. This background in primate evolution provides an introduction to hominid evolution, the focus of Chapter 5. Trends in hominid evolution and some of the more important hominid fossil finds are examined. The chapter then discusses different interpretations of the evolution of the hominids and the origins of *Homo sapiens*. Part II concludes with the study of modern human variation in Chapter 6. This chapter explores the different sources of human variation—genetic, environmental, and cultural—and how physical anthropologists examine this variation.

Part III presents archaeological perspectives on human culture spanning the earliest tool traditions through the appearance of complex societies and the state. Part III opens with an expanded discussion of Paleolithic cultures in Chapter 7. This chapter presents the archaeological evidence for early hominid and human behavior, dealing with the stone tools and technological developments of the Lower, Middle, and Upper Paleolithic. Chapter 8 concentrates on the origins of domestication and settled life. It includes a discussion of how archaeologists study the origins of domestication, as well as developments in different world areas. Chapter 9 presents a discussion of the rise of the state and complex societies. As in the preceding chapters, this discussion includes a substantive presentation of developments in different world areas, as well as the archaeological evidence that archaeologists use to evaluate the growth of political and social complexity in ancient societies.

In Part IV, the concluding section, Chapter 10 underscores the relevance of the fields of physical anthropology and archaeology to issues and concerns in modern life. It includes a review of the applied aspects of archaeology and physical anthropology, as well as the issues involved in preserving and interpreting the anthropological data in a modern context.

Anyone with comments, suggestions, or recommendations regarding this text is welcome to send e-mail (Internet) messages to the following address: *crdecors@maxwell.syr.edu*

BOXES

Boxed features offer an in-depth look at topics discussed in the text.

In Critical Perspectives boxes, which are designed to stimulate independent reasoning and judgment, students take the role of anthropologist by engaging in active, critical analysis of specific problems and issues that arise in anthropological research. These boxes encourage students to use rigorous standards of evidence when evaluating assumptions and hypotheses regarding scientific and philosophical issues that have no easy answers.

Anthropologists at Work boxes, which profile prominent anthropologists, humanize many of the issues covered in the chapters. These boxes go behind the scenes to trace the personal and professional development of some of today's leading anthropologists.

SUPPLEMENTS

Instructor's Manual with Tests This carefully constructed instructor's guide contains chapter outlines, chapter objectives, lecture/discussion topics, and classroom activities, as well as multiple-choice, true/false, and essay questions keyed to the text.

Companion Website™ In tandem with the text, students can now take full advantage of the World Wide Web to enrich their study of physical anthropology and archaeology through the DeCorse Web site. This resource correlates the text with related material available on the Internet. Features of the Web site include chapter objectives and study questions, as well as links to interesting material and information from other sites on the Web that can reinforce and enhance the content of each chapter. Address: *http://www.prenhall.com/decorse*

Anthropology on the Internet 1999–2000 This guide introduces students to the origin and innovations behind the Internet and provides clear strategies for navigating the complexity of the Internet and World Wide Web. Exercises within and at the end of the chapters allow students to practice searching for the myriad resources available to the student of anthropology. This 96-page supplementary book is free to students when shrink-wrapped as a package with *The Record of the Past: An Introduction to Physical Anthropology and Archaeology.*

The New York Times **Supplement,** *Themes of the Times,* **for Anthropology** *The New York Times* and Prentice Hall are sponsoring *Themes of the Times,* a program designed to enhance student access to current information relevant to the classroom. Through this program, the core subject matter provided in this text is supplemented by a collection of timely articles from one of the world's most distinguished newspapers, *The New York Times.* These articles demonstrate the vital, ongoing connection between what is learned in the classroom and what is happening in the world around us.

To enjoy the wealth of information of *The New York Times* daily, a reduced subscription rate is available. For information, call toll-free: 1-800-631-1222.

Prentice Hall and *The New York Times* are proud to cosponsor *Themes of the Times.* We hope it will make the reading of both textbooks and newspapers a more dynamic, involving process.

ACKNOWLEDGMENTS

A textbook such as this requires the support of many people. I appreciate the comments and suggestions from the many reviewers, including William Wedenoja, Southwest Missouri State University; Pamela Willoughby, University of Alberta; D. Tab Rasmussen, Washington University, St. Louis. At Syracuse University, Douglas V. Armstrong and Mark Fleishman were always at hand with helpful suggestions. Ray Scupin, at Lindenwood College, was also invaluable in providing recommendations, ideas, and references.

I also extend thanks to all of my colleagues who sent illustrations and information for use in the boxes. I would like to offer special thanks to Douglas V. Armstrong, George Fletcher Bass, Elizabeth Brumfiel, Glenn C. Conroy, Larry Kruckman, Merrick Posnansky, D. Tab Rasmussen, Stuart Tyson Smith, and Pamela Willoughby for supplying photos for the text.

I am very grateful for the continued support given to this project by Prentice Hall. Without the encouragement of Nancy Roberts, I would have never been able to complete the project. My thanks also go to the production editor, Barbara Reilly, and the copy editor, Mary Louise Byrd, who helped transform the manuscript into an effective teaching instrument. My warmest appreciation goes to Cindy, my wife, and my children, Christina and Nicholas, whose emotional support and patience throughout the publication process truly made this book possible.

About the Author

Photo courtesy of Mark Turney Bären Photography, 1997.

Christopher R. DeCorse received his bachelor of arts and master's degrees in anthropology and archaeology, completing his doctorate in archaeology at the University of California, Los Angeles. His theoretical interests include the interpretation of ethnicity, culture change, and variability in the archaeological record. Dr. DeCorse has excavated a variety of prehistoric and historic period sites in the United States, the Caribbean, and Africa, but his primary area of research has been in the archaeology, ethnohistory, and ethnography of Sierra Leone and Ghana. His most recent research has focused on culture contact and change at the African settlement of Elmina, Ghana, the site of the first European tradepost in sub-Saharan Africa. He is currently collaborating on several projects that examine connections between Africa and the Americas.

Dr. DeCorse has taught archaeology and general anthropology in various undergraduate and graduate programs, including the University of Ghana, Indiana University of Pennsylvania, and Syracuse University, where he is currently an associate professor in the Department of Anthropology. Dr. DeCorse is particularly interested in the interpretation and presentation of anthropology for undergraduates and the general public. In addition to *The Record of the Past*, Dr. DeCorse has co-authored *Anthropology: A Global Perspective*, a four-field anthropology text, and *Worldviews in Human Expression*, an introduction to the humanities from an anthropological perspective. He also serves on the advisory or editorial boards of *Annual Editions* in physical anthropology and archaeology, *International Journal of Historical Archaeology*, and *Beads: Journal of the Society of Bead Researchers*. He has participated on a number of committees and panels, including work as a consultant on human evolution and agricultural origins for the National Center for History in the Schools.

Dr. DeCorse has received several academic honors and awards, including Fulbright and Smithsonian fellowships. He has published more than thirty articles, reviews, and research notes in a variety of publications, including *The African Archaeological Review, Historical New Hampshire, Historical Archaeology,* and *Slavery and Abolition*. A volume on his work at Elmina, *Under the Castle Cannon*, and an edited volume, *Historical Archaeology in West Africa*, are forthcoming.

The Record of the Past

Chapter 1

INTRODUCTION
TO ANTHROPOLOGY

CHAPTER OUTLINE

THE FIELD OF ANTHROPOLOGY SEEKS TO understand humanity in the broadest sense, ranging from its biological origins in the distant past to the diversity of modern human experience. Anthropology's two overarching concerns are to understand the uniqueness and diversity of human societies around the world and to discover the fundamental similarities that link human beings—both in the past and the present. To accomplish these goals, anthropologists undertake studies the world over to assess the similarities and differences among societies. With these goals as a springboard, anthropology has forged distinctive objectives and propelled research that has broadened our understanding of humanity from the beginnings of human societies to the present.

Paleoanthropology and **archaeology,** the focus of this book, are the branches of anthropology that share a concern with the human past. They are also the fields of anthropology that include the study of tangible, physical remains that must be located and removed from the ground. The difference between the two subdisciplines lies with the questions each poses. *Physical anthropologists* are concerned with the biological aspects of humanity—the fossil remains of humans and human ancestors, the evolution of the human species, the biological diversity of modern humans, and the reasons for this diversity. In contrast, the *archaeologist* concentrates on the interpretation of past human cultures—their lifestyles, technology, and social systems—through the material remains they left behind. These different areas of concentration are seen in the division of the following chapters: Chapter 2 provides a background to paleoanthropological and archaeological investigation;

Chapters 3, 4, 5, and 6 concentrate on physical anthropology; and Chapters 7, 8, and 9 focus on archaeological research. Chapter 10 considers these subdisciplines in terms of their relevance to modern problems and concerns.

The subdisciplines of anthropology are connected and frequently overlap; indeed, many skills may be found in the same specialist. Paleoanthropologists may work with archaeologists to locate and excavate fossil sites. Archaeologists depend on physical anthropologists to interpret human skeletal remains uncovered in archaeological excavations. Researchers in both fields draw on scholars from other disciplines to help interpret their finds. For example, one project may include someone who specializes in geology and the identification of soils, whereas others may bring to the field expertise on the identification of particular fossil animals or plants. In studies of the more recent past, archaeologists may work closely with historians and cultural anthropologists. This interdisciplinary research allows for a more holistic, well-rounded interpretation of the past. This chapter introduces the distinctive approaches used by anthropologists to achieve these goals.

ANTHROPOLOGY: THE SUBFIELDS

The word *anthropology* stems from the Greek words *anthropo,* meaning "human beings" or "humankind," and *logia,* translated as "knowledge of" or "the study of." Thus, we can define **anthropology** as *the study of humankind.* This definition in itself, however, does

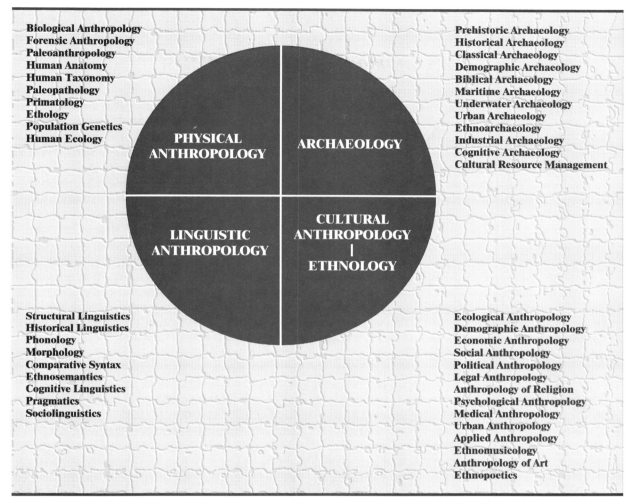

Biological Anthropology
Forensic Anthropology
Paleoanthropology
Human Anatomy
Human Taxonomy
Paleopathology
Primatology
Ethology
Population Genetics
Human Ecology

Prehistoric Archaeology
Historical Archaeology
Classical Archaeology
Demographic Archaeology
Biblical Archaeology
Maritime Archaeology
Underwater Archaeology
Urban Archaeology
Ethnoarchaeology
Industrial Archaeology
Cognitive Archaeology
Cultural Resource Management

PHYSICAL ANTHROPOLOGY

ARCHAEOLOGY

LINGUISTIC ANTHROPOLOGY

CULTURAL ANTHROPOLOGY | ETHNOLOGY

Structural Linguistics
Historical Linguistics
Phonology
Morphology
Comparative Syntax
Ethnosemantics
Cognitive Linguistics
Pragmatics
Sociolinguistics

Ecological Anthropology
Demographic Anthropology
Economic Anthropology
Social Anthropology
Political Anthropology
Legal Anthropology
Anthropology of Religion
Psychological Anthropology
Medical Anthropology
Urban Anthropology
Applied Anthropology
Ethnomusicology
Anthropology of Art
Ethnopoetics

FIGURE 1.1 This figure shows the four core subfields of anthropology. It also includes some of the various specializations that have developed within each of the subfields, which you will be introduced to in this text. Many of these specializations overlap one another in the actual studies carried out by anthropologists.

not really distinguish anthropology from other disciplines. After all, historians, psychologists, economists, geneticists, and sociologists, as well as scholars in many other fields, systematically study humankind in one way or another. Anthropology stands apart because of its holistic and comparative perspective that combines subdisciplines, or subfields, that bridge the natural sciences, the social sciences, and the humanities. The four traditional subdisciplines of anthropology—physical anthropology, archaeology, linguistic anthropology, and ethnology—give anthropologists a broad approach to the study of humanity the world over in both the past and present. Figure 1.1 shows

these subfields and the various specializations that make up each one. These traditional subdisciplines of anthropology are complemented by a fifth subfield—applied anthropology—that focuses on the relevance, the application of anthropological knowledge from the four other subfields to solve problems faced by modern society. A discussion of these subdisciplines and some of their principal specializations follows.

PHYSICAL ANTHROPOLOGY

Physical anthropology is the branch of anthropology concerned with humans as a biological species. As

ANTHROPOLOGISTS AT WORK
CLYDE SNOW: FORENSIC ANTHROPOLOGIST

Clyde Collins Snow obtained a master's degree in zoology from Texas Tech University and planned to pursue a Ph.D. in physiology, but his career plans were interrupted by military service. While stationed at Lackland Air Force Base near San Antonio, he was introduced to the field of archaeology and became fascinated with the ancient artifacts discovered in the surrounding area.

After leaving the military, Snow attended the University of Arizona, where his zoological training and archaeology interests led him to a Ph.D. in physical anthropology. He became skilled at identifying old bones and artifacts. With his doctoral degree completed, he joined the Federal Aviation Administration as a consulting forensic anthropologist, providing technical assistance in the identification of victims of aircraft accidents. Snow also lent his expertise to the design of safety equipment to prevent injuries in aircraft accidents.

As word of Snow's extraordinary skill in forensic anthropology spread, he was called to consult on and provide expert testimony in many criminal cases. His testimony was crucial at the sensational murder trial of John Wayne Gacy, accused of murdering more than thirty teenagers in the Chicago area. Snow also collaborated with experts in the reinvestigation of President John F. Kennedy's assassination. These experts built a full-scale model of Kennedy's head to determine whether Lee Harvey Oswald could have inflicted all of Kennedy's wounds. They did not uncover any scientific evidence to contradict the Warren Commission's conclusion that Oswald was the sole assassin.

such, it is the subdiscipline most closely related to the natural sciences. Physical anthropologists conduct research in two major areas: human evolution and modern human variation. The investigation of human evolution presents one of the most tantalizing areas of anthropological study. Research has now traced the African origins of humanity back over four million years. Fieldwork in other world areas has traced the expansion of early human ancestors throughout the world. Much of the evidence for human origins consists of **fossils,** the fragmentary remains of bones and living materials preserved from earlier periods. The study of human evolution through analysis of fossils is called **paleoanthropology** (the prefix *paleo* means "old" or "prehistoric"). Paleoanthropologists use a variety of sophisticated scientific techniques to date, classify, and compare fossil bones to determine the links between modern humans and their biological ancestors. They may work closely with archaeologists when studying ancient tools and activity areas to learn about the behavior of early human ancestors.

Other physical anthropologists focus their research on the range of physical variation within and among different human populations. These anthropologists study human variation by measuring physical characteristics, such as body size, variation in blood types, differences in skin color, or various genetic traits. Human osteology is the particular area of specialization within physical anthropology dealing with the study of the human skeleton. Such studies have wide-ranging applications, including the identification of murder victims from fragmentary skeletal remains to the design of ergonomic airplane cockpits (see the box "Clyde Snow: Forensic Anthropologist"). Physical anthropologists are also interested in evaluating how disparate physical characteristics reflect evolutionary adaptations to different environmental conditions, thus shedding light on why human populations vary.

An increasingly important area of research for some physical anthropologists is **genetics,** the study of the biological "blueprints" that dictate the inheritance of physical characteristics. Research on genetics examines a wide variety of questions. It has, for example, been important in identifying the genetic sources of some diseases such as sickle cell anemia, cystic fibrosis, and Tay-Sachs disease. Genetics has also become an increasingly important complement to paleoanthropological research. Through the study of the genetic makeup of modern humans, geneti-

More recently, Snow and his team have been recognized for their contributions to human rights issues. Snow served as a consultant to the Argentine government's National Commission on Disappeared Persons in its efforts to determine the fate of thousands of Argentineans who were abducted and murdered by military death squads between 1976 and 1983, when the country was under the rule of a military dictatorship. As a result of his investigations, Snow was asked to testify as an expert witness in the trial of the nine junta members who ruled Argentina during the period of military repression. He

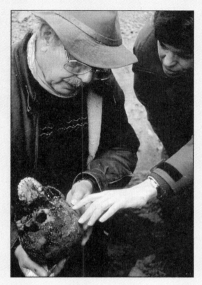

Dr. Clyde Snow.

also assisted people in locating their dead relatives.

Snow stresses that in his human rights investigative work he is functioning as an expert, not necessarily as an advocate. He must maintain an objective standpoint in interpreting his findings. The evidence he finds may then be presented by lawyers (as advocates) in the interests of justice. Snow's human rights work is supported by various agencies, such as the American Association of Advanced Sciences, the Ford Foundation, the J. Roderick MacArthur Foundation, Amnesty International, Physicians for Human Rights, and Human Rights Watch.

cists have been able to calculate the genetic distance between modern humans, thus providing a means of inferring the evolutionary relationships within the species. These data provide independent evidence for the African origins of the modern human species and human ancestors.

ARCHAEOLOGY

Through **archaeology,** the branch of anthropology that seeks out and examines the artifacts of past societies, we learn about the culture of those societies, the shared way of life of a group of people that includes their values, beliefs, and norms. **Artifacts,** the material products of former societies, provide clues to the past. Some archaeological sites reveal spectacular jewelry like that found by the movie character Indiana Jones, or the treasures of a Pharaoh's tomb. Most artifacts are not so glamorous, however. Despite the popular image of archaeology as an adventurous, even romantic pursuit, it usually consists of methodical, rigorous scientific research (see the box "Patty Jo Watson: Archaeologist"). For example, archaeologists often spend hours sorting through ancient garbage, or **middens,** to discover

how members of an early society ate their meals, what tools they used in their household and in their work, and what beliefs gave meaning to their lives. The broken fragments of pottery, stone, glass, and other materials are collected and carefully analyzed in a laboratory setting. It may take months or even years to fully complete the study of an excavation. Unlike the glorified adventures of fictional archaeologists, the real-world field of archaeology thrives on the intellectually challenging adventure of systematic, scientific research that enlarges our understanding of past societies.

Archaeologists have examined sites the world over, from campsites of the earliest humans to modern landfills. Some archaeologists investigate past societies whose history is not recorded in documentary sources or oral traditions. Known as *prehistoric archaeologists,* they study the artifacts of groups such as ancient inhabitants of Europe or the first humans to arrive in the Americas. These researchers have no written documents or oral traditions to help interpret the sites and artifacts that they recover—the archaeological record provides the primary source of information for their interpretations about the past. *Historical archaeologists,*

ANTHROPOLOGISTS AT WORK

PATTY JO WATSON: ARCHAEOLOGIST

After two years of premed studies at Iowa State University, Patty Jo Watson changed course and decided to pursue graduate study in archaeology at the University of Chicago. Working with Robert Braidwood, Watson became adept at using precise scientific methods to assess artifacts left behind by past societies. Early in her career, Watson participated in the excavation of prehistoric societies in the Middle East. Later, after marrying philosopher and geologist Richard (Red) Watson, she focused her energies on exploring the prehistory of caves.

Watson took a keen interest in Native American cave explorers in Kentucky, especially in the Mammoth Cave system, the longest network of caves in the world, with approximately three hundred miles of interconnected passageways. The system includes Salts Cave, Colossal Cave, Unknown

Dr. Patty Jo Watson.

Cave, Mammoth Cave, and several others. For more than thirty years, Watson and her colleagues have surveyed, mapped, and interpreted the prehistoric artifacts in the Mammoth Cave system. They uncovered evidence that Native American cave dwellers,

thought to be purely hunter-gatherers, actually cultivated plants in addition to eating game animals and wild plants. This discovery has yielded major insights concerning the development of agriculture in native North American societies.

Watson and two of her colleagues have written two major books on the philosophy of archaeology: *Explanation in Archaeology: An Explicitly Scientific Approach* (1971) and *Archaeological Explanation: The Scientific Method in Archaeology* (1984). As a result of her contributions to cave archaeology and the prehistory of Native American culture, made over a thirty-year career, Watson was inducted into the National Academy of Sciences, an honor the *New York Times* described as "second only to the Nobel Prize." She is one of a small number of American women selected for membership in this prestigious organization.

on the other hand, work with historians in investigating the artifacts of societies of the more recent past. For example, some historical archaeologists are probing the remains of plantations in the southern United States to gain an understanding of the lifestyles of enslaved Africans and slave owners during the nineteenth century. Other archaeologists, called *classical archaeologists,* conduct research on ancient civilizations such as in Egypt, Greece, and Rome. There are many more areas of specialization within archaeology that reflect area or topical specializations, or the time period on which the archaeologist works (see Figure 1.1). There are, for example, industrial archaeologists, Biblical archae-

ologists, Medieval and post-medieval archaeologists, and Islamic archaeologists. Underwater archaeologists work on a variety of places and time periods the world over; they are distinguished from other archaeologists by the distinctive equipment, methods and procedures needed to excavate under the water.

In a novel approach, still other archaeologists have turned their attention to the very recent past. For example, in 1972, William L. Rathje began a study of modern garbage as an assignment for the students in his introductory anthropology class. Even he was surprised at the number of people who took an interest in the findings. A careful study of garbage reveals insights about modern society that cannot be ferreted out in any other way. Whereas questionnaires and interviews depend on the cooperation and interpretation of respondents, garbage provides an unbiased physical record of human activity. Rathje's "garbology project" is still in progress and, combined with information from respondents, offers a unique look at patterns of waste management, consumption, and alcohol use in contemporary U.S. society (Rathje & Ritenbaugh, 1984, also see discussion in Chapter 10).

LINGUISTIC ANTHROPOLOGY

Linguistics, the study of language, has a long history that dovetails with the discipline of philosophy but is also one of the integral subfields of anthropology (Hickerson, 1980). Linguistic anthropology focuses on the relationship between language and culture. One of the wide-ranging areas of research in linguistic anthropology is **historical linguistics,** which concentrates on the comparison and classification of different languages to discern the historical links among languages. By examining and analyzing grammatical structures and sounds of languages, researchers discover connections among these grammatical and aural patterns and seek out sources of change among different language families and the relationship between different languages. This type of historical linguistic research is particularly useful in tracing the migration routes of societies through time. It may help confirm archaeological and paleoanthropological data gathered independently. For example, some researchers have integrated historical linguistic studies with archaeological and physical anthropological data to interpret the origins of Native American populations.

Structural linguistics, another area of specialization, deals with the structure of grammatical patterns and other linguistic elements to learn how contemporary languages mirror and differ from one another. Structural linguistics has also uncovered some intriguing relationships between language and thought patterns among different groups of people. Do people who speak different languages with different grammatical structures think and perceive the world differently from one another? For example, do native Chinese speakers think or view the world and their life experiences differently from native English speakers? These are some of the questions that structural linguists attempt to answer.

Linguistic anthropologists also examine the social usages of language in different cultures (see the box "Bambi B. Schieffelin: Linguistic Anthropologist"). This specialty, called **sociolinguistics,** is the systematic study of language use in various social settings to investigate the connections between language and social behavior. In most societies, people address others differently depending on social status. In Thailand, for example, there are thirteen forms of the pronoun *I*. One form is used with equals, other forms come into play with people of higher status, and some forms are used when males address females (Scupin, 1989).

ETHNOLOGY

Ethnology, more popularly known as **cultural anthropology,** is the subfield of anthropology that examines various contemporary societies throughout the world, from the tropical rainforests of Zaire and Brazil to the urban areas of the United States. Traditionally, most ethnologists conducted research on the non-Western, traditional cultures of Africa, Asia, the Middle East, Latin America, and the Pacific Islands, as well as on the Native American populations in the United States. Today, however, many anthropologists have turned to research on their own cultures to gain a better understanding of its institutions and cultural values.

Cultural anthropologists or ethnologists use a unique research strategy in conducting their fieldwork (Agar, 1980). This research strategy, referred to as **participant observation,** involves learning the language and culture of the group being studied by participating in the group's daily activities. Through participation, the ethnologist becomes deeply familiar with the group and can understand and explain the society and culture of the group as an insider (see the box "Napoleon Chagnon: Ethnologist").

An ethnologist's description of a particular society is based on what anthropologists call *ethnographic data,* and it is written up as an **ethnography.** A typical ethnography reports on the environmental setting, economic patterns, social organization, political system, and religious rituals and beliefs of the society under study. This information is then synthesized

ANTHROPOLOGISTS AT WORK
BAMBI B. SCHIEFFELIN: LINGUISTIC ANTHROPOLOGIST

As an undergraduate at Columbia University, Bambi B. Schieffelin was drawn to two fields: anthropology and comparative literature. After spending a summer on a field trip to rural Bolivia and a year in the southern highlands of Papua New Guinea, Schieffelin decided to pursue a doctorate in anthropology, with a specialty in linguistic anthropology. She combined the fields of developmental psychology, linguistics, and anthropology, which prepared her for fieldwork among the Kaluli people in Papua New Guinea. After completing her Ph.D., Schieffelin spent a year

teaching at the University of California at Berkeley and also teaching linguistics at Stanford University. Since 1986 she has been teaching at New York University in the department of anthropology.

Schieffelin's work focuses on language use and socialization. She studies how language is learned and acquired by children, and how it is used in various social contexts. She has collaborated with Elinor Ochs to develop innovative approaches to understanding how language use is influenced by socialization. Together they have edited several volumes, including

Language Socialization across Cultures (1987). In addition, Schieffelin has developed these topics in her own book, *The Give and Take of Everyday Life: Language Socialization of Kaluli Children* (1990).

Through their research on children's language socialization, Schieffelin and Ochs contributed to a cross-cultural understanding of this process. Until the early 1970s, most of the theories on language and socialization had been drawn from psychological research on middle-class Americans. Schieffelin and Ochs focused instead on language and socialization among

---

and evaluated in light of information and conclusions drawn on data from other area, the objective being to provide explanations of human society and behavior.

APPLIED ANTHROPOLOGY

The four preceding areas of study—physical anthropology, archaeology, linguistic anthropology, and ethnology—are well established and have long been regarded as the four primary subdisciplines of American anthropology. In recent years, however, a fifth major subfield has been recognized: **Applied anthropology** is the use of anthropological data to offer practical solutions to problems faced in modern society. While this subdiscipline draws on information gathered from the other subfields of anthropology, it is distinct in its applied, problem-solving orientation. These problems may be environmental, technological, economic, social, political, or cultural. Applied anthropologists are often referred to as *practicing*

anthropologists and are involved at various levels in the planning process for implementing change, planning development projects and addressing policy concerns. In this context, they frequently work as consultants to various government and private agencies.

Each of the other subdisciplines is represented in applied anthropology, and the topics of study are as far-ranging as the issues facing modern society. For example, an applied area of physical anthropology is a specialty known as **forensic anthropology,** which concentrates on the identification of human skeletal remains for legal purposes. Forensic anthropologists, trained to recognize subtle anatomical differences in human skeletal fragments, often assist police in solving homicides and missing persons cases. Archaeologists, on the other hand, grapple with the problems of balancing development and urban renewal efforts with the desire to preserve and study historic buildings and archaeological sites that are threatened with destruction. Archaeologists may be called to excavate—or

many different societies. They emphasized the importance of cultural practices in shaping verbal activities. For example, prior to their research, it was assumed that "baby talk" was the same all over the world. They found, however, that "baby talk" is not universal and is linked to ideas that people have about children.

In her ethnographic research, Schieffelin tape-records and transcribes everyday social interactions in different speech communities. She has carried out research in Papua New Guinea since 1967, focusing not only on language socialization but also on language change and the introduction of literacy into a nonliterate society. In addition to this work in a relatively traditional society, Schieffelin has worked in a number of urban speech communities in the United States, where linguistic diversity is apparent on every street corner. Her research in Philadelphia among Sino-Vietnamese focused on language socialization and literacy, and her studies of Haitians in New York analyzed language socialization and code-switching practices.

As a linguistic anthropologist, Schieffelin tries to integrate two perspectives. First, she focuses on how the study of language use can lead to insights into how culture is transmitted from generation to generation in everyday social interactions. Second, she analyzes the ways in which language expresses social relationships and cultural meanings across different social and political contexts. Her work represents the most current developments in linguistic anthropology today.

Linguistic anthropologists have broadened their vision of places in which to investigate language use—including legal, medical, scientific, educational, and political arenas—as these contexts are critical to an understanding of how power is produced and distributed. They are also studying all varieties of literacy, including television and radio, providing perspectives on these new forms of global communication.

"salvage"—an archaeological site to provide some record of it before it is destroyed. This applied type of archaeological research is referred to as *contract archaeology* or *cultural resource management.* Chapter 10 provides an in-depth look at applied anthropology with regard to the fields of physical anthropology and archaeology.

Cultural anthropology and linguistics also have applied components. They may address such varied concerns as cultural heritage preservation, helping native peoples conserve their indigenous languages, assist corporations and government agencies to understand better the impact of programs or development projects, or help resolve the difficulties faced in the resettlement of refugees. Applied anthropologists may serve as analysts, mediators, or facilitators among different private developers, government officials, and interest groups who are participating in a development project. The anthropologists try to reconcile differences among these groups, facilitating compromises that ideally will benefit all of the parties involved.

HOLISTIC ANTHROPOLOGY, INTERDISCIPLINARY RESEARCH, AND A COMPARATIVE PERSPECTIVE

Anthropological study is inherently interdisciplinary. Most anthropologists are trained in the four primary subfields of anthropology. Because of all the research being done in these different fields, however—more than three hundred anthropology journals and hundreds of books are published every year—no one individual can keep abreast of all the developments across the entire discipline, and anthropologists usually specialize in one of the subfields and focus their research on a particular area or topical questions. Nevertheless, most anthropologists are committed to a **holistic** approach to understanding humankind—a broad, comprehensive account that draws on all of the subfields. This holistic approach considers both the present conditions of a society and the past;

ANTHROPOLOGISTS AT WORK

NAPOLEON CHAGNON: ETHNOLOGIST

Napoleon Chagnon was born in Port Austin, Michigan, in 1938. As a physics major at the University of Michigan, he chose to take a few anthropology courses to fulfill some of his liberal arts requirements. These courses so captivated him that he changed his major and ultimately earned B.A., M.A., and Ph.D. degrees in anthropology at Michigan. After completing his Ph.D. in 1966, he joined the Department of Human Genetics at the University of Michigan Medical School and participated in a multidisciplinary study of the Yanomamö Indians of Venezuela and Brazil.

Chagnon has returned to do fieldwork among the Yanomamö almost every year since 1966, enabling him to conduct a long-term, systematic study of change within this population. His ethnographic studies, and the many excellent ed-

ucational films that he and colleague Timothy Asch have produced, have made the Yanomamö well known around the world. Chagnon's description of his first fieldwork experience with the Yanomamö demonstrates how "culture shock" can affect ethnologists who encounter a society radically different from their own. He describes his initial experience with the Yanomamö after a long voyage up the Orinoco River, arriving at a riverbank near the village:

I looked up and gasped when I saw a dozen burly, naked, sweaty, hideous men staring at us down the shafts of their arrows! Immense wads of green tobacco were stuck between their lower teeth and lips, making them look even more hideous, and strands of dark green slime dripped or hung from their nostrils—so long that

they clung to their [chests] or drizzled down their chins. (1997:11) [The dark green slime was the remains of a hallucinogenic drug that the Yanomamö had taken.]

Eventually Chagnon adjusted to his new environment and conducted a thorough, systematic ethnographic study. In a book about his fieldwork, *Studying the Yanomamö* (1974), he describes both his analytical techniques and his immersion into Yanomamö society. Over time, Chagnon became personally involved in the lives of the Yanomamö. His deep personal ties with the Yanomamö prompted him to help them grapple with problems stemming from their confrontation with new, external pressures. Chagnon has spearheaded an international effort to aid the Yanomamö in their adjustment to the modern world.

biological, as well as cultural phenomena. Through collaborative studies among the various specialists in the subfields, anthropologists can ask more broadly framed questions.

Nor is anthropology constrained by its own subfields to realize its research agenda. Although it stands as a distinct discipline, anthropology has strong links to other social and physical sciences. In physical anthropology, for example, there is substantial overlap with certain aspects of biology, zoology, and genetics. Study of fossil locations and archaeological sites may also draw heavily from geology and paleontology to interpret their formation, age, and content.

Anthropology also dovetails considerably with the field of history, which, like anthropology, encom-

passes study of a broad range of social phenomena. Every human event that has ever taken place in the world is a potential topic for both historians and anthropologists. Historians describe and explain human events that have occurred throughout the world; anthropologists place their biological, archaeological, linguistic, and ethnographic data in the context of these historical developments. In some instances these interests intersect in the same research topic, as, for example, in the field of historical archaeology, which deals with the study of archaeological sites that are known through documentary records and oral traditions.

Cultural anthropology, on the other hand, is closely related to sociology. In the past, cultural anthropologists examined the traditional societies of the world, whereas sociologists focused on industrial societies. Today cultural anthropologists and sociologists explore many of the same societies using similar research approaches. For example, both may rely on statistical and nonstatistical data whenever appropriate in their studies. Ethnology also overlaps with the fields of psychology, economics, and political science. Ethnologists draw on psychology when they assess the behavior of people in other societies. Psychological questions bearing on perception, learning, and motivation all figure in ethnographic fieldwork. Additionally, ethnologists probe the economic and political behavior and thought of people in various societies, using these data for comparative purposes.

In addition to its interconnections with the natural and social sciences, the discipline of anthropology is aligned with the humanities. Many anthropologists explore the creative dimensions of humanity, such as poetry, art, music, and mythology. *Ethnopoetics,* for example, is the study of poetry and how it relates to the experiences of people in different societies. For example, a provocative study of the poetry of a nomadic tribe of Bedouins in the Middle East has yielded new insights into the concepts of honor and shame in these societies (Abu-Lughod, 1987). Another related field, *ethnomusicology,* is devoted to the study of musical traditions in various societies throughout the world. Ethnomusicologists record and analyze music and the traditions that give rise to these musical expressions, exploring similarities and differences in musical performance and composition. Other anthropologists study the art of particular societies, such as pottery styles among Native American groups. As a result of these studies, we have a keener appreciation of the diverse creative abilities exhibited by humans throughout the world.

Another integral part of anthropology is its comparative perspective. Anthropologists working in all of the subfields emphasize a perspective that draws on examples of phenomena from throughout the world. By going beyond the specific or local conditions, anthropologists are able to evaluate better both the similar and distinct features present. Indeed, the objective of anthropology is not to simply describe a particular fossil find, archaeological site, or living society but rather to explain these phenomena. This perspective is used throughout this text to show how anthropologists place their findings in the interconnecting worldwide community of humanity.

ANTHROPOLOGICAL EXPLANATIONS

A fundamental question faced by all anthropologists is how to evaluate the interpretations they reach about the particular social, cultural, or biological phenomena they study. Human knowledge is rooted in personal experience, as well as in the beliefs, traditions, and norms maintained by the society they live in. This includes such knowledge as assumptions about putting on warm clothing in cold weather and bringing an umbrella if it is going to rain, for example. Yet it also includes notions about how food should be prepared, what constitutes "appropriate" behavior, and perceptions about the social roles of men and women. Religion constitutes another source of human knowledge. Religious beliefs and faith are most often derived from sacred texts, such as the Bible, Koran, and Talmud, but they are also based on intuitions, dreams, visions, and extrasensory perceptions. Most religious beliefs are cast in highly personal terms and, like personal knowledge, span a wide and diverse range. People who do not accept these culturally coded assumptions may be perceived as different, abnormal, or nonconformists by other members of their society. Yet ethnographic and cross-cultural research in anthropology has demonstrated that such culturally constituted knowledge is not as general as we might think. This research indicates that as humans, we are not born with this knowledge. Knowledge tends to vary both among different societies and among different groups within the same society.

Popular perceptions about other cultures have often been based on ethnocentric attitudes. *Ethnocentrism* is the practice of judging another society by the values and standards of one's own society. To some degree, ethnocentrism is a universal phenomenon. As humans learn the basic values, beliefs, and norms of their society, they tend to think of their own culture as preferable, ranking other cultures as less desirable. Members of a society may become so committed to particular cultural traditions that they cannot conceive of any other way of life. They often view other cultural traditions as strange or alien, perhaps even inferior, crazy, or immoral.

Such deeply ingrained perceptions are difficult to escape, even for anthropologists. Nineteenth-century anthropologists, for example, often reinforced ethnocentric beliefs about other societies. The twentieth century has also seen the co-opting of anthropological data to serve specific political and social ends (see discussions in Chapter 2 on archaeological interpretation). In the twentieth century, however, anthropologists increasingly began to recognize the influences that prevent the interpretation of other cultures from a more objective, systematic manner.

THE SCIENTIFIC METHOD

Given the preceding concerns, it is critical to understand how anthropological interpretations are evaluated. In contrast to personal knowledge and religious faith, anthropological knowledge is not based on traditional wisdom or revelations. Rather, anthropologists employ the scientific method, a logical system used to evaluate data derived from systematic observation. Researchers rely on the **scientific method** to investigate both the natural and the social worlds because the approach allows them to make claims about knowledge and to verify those claims with systematic, logical reasoning. Through critical thinking and skeptical thought, scientists strive to suspend their judgment about any claim for knowledge until it has been verified.

Testability and verifiability lie at the core of the scientific method. There are two ways of developing

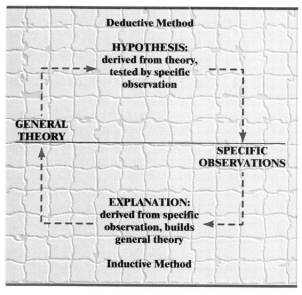

FIGURE 1.2 Deductive and inductive research methods.

testable propositions: the inductive method and the deductive method. In the **inductive method**, the scientist first makes observations and collects data (Figure 1.2). Much of the data collected are referred to as variables. A **variable** is any piece of data that changes from case to case. For example, a person's height, weight, age, and sex all constitute variables. Observations about different variables are then used by the researcher to develop hypotheses about the data. A **hypothesis** is a testable proposition concerning the relationship between particular sets of variables in the collected data. The practice of testing hypotheses is the major focus of the scientific method, as scientists test one another's hypotheses to confirm or refute them. If a hypothesis is found to be valid, it may be woven together with other hypotheses into a general theory. **Theories** are statements that explain hypotheses and observations about natural or social phenomena. Because of their explanatory nature, theories often encompass a variety of hypotheses and observations. One of the most comprehensive theories in anthropology is the theory of evolution (see Chapter 3). This theory helps explain a diversity of hypotheses about biological and natural phenomena, as well as discoveries by paleoanthropologists and geneticists.

In contrast to the inductive method, the **deductive method** of scientific research begins with a general theory from which scientists develop testable hypotheses. Data are then collected to evaluate these hypotheses. Initial hypotheses are sometimes referred to as "guesstimates" because they may be based on guesswork by the scientist. These hypotheses are tested through experimentation and replication. As with the inductive method, the testing and retesting of hypotheses and theories is used to ensure the reliability of observations made.

Through these methods researchers do not arrive at absolute truths. Theories may be invalidated or falsified by contradictory observations. Yet even if numerous observations and hypotheses suggest a particular theory is true, the theory always remains open to further testing and evaluation. The systematic evaluation of hypotheses and theories enables scientists to state their conclusions with a certainty that cannot be applied to personal and culturally construed knowledge.

Despite the thoroughness and verification that characterize the research, anthropological explanations have limitations. Unlike conclusions drawn in the nat-

Calvin and Hobbes   by Bill Watterson

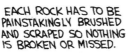

Despite the adventurous, exciting portrayal of archaeologists in movies, in real life archaeological research is meticulous and time consuming. Once dug up, an archaeological site is gone forever; it can be "reassembled" only through the notes kept by the archaeologist.

Source: "Calvin and Hobbes ©1988 Watterson. Reprinted with permission of Universal Press Syndicate. All rights reserved.

ural sciences, anthropological knowledge is frequently presented as tentative and hypothetical because the conclusions are based on assumptions and presuppositions about the myriad variables that affect human thought and behavior. The complexities of the phenomena being studied make it difficult to assess all of the potential variables, and disagreements about interpretations are common. The point here, however, is not that progress is impossible. Anthropological evidence can be verified or discarded by making assumptions explicit and weeding out contradictory, subjective knowledge. Poor hypotheses are rejected and replaced by better explanations. Interpretations can be made stronger by drawing on independent lines of evidence to support and evaluate theories. This process makes the scientific method much more effective than other means of acquiring knowledge.

WHY STUDY ANTHROPOLOGY?

Why study anthropology? Many students would find the answer to this question straightforward if not particularly thought-provoking. Anthropology courses are frequently found in the core requirements for a general, liberal arts education. The rational for such inclusion is well thought out. Studies indicate that a well-rounded education contributes to a person's success in any chosen career, regardless of his or her

major. Because of its broad interdisciplinary nature, anthropology is especially well suited to this purpose. By being exposed to the cultures and lifestyles of unfamiliar societies, students develop a more critical and analytical stance toward conditions in their own society, as well as a more open-minded perspective toward others. Anthropology promotes a cross-cultural perspective that allows us to see ourselves as part of one human family in the midst of tremendous diversity.

While these general goals are laudable, the study of anthropology offers more pragmatic applications (Omohundro, 1998). As seen in the discussion of applied anthropology, all of the traditional subfields of anthropology have areas of study with direct relevance to modern life. Many students have found it useful to combine an anthropology minor or major with another degree. For example, given the increasingly multicultural and international focus of today's world, students who go on to careers in business, management, marketing, or public service may find it advantageous to have some anthropology courses on their résumés. The concepts and knowledge gleaned from anthropology may find practical application in dealing with issues of cultural and ethnic diversity, multiculturalism, on a daily basis. Similarly, policy makers in federal, state, and local governments may find it useful to have an understanding of historic preservation issues and cultural resource management concerns. In education, various aspects of anthropology, including evolution, study of the human past, non-European cultures, and the interpretation of cultural and social phenomena, are increasingly being integrated into elementary and secondary school curricula. Education majors preparing for the classroom can draw on their background in anthropology to provide a more insightful context for some of these issues.

 ## SUMMARY

Anthropology consists of four traditional subfields: physical anthropology, archaeology, linguistic anthropology, and ethnology. Each of these subfields uses distinctive methods for examining humanity the world over in both the past and the present. Physical anthropologists investigate human evolution and the physical variation of human populations across many geographic regions. Archaeologists study the past by analyzing artifacts, the material remains of past societies. Linguistic anthropologists focus their studies on languages, seeking out historical relationships among languages, pursuing clues to the evolution of particular languages, and comparing one language with another to determine differences and similarities. Ethnologists conduct fieldwork in various societies to examine people's lifestyles. They describe these societies in written studies, called ethnographies, which highlight behavior and thought patterns characteristic of the people studied. In examining societies, ethnologists use systematic research methods and strategies, primarily participant observation, which involves participating in the daily activities of the people they are studying. These four fields are complemented by a fifth subfield—applied anthropology—which uses the research methods and data from the other subdisciplines to address problems and concerns facing the modern world.

Anthropological research is inherently interdisciplinary and it frequently intersects with other disciplines. In the natural sciences, certain aspects of anthropology intersect with biology, geology, paleontology, and genetics. On the other hand, in social sciences, the fields of history, sociology, and psychology overlap with certain aspects of anthropological inquiry. Yet anthropology is also a humanistic discipline that focuses on such cultural elements as art, music, and religion. By bridging these diverse fields, anthropology enables us to look at humanity's biological and cultural heritage with a broad perspective.

Anthropologists draw on the scientific method to help evaluate their interpretations about the natural or social phenomena. By making their observations in a careful and systematic way, they generate theories—testable explanations of why certain biological characteristics, behaviors, or cultural characteristics occur.

For students, anthropology creates a broader awareness and a deep appreciation of humanity in both the past and the present. By evaluating anthropological data, students develop critical-thinking skills. And the process of anthropological inquiry—exploring other cultures and comparing them to one's own—helps students appreciate the advantages and difficulties of both their own society and others. Such understanding may find practical application in business, marketing, education, and public service in an increasingly multicultural and interconnected world.

 ## QUESTIONS TO THINK ABOUT

1. What are the distinctive features that distinguish the field of anthropology from other disciplines?

2. How does the field of anthropology bridge the sciences and the humanities?

3. How do anthropologists utilize the scientific method in their studies? What are the limitations of the scientific method in anthropological studies?

4. Why should a student study anthropology?

 KEY TERMS

anthropology
applied anthropology
archaeology
artifacts
cultural anthropology
deductive method
ethnography
ethnology

forensic anthropology
fossils
genetics
historical linguistics
holistic
hypothesis
inductive method
linguistics

middens
paleoanthropology
participant observation
scientific method
sociolinguistics
structural linguistics
theories
variable

 SUGGESTED READINGS

DEETZ, JAMES. 1996. *In Small Things Forgotten: The Archaeology of Early American Life*. New York: Bantam, Doubleday Dell. A popular, highly readable overview of historical archaeology in the United States by one of the leading scholars in the field. It discusses the sources of information and special dating methods that can be used by archaeologists studying the recent past, as well as how archaeologists have addressed specific research questions.

FEDER, KENNETH L. 1996. *Frauds, Myths and Mysteries: Science and Pseudoscience in Archaeology,* 2nd edition. Mountain View, CA: Mayfield. An excellent introduction to how archaeologists evaluate their finding, highlighted with entertaining discussions of several frauds and myths, as well as genuine mysteries about the past.

SCHICK, KATHY D., and NICHOLAS TOTH. 1993. *Making Silent Stones Speak: Human Evolution and the Dawn of Technology*. New York: Simon and Schuster. An interesting and readable account of more than two decades of research on the origins of human tool manufacture and use, the volume provides a clear illustration of the interdisciplinary, collaborative nature of paleoanthropological research. A book that can be read by both the novice and experienced researcher.

SCUPIN. RAYMOND, and CHRISTOPHER R. DeCORSE. 1998. *Anthropology: A Global Perspective,* 3rd ed. Upper Saddle River, NJ: Prentice Hall. A general introduction to anthropology, including expanded discussions on different anthropological concepts, theories, and conclusions.

Chapter 2

THE RECORD
OF THE PAST

CHAPTER OUTLINE

Answering Questions

Fossils and Fossil Localities
Paleoanthropological Study

The Archaeological Record
Locating Sites / Archaeological Excavation

Dating Methods
Relative Dating / Numerical, or Absolute, Dating / Seriation

Interpretations about the Past

WHY STUDY THE HUMAN PAST? DURING the early history of anthropology, the answer to this question was straightforward. The study of fossils and artifacts of the past sprang out of a curiosity about the world and the desire to collect and organize objects. This curiosity was, in part, a reflection of the increasing interest in the natural world that arose with the scientific revolution in the Western world beginning in the fifteenth century (see Chapter 3). For early collectors, however, the object was often an end in itself—items were placed on shelves to look at, with little or no interest expressed in where the fossil might have come from or what the artifact and associated materials might tell about the people that produced them. Collectors of this kind are called **antiquaries.**

Early antiquarian collections often incorporated many different items in addition to fossils and archaeological materials. For example, the museum of Olaus Wormius, a seventeenth-century Danish scholar, included uniquely shaped stones, seashells, ethnographic objects, and curiosities from around the world, in addition to fossils and ancient stone tools (Daniel, 1981). While these objects were sometimes described and illustrated with great care, they were not analyzed or interpreted to shed light on the evolution of life or the lifeways of ancient humans. Of course, ancient coins, metal artifacts, and jewelry were recognized for what they were, but stone tools and even ancient pottery were generally regarded as naturally occurring objects or the work of trolls, elves, and fairies.

By the late eighteenth century, scholars started to move beyond the simple description of objects to

an increasing appreciation of the significance of fossil remains and the material traces of ancient human societies. This appreciation could be examined within the context of a host of new observations in the natural sciences, including many about the geological record and the age of the earth. In 1797, an English country gentleman named John Frere published an account of some stone tools he had found in a gravel quarry in Suffolk. While brief, the description is tantalizing in terms of the changing attitude toward traces of the past. The stone tools—actually Paleolithic hand axes—were found at a depth of more than twelve feet in association with the bones of extinct animals, in a layer of soil that appeared undisturbed by more recent materials. Frere correctly surmised that the tools were "from a very remote period indeed, even beyond that of the present world" (Daniel, 1981, p. 39).

The nineteenth century saw the first fossil finds of ancient human ancestors. These included the bones found in the Neander Valley of Germany in 1857, now recognized as an archaic human species, *Homo neanderthalensis,* or Neandertal man (see Chapter 5). Although this was a historic discovery, the significance of the fossil was not realized at the time. Interpretations were diverse. Some scholars correctly interpreted the finds as an early human ancestor, but others variously dismissed the bones as those of a Cossack soldier, an elderly Dutchman, a powerfully built Celt, or a pathological idiot (Johanson & Edey, 1981, pp. 28–29). Information continued to accumulate, however, and by the end of the century, the roots of modern archaeological and paleoanthropological study were well established.

CRITICAL PERSPECTIVES

ENGENDERING ARCHAEOLOGY: THE ROLE OF WOMEN IN AZTEC MEXICO

The interpretation of the material record poses a challenge to archaeologists. It provides excellent evidence on some subjects—researchers can readily discuss ancient technology, diet, hunting techniques, and the plan of an ancient settlement—but some topics are more difficult to address. What were the marriage customs, the political system, or the religious beliefs of the ancient inhabitants of a site? These factors are by nature nonmaterial and are not preserved archaeologically. Even documentary records may offer only limited insight on some topics.

In a fascinating study of gender among the Aztec of ancient Mexico, archaeologist Elizabeth Brumfiel utilized both the archaeological and the documentary record to provide new insights into the past (Brumfiel, 1983, 1991). The Aztec civilization was flourishing in central Mexico when the Spanish reached the Americas. It had emerged as the principal state in the region by the fifteenth century, eventually dominating an area stretching from the Valley of Mexico to modern-day Guatemala, some 500 miles to the southwest. The capital, Tenochtitlan, was an im-pressive religious center built on an island in Lake Texcoco. The city's population numbered tens of thousands when the Aztec leader, Montezuma, was killed during fighting with Spanish conquistadors led by Hernan Cortés in 1520. Within decades of the first Spanish contact, the traces of the Aztec Empire had crumbled and been swept aside by European colonization.

Records of the Aztec civilization survive in documentary accounts recorded by the Spanish. The most comprehensive is a monumental treatise on Aztec life, from the raising of children to religious beliefs, written by Fray Bernardino de Sahagun. It is the most exhaustive record of a Native American culture from the earliest years of European contact. For this reason it has been a primary source of information about Aztec life and culture.

Sahagun's description of the role of women in Aztec society focuses on weaving and food preparation. Regrettably, as Brumfiel points out, his work offers little insight into how these endeavors were tied to other economic, political, and religious activities. In addition, Sahagun does not comment on some of his own illustra-tions, which show women involved in such undertakings as healing and marketing. Interpretations based solely on Sahagun's descriptions marginalize women's roles in production as nondynamic and of no importance in the study of culture change.

To obtain a more holistic view of women in Aztec society, Brumfiel turned to other sources. The Aztecs also possessed their own records. Although most of these were sought out and burned by the zealous Spanish priests, some Aztec codices survive. These sources indicate that textiles were essential as tribute, religious offerings, and exchange. Many illustrations also depict women in food production activities. In addition to various categories of food, the codices show the griddles, pots, and implements used in food preparation.

Independent information on these activities is provided by the archaeological record. For example, the relative importance of weaving can be assessed by the number and types of spindle whorls, the perforated ceramic disks used to weight the spindle during spinning, found on sites. Archaeological indications of dietary

ANSWERING QUESTIONS

Few modern archaeologists or paleoanthropologists would deny the thrill of finding a well-preserved fossil, an intact arrow point, or the sealed tomb of a king, but the romance of discovery is not the primary driving force for these scientists. In contrast to popular movie images, the modern researcher is likely to spend more time in a laboratory or in front of a word processor than looking for fossils or exploring lost cities. Perhaps their most fundamental desire is to reach back in time to understand more fully our past.

practices can be inferred from ceramic griddles, cooking pots, jars, and stone tools used in the gathering and preparation of food.

Brumfiel notes that the most interesting aspect of archaeological data on both weaving and food preparation is the variation. Given the static model of women's roles seen in the documentary records, a uniform pattern might be expected in the archaeological data. In fact, precisely the opposite is true. Evidence for weaving and cooking activities varies in different sites and over time. Brumfiel suggests that the performance of these activities was influenced by a number of variables, including environmental zones, proximity to urban markets, social status, and intensified agricultural production.

Food preparation, essential to the household, was also integral to the tenfold increase in the population of the Valley of Mexico during the four centuries preceding Spanish rule. As population expanded during the later Aztec period, archaeological evidence indicates that there was intensified food production in the immediate hinterland of Tenochtitlan. Conversely, the evidence for weaving decreases, indicating that women shifted from weaving to market-oriented food production. These

Aztec codex showing women weaving.

observations are not borne out at sites further away from the Aztec capital. In more distant sites women intensified the production of tribute cloth with which the Aztec empire transacted business.

This model provides a much more dynamic view of women's roles in Aztec society. The observations are consistent with the identification of the household as a flexible social institution that varies with the presented opportunities and constraints. It also underscores the importance of considering both women's and men's roles as part of an interconnected, dynamic system.

Brumfiel's research provides insights into the past that neither

archaeological nor documentary information can supply on its own. She was fortunate to have independent sources of information that she could draw on to interpret and evaluate her conclusions.

Points to Ponder

1. In the absence of any documentary or ethnographic information, how can archaeologists examine the gender of past societies?

2. Can we automatically associate some artifacts with men or with women?

3. Would interpretations vary in different cultural settings?

This book deals with some of the major questions that have been addressed by paleoanthropologists and archaeologists: the evolution of the human species, the human settlement of the world, the origins of agriculture, and the rise of complex societies and the state.

Although anthropologists make an effort to document the record of bygone ages as fully as possible,

they clearly cannot locate every fossil, document every archaeological site, or even record every piece of information about each artifact recovered. Despite decades of research only a minute portion of such important fossil localities as those in the Fayum Depression in Egypt and Olduvai Gorge in Tanzania have been studied. In examining an archaeological site or

CRITICAL PERSPECTIVES

HISTORICAL ARCHAEOLOGY

Some archaeologists have the luxury of written records and oral histories to help them locate and interpret their finds. Researchers delving into ancient Egyptian sites, the ancient Near East, Greek and Roman sites, Chinese civilization, Mayan temples, Aztec cities, Islamic sites, biblical archaeology, and the settlements of medieval Europe can all refer to written sources, ranging from religious texts to explorers' accounts and tax records.

Why dig for archaeological materials if written records or oral traditions can tell the story? Although such sources may provide a tremendous amount of information, they do not furnish a complete record (Beaudry, 1988; Deetz, 1996). Whereas the life story of a head of state, records of trade contacts, or the date of a temple's construction may be pre-

served, the lives of many people and the minutia of everyday life were seldom noted. In addition, the written record is often biased by the writer's personal or cultural perspective. For example, much of the written history of Native Americans, sub-Saharan Africans, Australian Aborigines, and many other indigenous peoples were recorded by European missionaries, traders and administrators, who frequently provided only incomplete accounts viewed in terms of their own interests and beliefs.

Information from living informants may also provide important information about some populations, particularly societies with limited written records. In recognizing the significance of such nonwritten sources, however, it is also necessary to recognize their own distinct limitations.

The specific roles oral traditions played (and continue to play) varied in different cultural settings. Just as early European chroniclers viewed events with reference to their own cultures' traditions, so oral histories are shaped by the worldviews, histories, and beliefs of the various cultures that employ them. Interpreting such material may be challenging for individuals outside the originating cultures. Study of the archaeological record may provide a great deal of information not found in other sources and provide an independent means of evaluating conclusions drawn on the basis of other sources of information (see the box "Engendering Archaeology: The Role of Women in Aztec Mexico"). For example, it has proven particularly useful in assessing change and continuity in indigenous populations during

even a particular artifact, many different avenues of research might be pursued (see the box "Engendering Archaeology: The Role of Women in Aztec Mexico" on pages 18–19). For example when investigating pottery from a particular archaeological site, some archaeologists might concentrate on the technical attributes of the clay and the manufacturing process (Rice, 1987). Others might focus on the decorative motifs on the pottery and how these relate to the myths and religious beliefs of the people who created them. Still other researchers might be most interested in the pottery's distribution (where it was found) and what this conveys about ancient trade patterns.

Research is guided by the questions about the past that the anthropologist wants to answer. To en-

sure that appropriate data are recovered to address these questions, the paleoanthropologist or archaeologist begins a project by preparing a **research design,** in which the objectives of the project are set out and the strategy for recovering the relevant data is outlined. The research design must take into account the types of data that will be collected and how those data relate to existing anthropological knowledge. Within the research design, the anthropologist specifies what types of methods will be used for the investigation. Different topical or area specializations require specific background knowledge or familiarity with specialized techniques (see the boxes "Historical Archaeology" and "Underwater Archaeology"). The anthropologist must also be

the past 500 years (DeCorse, 1992, 1998; Rogers & Wilson, 1988).

In North America, during the past several decades, an increasing amount of work has concentrated on the history of immigrants from Europe, Asia, Africa, and other world areas who arrived in the last 500 years. Archaeological studies have proven of great help in interpreting historical sites and past lifeways, as well as culture change, sociopolitical developments, and past economic systems (Leone & Potter, 1988; Noel Hume, 1983). Among the most significant areas of study is the archaeology of slavery (Singleton, 1985, 1999). Although living in literate societies,

slaves were prohibited from writing and left a very limited documentary record of their own. Archaeological data have been used to provide a much more

complete picture of plantation life and slave society.

Points to Ponder

1. What are some different sources of "historical" information—written and orally preserved accounts—that you can think of? How are these different from one another in terms of the details they might provide?

2. Consider a particular activity or behavior important to you (for example, going to school, participating in a sport, or pursuing a hobby). How would evidence of the activity be presented in written accounts, oral histories, and the archaeological record?

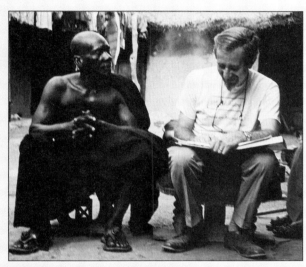

Archaeologist Merrick Posnansky interviewing the chief of the town of Hani, Ghana in 1983. Researchers can use knowledge gathered from living informants to help interpret archaeological finds.
Source: Courtesy of Merrick Posnansky, UCLA.

well grounded in the different theoretical perspectives of anthropology relevant to the research questions. Before going into the field, the researcher also analyzes the existing data, which might include geological surveys, archaeological research reports, descriptions by travelers to the region, and interviews with local inhabitants. These data help place the fossils or the archaeological sites to be studied in a broad context. Generally, the research design is then reviewed by other anthropologists, who recommend it for funding by various government or private research foundations.

Although this book attempts to provide an overview of some of the techniques used and some current interpretations, new finds and improved

methods are constantly changing the amount and kind of information available. New fossils are constantly uncovered and archaeological sites exposed. Each of these discoveries adds to the amount of information available to interpret the past—and to evaluate and revise existing interpretations.

FOSSILS AND FOSSIL LOCALITIES

Paleoanthropologists are particularly concerned with the study of ancient humans and human ancestors. Central to this research is the recovery of fossilized remains of these early ancestors. **Fossils**

CRITICAL PERSPECTIVES
UNDERWATER ARCHAEOLOGY

Sunken ships, submerged settlements, and flooded towns: This wide variety of sites of different time periods, in different world areas, shares the need for specialized techniques to locate, excavate, and study them (Greene, 1990; Throckmorton, 1987). Although efforts were occasionally made in the past to recover cargoes from sunken ships, it was only with the invention and increasing accessibility of underwater breathing equipment during the twentieth century that the systematic investigation of underwater sites became feasible.

A tantalizing example of an underwater archaeological project is the excavation and raising of the preserved remains of the *Mary Rose,* the pride of the young English navy and the flower of King Henry VIII's fleet. The 700-ton warship, which was probably the first English warship designed to carry a battery of guns between its decks, foundered and sank in Portsmouth harbor on a warm July afternoon in 1545. Henry VIII, camped with his army at Southsea Castle, is said to have witnessed the disaster and heard the cries of the crew. In the 1970s, the site of the *Mary Rose* rediscovered and was systematically explored by

volunteer divers from around the world. The ship produced a spectacular array of over 14,000 artifacts ranging from the massive cannon to musical instruments, famed English longbows, and navigational equipment. Finds from the *Mary Rose* and the preserved portions of the hull can be seen at the Mary Rose Ship Hall and Exhibition at the H.M. Naval Base, Portsmouth, England.

Most people associate underwater archaeology with sunken ships, and this in fact represents an important part of the subdiscipline. However, natural disasters may also submerge cities and towns. Such was the case of Port Royal, Jamaica, a flourishing trade center and gathering place for pirates during the seventeenth century. In 1692, a violent earthquake and tidal wave submerged or buried portions of the city, preserving a record for future archaeologists. Excavations at the site spanning the last three decades have recovered a wealth of materials from seventeenth-century life (Hamilton & Woodward, 1984).

Finds from beneath the waves have a great deal of relevance for archaeological interpretation. They may provide a record of human

settlement during periods of lower sea level. Often artifacts from underwater sites are better preserved and so present a wider range of materials than those from land. Even more important, underwater sites are immune to the continued disturbances associated with human activity typical of most land sites. Shipwrecks and sunken cities like Port Royal can be compared to time capsules, containing a selection of artifacts that were in use in a certain context at a specific time. Archaeologists working on land seldom have such clearly sealed archaeological deposits.

Points to Ponder

1. Archaeological excavation on land is a meticulous and careful process. Discuss how excavation and recording methods would have to be modified to conduct archaeological research underneath the water.

2. Given the unique location and preservation found at underwater sites, why might they be more appropriate or important for considering certain types of research questions than land sites?

are the preserved remains, impressions, or traces of living creatures from past ages. They form when an organism dies and is buried by soft mud, sand, or silt (Figure 2.1). Over time this sediment hardens, preserving the remains of the creature within. Occasionally, conditions may be such that actual por-

tions of an organism are preserved—fragments of shell, teeth, or bone. But most fossils have been altered in some way, the decayed parts of bone or shell having been replaced by minerals or surrounding sediment. Even in cases in which fragments of bone or shell are present, they have often

FIGURE 2.1   How Fossils Form   Only a small number of the creatures that have lived are preserved as fossils. After death, predators, scavengers, and natural processes destroy many remains, frequently leaving only fragmentary traces for researchers to uncover.

been broken or deformed and need to be carefully reconstructed.

Paleoanthropologists refer to places where fossils are found as **fossil localities.** These are spots
where predators dropped animals they had killed,
places where creatures were naturally covered by
sediments, or sites where early humans actually
lived. The formation of these fossil beds is complex indeed. Only a small number of the once-liv-
ing creatures are preserved in the fossil record. After
death, few animals are left to lie peacefully, waiting to be covered by layers of sediment and preserved as fossils. Many are killed by predators that
scatter the bones. Scavengers may carry away parts
of the carcass, and insects, bacteria, and weather
quickly destroy many of the remains that are left. As
a result, individual fossil finds are often very incomplete.

Despite the imperfection of the fossil record, a
striking history of the earth's past has survived. Careful study of fine-grained sediments sometimes reveals the preservation of minute fossils of shellfish,
algae, and pollen. Improved techniques, such as
computer and electronic scanning equipment, have
revealed that images of the delicate structure in
bones or the interior of a skull may be preserved in
a fossil (see the Critical Perspectives box, "New Perspectives on the Taung Child," in Chapter 5). Scientists have identified some early human ancestors on
the basis of very fragmentary remains. Paleoanthropologists working with these finds may report the
discovery of a new fossil species, describing the fossils and noting the tenuous nature of their conclusions. As more evidence is uncovered, the original
interpretation may be confirmed, reinterpreted, or
declared false in light of the new findings.

PALEOANTHROPOLOGICAL STUDY

While much of paleoanthropological research focuses on the locating and study of fossil remains, the
overarching concern is to provide a broad understanding of early humans and human ancestors. As
will be discussed in Chapter 5, the behavior, diet,
and activities of these early humans were very different from those of modern humans. Determining
their behavior, as well as the age of the finds and
the environment in which early humans lived, is
challenging and dependent on an array of specialized skills and techniques. Understanding depends

on the holistic, interdisciplinary approach that characterizes anthropology.

As in all anthropological research, a paleoanthropological project begins with a research design out-
lining the objectives of the project and the
methodology to be employed. This would include
description of the region and the time period to be
examined and an explanation of how the proposed
research would contribute to existing knowledge.
The initial survey work for a paleoanthropological
project often relies on paleontologists and geologists,
who provide an assessment of the age of the deposits
within the region to be studied and the likely conditions that contributed to their formation. Clues about
the age may be determined through the identification of distinctive geological deposits and associated
flora and faunal remains (see the discussion of dating methods and faunal correlation later in the chapter). Such information also helps in the reconstruction
of the paleoecology of the region. **Paleoecology**
(*Paleo,* from the Greek, meaning old, and *ecology,*
meaning "study of environment") is the study of ancient environments. Some areas might be of the appropriate age but not have had the right conditions
to fossilize and preserve remains—the remains of
early humans ancestors that may be present might
be so fragmentary and mixed with deposits of other
ages to be of limited use. Another consideration is
the accessibility of fossil deposits. Fossils may be
found in many areas, but they often lie buried under
deep deposits that make it impossible for researchers
to study them and assess their age and condition. On
other instances, however, erosion by wind or water
exposes underlying layers of rock that contain fossils.

Based on the information provided by paleontologists and geologists, more detailed survey work is
undertaken to locate traces of early humans. This
stage of the research may draw on the skills of the
archaeologist, who is trained to examine the study of
material remains of past societies (see discussion
below). Once a fossil locality is located, systematic
excavations are undertaken to reveal buried deposits.
In excavating, paleoanthropologists take great pains
to record a fossil's exact position in relation to the
surrounding sediments; only if the precise location
is known can a fossil be accurately dated and any associated archaeological and paleontological materials be accurately interpreted.

Of particular importance in interpreting fossil localities is the **taphonomy** of the site—the study of

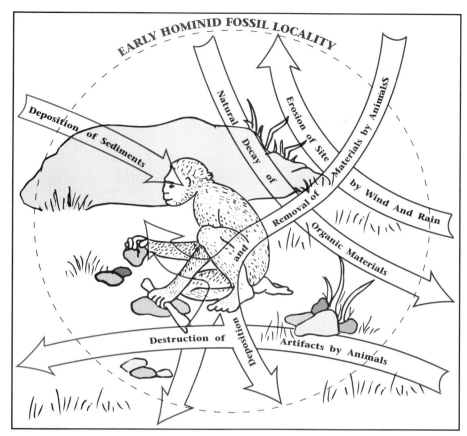

EARLY HOMINID FOSSIL LOCALITY

Deposition of Sediments

Natural Decay of

Erosion of Site

Removal of Materials by Animalss

by Wind And Rain

Organic Materials

and

Destruction of

Deposition

Artifacts by Animals

FIGURE 2.2   The Knotty Problem of Taphonomy
A variety of different activities and events contributes to the formation of an individual fossil locality. These include activities by the early human ancestors, but also such natural processes as decomposition and decay, erosion by wind and rain, and movement of bones and artifacts by animals. Paleoanthropologists must try and decipher these different factors in interpreting the behavior of early human ancestors.

the variety of natural and behavioral processes that led to the formation of the deposits uncovered. As seen in Figure 2.2, the taphonomy of an individual fossil locality may be complex and the unraveling of the history that contributed to its formation very complex indeed. The fossil locality may include traces of the activities of early humans—the artifacts resulting from their behavior, tool manufacture, and discarded food remains, as well as the remains of the early humans themselves. On the other hand, these traces may be altered by a host of disturbances, including erosion by wind and rain and destruction and movement by wild animals.

Once removed from the ground, the detailed analysis of the finds begins. This starts with the careful cleaning of fossil remains and associated materials. Fossils are generally preserved in a hardened mineralized deposit, and cleaning may be tedious and time-consuming. Artifacts and faunal remains from the excavations will be labeled and carefully described, and any fossil remains of early humans re-

constructed. Drawing on all of the geological, paleontological, archaeological, and physical anthropological information, the paleoanthropologist then attempts to place the discoveries in the context of other discoveries and interpretations. The anatomical characteristics of the fossils of the early humans will be compared to other fossils to try and assess their evolutionary relationship, and the other data will be brought to bear on the reconstruction of the ancient environment and models of the way they lived.

THE ARCHAEOLOGICAL RECORD

Culture is a fundamental concept within the discipline of anthropology. In popular use, most people use the word *culture* to refer to "high culture"—Shakespeare's works, Beethoven's symphonies, Michelangelo's sculptures, gourmet cooking, imported wines, and so on. Anthropologists, however,

Bones and other archaeological remains are often fragmentary and must be excavated with great care. Here a researcher carefully extracts fossilized bone from sediments at the early hominid site of Sterkfontein, South Africa.
Source: Courtesy of C.R. DeCorse.

use the term in a much broader sense to refer to a shared way of life that includes values, beliefs, and norms transmitted within a particular society from generation to generation. This view of culture includes agricultural practices, social organization, religion, political systems, science, and sports. **Culture** thus encompasses all aspects of human activity from the fine arts to popular entertainment, from everyday behavior to the most deeply rooted religious beliefs. It contains the plans, rules, techniques, and designs for living.

Archaeologists are concerned with the study and interpretation of the cultures of past human societies. This presents an inherent difficulty. By its very nature, culture is *nonmaterial*—that is, it refers to the intangible products of human society (such as values, beliefs, religion, and norms) that are not preserved archaeologically. Hence, archaeologists must rely on the artifacts—the physical remains of past societies. This residue of the past is called **material culture.** Material culture consists of the physical products of human society (ranging from weapons to clothing). The earliest traces of material culture are stone tools

dating back more than two and a half million years: simple choppers, scrapers, and flakes. Modern material culture consists of all the physical objects that a contemporary society produces or retains from the past, such as tools, streets, buildings, homes, toys, medicines, and automobiles. Archaeologists investigate the material culture of the societies they study, and they also examine the relationship between the material culture and the nonmaterial culture: the values, beliefs, and norms that represent the patterned ways of thinking and acting within a society.

The preservation of archaeological materials varies (Schiffer, 1987). Look at the objects that surround you. How long would these artifacts survive if left uncared for and exposed to the elements? As is the case with the fossil record, the archaeological past is a well-worn and fragmentary cloth rather than a complete tapestry. Stone artifacts endure very well, and thus it is not surprising that much of our knowledge of early human lifeways is based on stone tools. Ceramics and glass may also survive very well; but iron and copper corrode, and organic materials, such as bone, cloth, paper, and wood, generally disappear quickly.

In some cases environmental conditions that limit insect and microbial action and protect a site from exposure to the elements may allow for the striking preservation of archaeological materials. Some of the most amazing cases are those in which items have been rapidly frozen. An illustration of this kind of preservation is provided by the discovery in 1991 of the 5,300-year-old frozen remains of a Bronze Age man by hikers in Italy's Tyrol mountains. With the body were a wooden backpack, a wooden bow, fourteen bone-tipped arrows, and fragments of clothing. In other instances, a waterlogged environment, very dry climate, or rapid burial may create conditions for preservation. Such unique instances provide archaeologists with a much more complete record than is usually found.

Places of past human activity that are preserved in the ground are called **archaeological sites.** Sites reflect the breadth of human endeavor. Some are settlements that may have been occupied for a considerable time, for example, a Native American village or an abandoned gold-mining town in the American West. Other sites reflect specialized activities, for instance, a ceremonial center, a burial ground, or a place where ancient hunters killed and butchered an animal.

In some cases, environmental conditions may allow for amazing preservation, as illustrated by the 4,000-year-old naturally mummified remains of a woman from arid hills near the Chinese city of Urumqi (left); and the 5,300-year-old frozen remains of a man found in Italy's Tyrol mountains (above).

Much of the archaeologist's time is devoted to the study of **artifacts,** any objects made or modified by humans. These include everything from chipped stone tools and pottery to plastic bottles and computers. Nonmovable artifacts, such as an ancient fire hearth, a pit dug in the ground, or a wall, are called **features.** In addition to artifacts and features, archaeologists examine items recovered from archaeological sites that were not produced by humans but nevertheless provide important insights into the past. Animal bones, shells, and plant remains recovered from an archaeological site furnish information on both the past climatic conditions and the diet of the early inhabitants. The soil of a site is also an important record of past activities and the natural processes that affected a site's formation. Fires, floods, and erosion all leave traces in the earth for the archaeologist to discover. All of these data may yield important information about the age, organization, and function of the site being examined. These non-artifactual organic and environmental remains are referred to as **ecofacts.**

Of crucial importance to the archaeologist is the **context** in which archaeological materials are found. An artifact's context is its specific location in the ground and associated archaeological materials. Without a context, an artifact offers only a limited amount of potential information. By itself, a pot may be identified as something similar to other finds from a specific area and time period, but it provides no new information. If, however, it and similar pots are found to contain offerings of a particular kind and are associated with female burials, a whole range of other inferences may be made about the past. By removing artifacts from sites, laypersons unwittingly cause irreparable damage to the archaeological record.

Archaeological interpretation has historically been strongly influenced by cultural anthropology theory (Lamberg-Karlovsky, 1989; Trigger, 1989). Archaeologists don't dig up cultural systems. Social systems, marriage patterns, and religious beliefs are not preserved archaeologically, but material remains that might reflect them are. Cultural anthropology helps archaeologists understand how cultural systems work and how the archaeological record might reflect portions of these systems. On the other hand, archaeology offers cultural anthropology a time depth that cannot be obtained through observations of living populations. The archaeological record provides a record of past human behavior. Clearly it furnishes important insights into past technology, providing answers to such questions as "When did people learn to make pottery?" and "How was iron smelted?" However, artifacts also offer clues to past ideals and belief systems. Consider, for example, what meanings and beliefs are conveyed by such artifacts as a Christian cross, a Jewish menorah, or a Hopi kachina figure. Other artifacts convey cultural beliefs in more subtle ways. Everyday items such as the knife, fork, spoon, and plate that Americans eat with are not the only utensils suitable for the task; indeed, food preference itself is a culturally influenced choice.

LOCATING SITES

In 1940, schoolboys retrieving their dog from a hole in a hillside near Montignac, France, found themselves in an underground cavern. The walls were covered with delicate black and red paintings of bison, horses, and deer. The boys had discovered Lascaux cave, one of the finest known examples of Paleolithic cave art.

Chance findings such as this sometimes play a role in the discovery of archaeological remains, but researchers generally have to undertake a systematic examination, or **survey,** of a particular area, region, or country to locate sites. They will usually begin by examining previous descriptions, maps, and reports of the area for references to archaeological sites. Informants who live and work in the area may also be of great help in directing archaeologists to discoveries.

Of course, some archaeological sites are more easily located than others; the great pyramids near Cairo, Egypt; Stonehenge in southern England; and the Parthenon of Athens have never been lost. Though interpretations of their precise use may differ, their impressive remains are difficult to miss. Unfortunately, many sites, particularly some of the more ancient, are more difficult to locate. The settlements occupied by early humans were usually small, and only ephemeral traces are preserved in the ground. In many instances these may be covered under many feet of sediment. Examination of the ground surface may reveal scatters of artifacts or discolorations in the soil, which provide clues to buried deposits. Sometimes nature inadvertently helps archaeologists, as erosion may expose sites. Archaeologists can also examine road cuts, building projects, and freshly plowed land for archaeological materials.

In the field, an archaeologist defines what areas will be targeted for survey. These will be determined by the research design but also by environmental and topographical considerations, as well as the practical constraints of time and money. Archaeological surveys can be divided into *systematic* and *unsystematic* approaches (Renfrew & Bahn, 1996). The latter is simpler, the researcher simply walking over trails, riverbanks, and plowed fields in the survey area and making notes of any archaeological material observed. This approach avoids the problem of climbing through thick vegetation or rugged terrain. Unfortunately, it may also produce a biased sample of the archaeological remains present—ancient land uses might have little correspondence with modern trails or plowed fields.

Researchers use many different methods to ensure more systematic results. In some instances, a region, valley, or site is divided into a grid, which is then walked systematically. In other instances, *transects* may provide useful information, particularly where vegetation is very thick. In this case, a straight line, or transect, is laid out through the area to be surveyed. Fieldworkers then walk along this line, noting changes in topography, vegetation, and artifacts.

SUBSURFACE TESTING AND SURVEY Because many archaeological sites are buried in the ground, many surveys incorporate some kind of subsurface testing. This may involve digging auger holes or shovel test pits at regular intervals in the survey, the soil from which is examined for any traces of archaeological material. This technique may provide important information on the location of an archaeological site, its extent, and the type of material represented.

Today many different technological innovations allow the archaeologist to prospect for buried sites without lifting a spade. The utility of these tools can be illustrated by the magnetometer and resistivity meter. The **proton magnetometer** is a sensor that can detect differences in the soil's magnetic field caused by buried features and artifacts. A buried foundation will give a different reading than an ancient road, both being different from the surrounding undisturbed soil. As the magnetometer is systematically moved over an area, a plan of buried features can be created.

Electrical **resistivity** provides similar information, though it is based on a different concept. A resistivity meter is used to measure the electrical current passing between electrodes that are placed in the ground. Variation in electrical current indicates differences in the soil's moisture content, which in turn reflects buried ditches, foundations, or walls, which retain moisture to varying degrees.

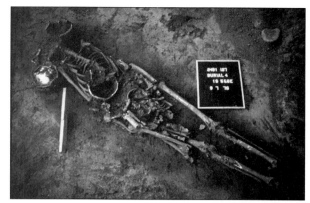

Aerial photography often allows the identification of archaeological sites that may be invisible on the ground. This aerial photograph of a recently plowed cornfield in Perry County, southern Illinois, led to the discovery of the Grier Site. Subsequent excavation revealed that the site had been occupied from the Archaic to the Mississippian; the burials date to about 1000 B.C.
Source: Courtesy of Larry Kruckman, Indiana University, Pennsylvania.

Although at times yielding spectacular results, techniques such as magnetometer and resistivity surveys are not without their limitations. Buried metal at a site may confuse the magnetic readings of other materials, and a leaking hose wreaks havoc with a resistivity meter. Both techniques may produce confusing patterns as a result of shallowly buried geological features such as bedrock.

REMOTE SENSING An archaeologist was once heard to say that "one ought to be a bird to be a field archaeologist," and indeed, the perspective provided by **aerial photography** is a boon to archaeologists (Daniel, 1981: 165). Experiments with aerial photography occurred prior to World War I, but it was during the war that its potential importance to archaeological surveys was recognized. Pilots noticed that some sites, invisible on the ground, were dramatically illustrated from the air. Aerial photography is important in locating sites, but it is also of particular use in mapping and interpretation (Kruckman, 1987).

Of less use to archaeologists are photographs taken from extremely high altitudes by satellites or space shuttles (Ebert, 1984). Often the scale of these pictures, and their cost, make them of limited immediate use. A striking application of such sophisticated techniques, however, is illustrated by some of the research in Mesoamerica. In 1983, National Aeronautics and Space Administration scientists, working with archaeologists, were able to identify ancient Mesopotamian and Mayan settlements and farmlands that had not been located with other techniques. Space imaging radar, which can detect features buried under six feet of sand, proved helpful in identifying ancient caravan routes on the Arabian Peninsula. These routes enabled researchers to locate the lost city of Ubar, a trade center that was destroyed around A.D. 100, and the city of Saffara on the Indian Ocean. As space age technology becomes both more refined and more affordable, it may provide an increasingly important resource for archaeologists.

ARCHAEOLOGICAL EXCAVATION

Archaeological surveys provide invaluable information about the past. The distribution of sites on the landscape offers knowledge about the use of natural resources, trade patterns, and political organization. Surveys also help define the extent of specific sites and allow for a preliminary assessment of their age and function. These data are invaluable in interpreting regional developments and how individual sites form part of a larger picture. However, depending on the project's research objectives, an archaeologist may want more detailed information about individual sites. Once archaeological sites in a

Satellite photo of the Nile River in Egypt illustrates the stark contrast between the river's floodplain and the surrounding desert. At the southern edge of the image is Luxor, which includes the ruins of the ancient Egyptian city of Thebes. Archaeologists are increasingly able to use space-age technology to locate archaeological features.

region have been located, they may be targeted for systematic archaeological excavation (Figure 2.3).

Excavation is costly, time-consuming, and also destructive. Once dug up, an archaeological site is gone forever; it can be "reassembled" only through the notes kept by the archaeologist. For this reason, archaeological excavation is undertaken with great care. Although picks and shovels may occasionally come into play, the tools used most commonly are the trowel, whisk broom, and dustpan. Different techniques may be required for different kinds of sites. For example, more care might be taken in excavating the remains of a small hunting camp than a nineteenth-century house in an urban setting covered with tons of fill. On underwater sites, researchers must contend with recording finds using specialized techniques, while wearing special breathing apparatus (see the box "George Fletcher Bass: Underwater Archaeologist"). Nevertheless, whatever the site, the archaeologist carefully records the context of each artifact uncovered, each feature exposed, and any changes in surrounding soil.

Work usually begins with the clearing of the site and the preparation of a detailed site plan. A grid is placed over the site. This is usually fixed to a **datum point,** some permanent feature or marker that can be used as a reference point and will allow the excavation's exact position to be relocated. As in the case of other facets of the research project, the areas to be excavated are determined by the research design. Excavations of *midden* deposits, or ancient trash piles, often provide insights into the range of artifacts at a site, but excavation of dwellings might provide more information into past social organization, political organization, and socioeconomic status.

A question often asked archaeologists is how deep they have to dig to "find something." The answer is, "Well, that depends." The depth of any given archaeological deposit is contingent on a wide range of variables, including the type of site, how long it was occupied, the types of soil represented, and the environmental history of the area. In some cases artifacts thousands or even hundreds of thousands of years old may lie exposed on the surface. In other cases flooding, burial, or cultural activities may cover sites with thick layers of soil. A clear illustration of this is seen in *tells,* or settlement mounds, in the Near East, which sometimes consist of archaeological deposits covering more than 100 square acres many feet deep.

DATING METHODS

How old is it? This simple question is fundamental to the study of the past. Without the ability to order temporally the developments and events that occurred in the past, there is no way of assessing evolutionary changes, cultural developments, or technological improvements. Paleoanthropologists and archaeologists employ many different dating techniques. Some of these are basic to the interpretation of both fossil localities and archaeological sites. Others are more appropriate for objects of certain ages or for particular kinds of materials (for example volcanic stone, as opposed to organic material). Hence, certain techniques are more typically associated with archaeological research than paleoanthropological research, and vice versa. In any given project, several different dating techniques are typically used in conjunction with one another to independently validate the age of the materials being

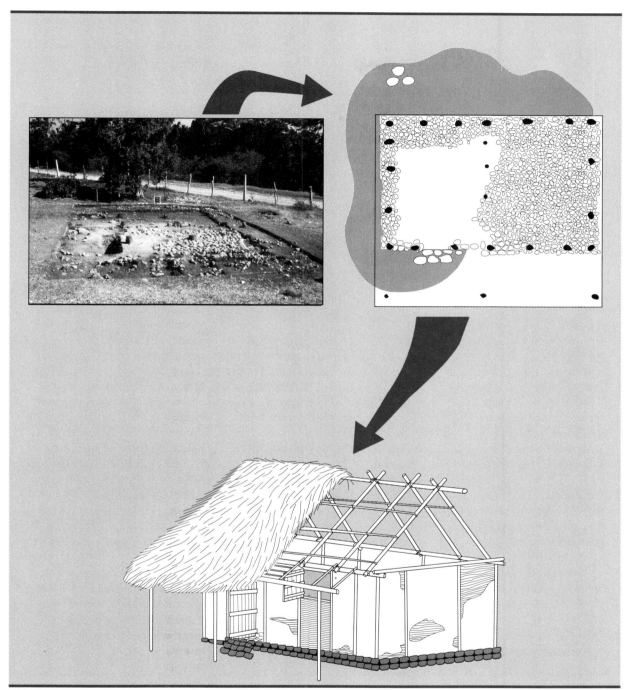

FIGURE 2.3   Excavation, archaeological plan, and artist's reconstruction of an eighteenth-century slave cabin at Seville Plantation, St. Annes, Jamaica. The meticulous recording of excavated artifacts and features allows archaeologists to reconstruct the appearance of past settlements. In this case, eighteenth-century illustrations and written descriptions helped the artist add features, such as the roof, that were not preserved archaeologically.

Source: Courtesy of Douglas V. Armstrong, Syracuse University.

ANTHROPOLOGISTS AT WORK
GEORGE FLETCHER BASS: UNDERWATER ARCHAEOLOGIST

George Fletcher Bass is one of the pioneers of underwater archaeology, a field that he actually didn't set out to study and, indeed, a field that was virtually unrecognized as a discipline when he entered it. Bass began his career with a master's degree in Near Eastern archaeology at Johns Hopkins University, which he completed in 1955. He then attended the American School of Classical Studies at Athens, and excavated at the sites of Lerna, Greece, and Gordion, Turkey. Following military service in Korea, Bass began his doctoral studies in classical archaeology at the University of Pennsylvania. It was there, in 1960, that he was asked by Professor Rodney S. Young if he would learn to scuba dive in order to direct the excavation of a Bronze Age shipwreck discovered off Cape Geldonya, Turkey. Bass's excavations of this site were the first time an ancient shipwreck was excavated in its entirety under the water.

During the 1960s, Bass went on to excavate two Byzantine shipwrecks off Yassi Ada, Turkey. At these sites he developed a variety of specialized methods for underwater excavation, including new mapping techniques, a sub-

Dr. George Bass, after a dive.

mersible decompression chamber, and a two-person submarine. In 1967, his team was the first to locate an ancient shipwreck with side-scanning sonar. In addition to setting standards for underwater archaeological research, these excavations captured popular imagination and revealed shipwrecks as time capsules containing a spectacular array of artifacts, many unrecovered from terrestrial sites (Bass, 1963, 1973; Throckmorton, 1962).

After completing his doctorate in 1964, Bass joined the faculty at the University of Pennsylvania. He remained there until 1973, when he left to found the Institute of

Nautical Archaeology (INA), which has been affiliated with Texas A&M University since 1976. Under his guidance, the INA has become one of the world's premier programs in underwater archaeology. The institute has conducted research throughout the world on shipwrecks and sites of a diversity of time periods. Bass has continued to focus on shipwrecks in Turkey, where he is an honorary citizen of the town of Bodrum. Some of his more recent projects include a fourteenth century B.C. wreck with a cargo of copper, ivory, tin, glass, and ebony, and a medieval ship with a large cargo of Islamic glass.

George Bass is currently the George T. and Gladys H. Abell Distinguished Professor of Nautical Archaeology and holder of the George O. Yamini Family Chair in Nautical Archaeology at Texas A&M University. He has written or edited seven books and is the author of more than 100 articles. Because of his unique contribution to underwater archaeology, Bass has been widely recognized and has received awards from the National Geographic Society, the Archaeological Institute of America, and the Society for Historical Archaeology.

examined. Dating methods can be divided into two broad categories that incorporate a variety of specific dating techniques. These broad categories are relative and absolute dating. Accurate dating of discoveries depends on both methods.

RELATIVE DATING

Relative dating refers to dating methods that determine whether one particular event occurred before or after another. The most basic relative dating method is *stratigraphic dating,* a technique pioneered by the seventeenth-century Danish scientist Niels Stensen (1638–1687). Today Stenson is better known by the latinized version of his name, Nicholas Steno. Steno was the first person to suggest that the hard rock where fossils are found had once been soft sediments that had gradually solidified. Because sediments had been deposited in layers, or *strata,* Steno argued that each successive layer was younger than the layers underneath. Steno's **law of supraposition** states that in any succession of rock layers, the lowest rocks have been there the longest and the upper rocks have been in place for progressively shorter periods. This assumption forms the basis of stratigraphic dating.

Steno was concerned with the study of geological deposits, but stratigraphic dating is also key in dating archaeological materials (Figure 2.4). Cultural materials and the trash associated with human occupation often accumulate to striking depths. Like all relative dating methods, stratigraphic dating does not allow researchers to assign an actual numerical age to a fossil or artifact. Rather, it indicates only whether one fossil is older or younger than another within the same stratigraphic sequence. This technique is essential to paleoanthropological and archaeological interpretation because it allows researchers to evaluate change through time. However, researchers must take notice of any disturbances that may have destroyed the order of geological or archaeological deposits. Disturbances in the earth's crust, such as earthquakes and volcanoes, can shift or devastate stratigraphic layers. Archaeological sites may be ravaged by erosion, burrowing animals, and human activity.

FAUNAL SUCCESSION   One of the first people to record the location of fossils systematically was William Smith (1769–1839), the "father" of English geology. An engineer at a time when England was being transformed by the construction of railway lines and canals, Smith noticed that as rock layers were exposed by the construction, distinct fossils occurred in the same relative order again and again. He soon found that he could arrange rock samples from different areas in the correct stratigraphic order solely on the basis of the fossils they contained. Smith had discovered the principle of **faunal succession** (literally, "animal" succession). A significant scientific milestone, Smith's observations were made sixty years before Darwin proposed his evolution-

ary theories to explain how and why life forms changed through time.

Since Smith's era, paleontologists have studied thousands of fossil localities around the world. Information on the relative ages of fossils from these sites casts light on the relative ages of fossils that are not found in stratigraphic context. Placing fossils in a relative time frame in this way is known as **faunal correlation.**

PALYNOLOGY   Remains of plant species, which have also evolved over time, can be used for relative dating as well. **Palynology** is the study of pollen grains, the minute male reproductive parts of plants. By examining preserved pollen grains, we can trace the adaptations vegetation underwent in a region from one period to another. In addition to helping scientists establish the relative ages of strata, studies of both plant and animal fossils offer crucial clues to the reconstruction of the environments where humans and human ancestors lived.

THE FUN TRIO   Scientists can determine the relative age of bones by measuring the elements of fluorine, uranium, and nitrogen in the fossil specimens. These tests, which can be used together, make up the *FUN trio.* Fluorine and uranium occur naturally in groundwater and gradually collect in bones after they are buried. Once absorbed, the fluorine and uranium remain in the bones, steadily accumulating over time. By measuring the amounts of these absorbed elements, scientists can estimate the length of time the bones have been buried. Nitrogen works in the opposite way. The bones of living animals contain approximately 4 percent nitrogen, and when the bones start to decay the concentration of nitrogen steadily decreases. By calculating the percentage of nitrogen remaining in a fossilized bone, scientists can calculate its approximate age.

The FUN trio techniques all constitute relative dating methods because they are influenced by local environmental factors. The amounts of fluorine and uranium in groundwater differ from region to region, and variables such as temperature and the chemicals present in the surrounding soil affect the rate at which nitrogen dissipates. Because of this variation, relative concentrations of fluorine, uranium, and nitrogen in fossils from two world areas may be similar yet differ significantly in age. The techniques are thus of greatest value in establishing the relative age of fossils from the same deposit.

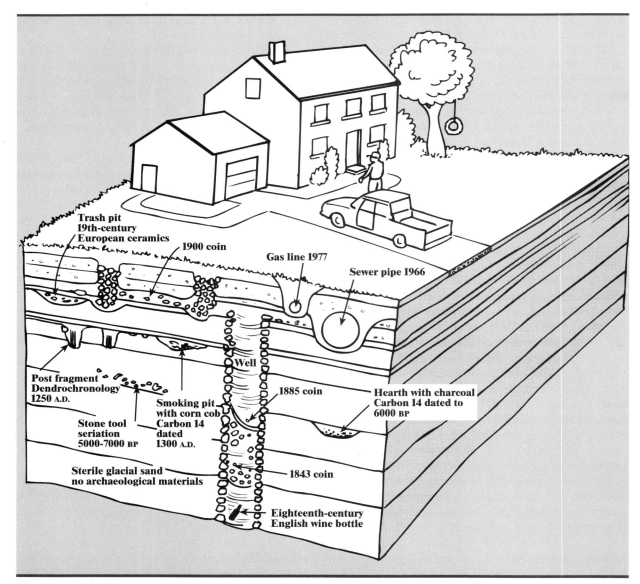

**Trash pit
19th-century
European ceramics**

1900 coin

Gas line 1977

Sewer pipe 1966

Well

**Post fragment
Dendrochronology
1250 A.D.**

1885 coin

**Hearth with charcoal
Carbon 14 dated to
6000 BP**

**Smoking pit
with corn cob
Carbon 14
dated
1300 A.D.**

**Stone tool
seriation
5000-7000 BP**

**Sterile glacial sand
no archaeological materials**

1843 coin

**Eighteenth-century
English wine bottle**

FIGURE 2.4 Stratigraphy Archaeological materials and the remnants of human occupation often accumulate to striking depth. This hypothetical profile illustrates the potentially complex nature of the archaeological record and how different techniques might be combined to date discoveries.

NUMERICAL, OR ABSOLUTE, DATING

During the nineteenth century, scientists experimented with many methods designed to pinpoint the numerical age of the earth itself and the geological epochs. Many of these methods were based on observations of the physical world. Studies of erosion rates, for example, indicated that it had taken millions of years to cut clefts in the earth like the Grand Canyon in the United States. Other strategies were based on the rates at which salt had accumulated in the oceans, the earth had cooled, and geological sediments had formed (Prothero, 1989).

These early approaches were flawed by a limited understanding of the complexity of such natural processes and therefore greatly underestimated the earth's age. For example, Sir Arthur Keith, a prominent English paleoanthropologist of the early twentieth century, posited that the Eocene epoch began approximately 2 million years ago. Yet modern estimates place the Eocene between 55 million and 34 million years ago. In contrast to these early researchers, today's scientists have a wide variety of highly precise methods of dating paleontological and archaeological finds.

Several of the most important numerical dating techniques used today are based on *radioactive decay,* a process in which *radioisotopes,* unstable atoms of certain elements, break down, or decay, by throwing off subatomic particles and energy over time. These changes can produce either a different isotope of the same element or another element entirely. Radioactive decay, which occurs at a set rate regardless of environmental conditions, can be measured with a device called a *mass spectrometer.* By calculating how much decay has occurred in a geological specimen or an artifact, scientists can assign to it a numerical age.

RADIOCARBON DATING The technique of using radioactive decay for archaeological dating was pioneered by Willard Libby, who received the 1960 Nobel Prize in chemistry for his work on radiocarbon dating. **Radiocarbon dating,** as its name implies, is based on the decay of carbon 14 ($_{14}C$), a radioactive (unstable) isotope of carbon that eventually decays into nitrogen. The concentration of carbon 14 in a living organism is comparable to that of the surrounding atmosphere and is absorbed by the organism as carbon dioxide (CO_2). When the organism dies, the intake of CO_2 ends. Thus, as the carbon 14 in the organism begins to decay, it is not replaced by additional radiocarbon from the atmosphere.

Like other radioisotopes, carbon 14 decays at a known rate that can be expressed in terms of its *half-life,* the interval of time required for half of the radioisotope to decay. The half-life of carbon 14 is 5,730 years. By measuring the quantity of carbon 14 in a specimen, scientists can determine the amount of time that has elapsed since the organism died.

Radiocarbon dating is of particular importance to archaeologists because it can be used to date organic matter, including fragments of ancient wooden tools,

charcoal from ancient fires, and skeletal material. The technique has generally been used to date materials less than 50,000 years old. The minuscule amounts of radiocarbon remaining in materials older than this make measurement difficult. However, refined techniques have produced dates of up to 80,000 years old (Prothero, 1989).

POTASSIUM-ARGON AND FISSION-TRACK DATING
Several isotopes that exhibit radioactive decay are present in rocks of volcanic origin. Some of these isotopes decay at very slow rates over billions of years. Two radiometric techniques that have proven of particular help to paleoanthropologists and archaeologists studying early human ancestors are potassium-argon and fission-track dating. These methods do not date fossil material itself. Rather, they can be used to date volcanic ash and lava flows, which are associated with fossil finds. Fortunately, many areas that have produced fossil discoveries were volcanically active in the past and can be dated by using these techniques. These methods have been employed at such fossil localities as the Fayum Depression in Egypt (see Chapter 4) and Olduvai Gorge in Tanzania (see Chapter 5).

In **potassium-argon dating,** scientists measure the decay of a radioisotope of potassium, known as potassium 40 ($_{40}K$), into an inert gas, argon (Ar). During the intense heat of a volcanic eruption, any argon present in a mineral is released, leaving only the potassium. As the rock cools, the potassium 40 begins to decay into argon. Because the half-life of $_{40}K$ is 1.3 billion years, the potassium-argon method can be used to date very ancient finds as well as more recent deposits. Although this technique has been used successfully to date rocks as young as 10,000 years old, it is most effective on samples dating between 1 million and 4.5 billion years old (Prothero, 1989).

Fission-track dating is based on the decay of a radioactive isotope of uranium ($_{238}U$), which releases energy at a regular rate. In certain minerals, microscopic scars, or tracks, from the decay process are produced. By counting the number of tracks in a sample, scientists can estimate fairly accurately when the rocks were formed. Fission-track dating specifies the age of rocks between 10 million and 4.5 billion years old.

DENDROCHRONOLOGY **Dendrochronology** is a unique type of numerical dating based on growth

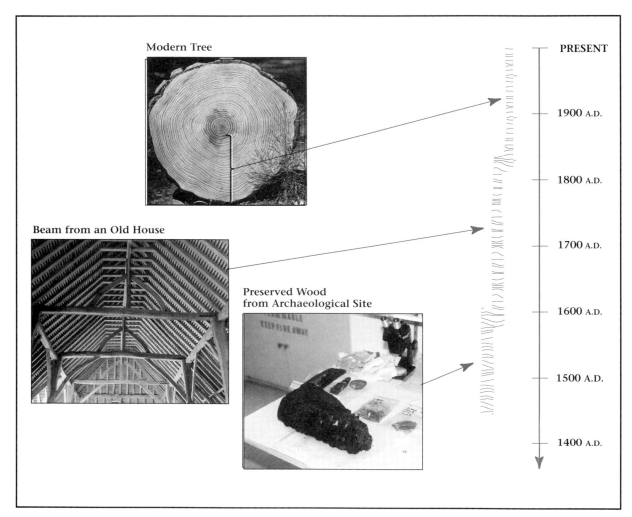

FIGURE 2.5 Dendrochronology   Dendrochronology is based on the careful examination of distinct patterns of thick and thin rings which preserve a record of a region's environmental history. As illustrated here, samples of wood from different contexts may be pieced together to provide a master dendrochronology. Fragments of wood from archaeological sites can then be compared to this dendrochronology to determine the period in which the tree lived.

rings in trees (Figure 2.5). Because a ring corresponds to a single year, the age of a tree can be determined by counting the number of rings. This principle was recognized as early as the late eighteenth century by the Reverend Manasseh Cutler, who used it to infer that a Native American mound site in Ohio was at least 463 years old.

Today tree-ring dating has become a great deal more sophisticated. In addition to recording annual growth, tree rings also preserve a record of environmental history—thick rings represent years when the tree received ample rain; thin rings denote dry spells. In more temperate regions, the temperature and the amount of sunlight may affect the thickness of the rings. Trees of the same species in a localized area will generally show a similar pattern of thick and thin rings. This pattern can then be overlapped with patterns from successively older trees to build up a master dendrochronology sequence. In the American Southwest, a sequence using the bristlecone pine has now been extended to almost 9,000 years ago. Work on oak in European sites has been extended even farther.

The importance of the method is manifest. Dendrochronology has proven of great significance in areas such as the American Southwest, where the dry conditions often preserve wood. The growth rings in fragments of wood from archaeological sites can be compared to the master dendrochronology sequence, and the date the tree was cut down can be calculated. Even more important, dendrochronology provides an independent means of evaluating radiocarbon dating. Fragments dated by both techniques confirm the importance of radiocarbon as a dating method. However, wood dated by both techniques indicates that carbon 14 dates more than 3,000 years old are increasingly younger than their actual age. The reason for this lies in the amount of carbon 14 in the earth's atmosphere. Libby's initial calculations were based on the assumption that the concentration was constant over time, but we now know that it has varied. Dendrochronologies have allowed scientists to correct, or calibrate, radiocarbon dates, rendering them more accurate.

SERIATION

Unlike the dating techniques discussed thus far, which utilize geological, chemical, or paleontological principles, seriation is based on the study of archaeological materials. Simply stated, **seriation** is a dating technique based on the assumption that any particular artifact, attribute, or style will appear, gradually increase in popularity until it reaches a peak, and then progressively decrease. Archaeologists measure these changes by comparing the relative percentages of certain attributes or styles in different stratigraphic levels in a site or in different sites. Using the principle of increasing and decreasing popularity of attributes, archaeologists are able to place categories of artifacts in a relative chronological order.

The principles of seriation can be illustrated by examining stylistic changes in New England gravestones of the seventeenth, eighteenth, and nineteenth

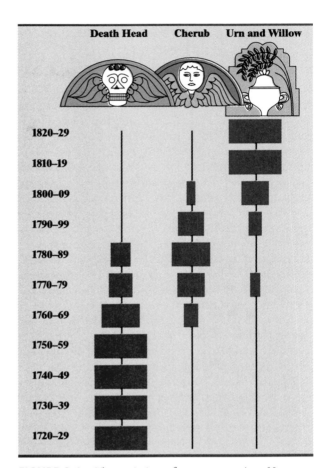

FIGURE 2.6   The seriation of gravestones in a New England cemetery by archaeologist James Deetz illustrates the growth and gradual decline in popularity of a closely dated series of decorative motifs.

Source: From *In Small Things Forgotten* by James Deetz. Copyright © 1977 by James Deetz. Used by permission of Doubleday, a division of Bantam Doubleday Dell Publishing Group, Inc.

centuries. Unlike many artifacts, gravestones can be closely dated. To validate the principle of seriation, archaeologist James Deetz charted how gravestone designs changed through time. His study of gravestones in Stoneham Cemetery, Massachusetts, as illustrated in Figure 2.6, demonstrates the validity of the method. In the course of a century, death's-head motifs were gradually replaced by cherub designs, which in turn were replaced by urn and willow decorations. The study also illustrates how local variation in beliefs and trade patterns may influence the popularity of an attribute.

INTERPRETATIONS ABOUT THE PAST

Views of the past are, unavoidably, tied to the present. As we discussed in Chapter 1, anthropologists try to validate their observations by being explicit about their assumptions. Prevailing social and economic conditions, political pressures, and theoretical perspectives all may affect interpretation. During the early twentieth century, bits and pieces of physical anthropology, archaeology, and linguistic information were muddled together to support the myth of a superior German race (Daniel, 1981). For example, Gustav Kossina, initially trained as a philologist, distorted archaeological interpretations to bolster chronologies that showed development starting in Germany and spreading outward to other parts of Europe.

Archaeological and historical information was also used to validate racist apartheid rule in South Africa. South African textbooks often proffered the idea that black, Bantu-speaking farmers migrating from the north and white, Dutch-speaking settlers coming from the southwest arrived in the South African hinterland at the same time. This interpretation had clear relevance to the present: Both groups had equal claim to the land. However, over the past two decades a new generation of archaeologists has knocked the underpinning from this contrived history (Hall, 1988). Archaeological evidence indicates that the ancestors of the black South Africans had moved into the region by A.D. 200, 1,500 years before the initial European settlement.

In these cases, versions of the past were constructed with dangerous effects on the present. More commonly, errors in interpretation are less intentional and more subtle. All researchers carry their own personal and cultural bias with them. Human societies are complex, and how this complexity is manifested archaeologically varies. These factors make the evaluation of interpretations challenging, and differences of opinion frequently occur.

Although there is no formula that can be used to evaluate all paleoanthropological and archaeological materials, there are useful guidelines. As seen in the preceding chapter, a key aspect of anthropological research is a systematic, scientific approach to data. Outmoded, incorrect interpretations can be revealed through the testing of hypotheses and replaced by more convincing observations. The validity of a particular interpretation can be strengthened by the use of independent lines of evidence; if these lead to similar conclusions, the validity of the interpretation is strengthened. Academic books and articles submitted for publication are reviewed by other researchers, and authors are challenged to clarify points and strengthen observations. In many cases the evaluation of a particular theory or hypothesis must await the accumulation of data. Many regions of the world and different aspects of the past are virtually unstudied. Therefore, any theories about these areas or developments must remain tentative and subject to reevaluation.

 SUMMARY

Paleoanthropologists and archaeologists examine different aspects of the human past. Paleoanthropologists concentrate on the evolution of humans as a biological species and the behavior of early human ancestors, whereas archaeologists are concerned with past human cultures—their lifestyles, technology, and social systems—through the material remains they left behind. Both subdisciplines overlap and utilize experts from other fields to provide a holistic interpretation of the past.

Paleoanthropologists work with fossils, the preserved traces of past life. Places where fossils are found are termed fossil localities. The fossil record is far from complete; only a small portion of the creatures that have lived are preserved. Nevertheless, an impressive record of past life has survived. Careful study and improved technology reveal minute fossils of shellfish, algae, and pollen and images of the delicate structure in bones. Paleoanthropology integrates the fields of geology, paleontology, and archaeology, as well as physical anthropology, to provide a more holistic interpretation of the emergence and the behavior of early human ancestors.

The archaeological record encompasses all the material traces of past cultures. Places of past human activity that are preserved in the ground are called archaeological sites. Sites contain artifacts, objects made or modified by humans, as well as other traces of past human activity and environmental conditions that affected the site. In studying archaeological materials, archaeologists are particularly interested in the context, the specific location of finds and associated materials. Understanding the context is of key importance in determining the age, uses, and meanings of archaeological materials.

Archaeologists concentrate on many different aspects of the past. To ensure that data relevant to the archaeologists' questions are recovered, projects begin with a research design, which sets out the objectives and formulates the strategy for recovering the pertinent information. Specialized fields of study in archaeology may require special approaches or techniques. For example, historical archaeologists draw on written records and oral traditions to help interpret archaeological remains. Underwater archaeologists require special equipment to locate and excavate sites.

Archaeological sites provide important information about the past, for example, the use of natural resources, trade patterns, and political organization. Sites can be located in many different ways. Often traces of a site may survive on the ground; and local informants, maps, and previous archaeological reports may be of help. To discover sites, archaeologists may survey areas, looking for any indications of archaeological remains. Surface examinations may be supplemented by subsurface testing to identify buried deposits. Technological aids such as the magnetometer or resistivity meter may also help archaeologists identify artifacts and features beneath the ground.

Depending on a project's objectives, archaeological sites may be targeted for excavation. Digging is always undertaken with great care, and information about the work is carefully recorded. Before excavation, a site is divided into a grid, which allows each artifact to be carefully located. The depth of an excavation depends on a number of variables, including the type of site, the length of occupation, the soils represented, and the area's environmental history.

Dating of fossils and archaeological materials is of key importance in the interpretation of the past. Without the ability to order finds temporally, there is no way of assessing technological changes, cultural developments, or evolutionary developments. Paleoanthropologists and archaeologists use many different dating techniques, which can be classified as either relative or absolute dating methods. Methods such as stratigraphic dating, faunal succession, and fluorine, uranium, and nitrogen analyses provide only relative ages for finds in the same deposits. Absolute techniques like radiocarbon dating, potassium-argon dating, and dendrochronology can be used to assign actual numerical ages to finds.

Interpretations of the past are inevitably influenced by the present. At times theories have been used to support political ends, as seen in Nazi Germany and the apartheid policies of South Africa. Researchers try to avoid bias by employing systematic, scientific methodology. Theories can be revealed as false through testing and replaced by more convincing arguments. These in turn can be negated or strengthened by exploring new lines of evidence. Archaeological theories, often derived from cultural anthropology, help archaeologists conceptualize how cultures work and what aspects of a past culture might be preserved archaeologically. Ultimately, this reflection provides a more complete explanation of the dynamics of past cultures and culture change.

QUESTIONS TO THINK ABOUT

1. What are the distinct issues, concerns, and methods that distinguish paleoanthropology and archaeology?

2. How do the archaeological and fossil records differ in terms of their content?

3. What is meant by the term *historical archaeology*? What are some ways in which archaeological research can improve our understanding of history?

4. What are the principal issues that you would address in preparing a research design for an archaeological or paleoanthropological project? What concepts, activities, and logistics would you consider?

5. A great deal of archaeological information can be obtained without moving a single shovelful of dirt. Describe three different methods used by archaeologists to explore sites in the field that are not dependent upon excavation.

6. What are the principal differences between relative and absolute dating? Discuss two methods apiece of relative dating and absolute dating, describing the advantages and limitations of each.

 KEY TERMS _____

aerial photography
antiquaries
archaeological sites
artifacts
context
culture
datum point
dendrochronology
ecofacts
faunal correlation

faunal succession
features
fission-track dating
fossils
fossil localities
law of supraposition
material culture
paleoecology
palynology
potassium-argon dating

proton magnetometer
radiocarbon dating
relative dating
research design
resistivity
seriation
survey
taphonomy

 SUGGESTED READINGS _____

DANIEL, GLYN. 1981. *A Short History of Archaeology*. New York: Thames & Hudson. A brief introduction to the history of archaeology by one of the leading scholars on the subject. This volume concentrates on developments in Europe, Africa, and Asia and makes a nice contrast with the Willey and Sabloff volume also listed here.

FERGUSON, LELAND. 1992. *Uncommon Ground: Archaeology and Early African America, 1650–1800*. Washington, DC: Smithsonian Institution Press. A highly readable and interesting account synthesizing archaeological and documentary sources to provide a fuller account of the early history of enslaved Africans in the Americas.

PROTHERO, DONALD R. 1989. *Interpreting the Stratigraphic Record*. New York: W. H. Freeman. An authoritative work on the principle, techniques, and methods used in dating and interpreting the geological record.

RENFREW, COLIN, and PAUL BAHN. 1996. *Archaeology: Theories, Methods and Practice*. New York: Thames & Hudson. An introductory text that provides a brief overview of the history of archaeology and a first-rate, understandable survey of the methods and techniques archaeologists use in the field and in the laboratory.

THROCKMORTON, PETER, ed. 1987. *The Sea Remembers: Shipwrecks and Archaeology from Homer's Greece to the Rediscovery of the Titanic*. New York: Weidenfeld & Nicholson. A compilation on nautical archaeology by one of the pioneers in the field. It provides an enjoyable, broad survey of the variety of sites and discoveries that have been recovered from shipwreck sites.

WILLEY, GORDON R., and JEREMY A. SABLOFF. 1993. *A History of American Archaeology*, 3d ed. New York: W. H. Freeman. A comprehensive and very readable overview of the origins and development of American archaeology, beginning with the first descriptions of monuments by European explorers to the present theoretical approaches.

Chapter 3

EVOLUTION

CHAPTER OUTLINE

ONE OF THE CHALLENGES OF PHYSICAL anthropology is to provide insights into the origins of humankind. The fossil record preserves traces of past life on earth, clearly charting a progression of simple one-celled organisms to increasingly diverse forms. A small portion of the fossil evidence relevant to human evolution is presented in Chapters 4 and 5. How did different forms of life appear and new species arise? The biological explanations for this change are the focus of this chapter.

Theories concerning the evolution of life date back to the ancient Greeks, but it was only during the nineteenth century that the first comprehensive theories of evolution were developed. They were made possible through discoveries in many different areas. The acceptance of evolutionary theory is based on research in many different fields. Indeed, the value of evolutionary theory is its utility as a consistent explanation for a wide variety of phenomena. Before examining the scientific basis for our understanding of evolution it is useful to consider other explanations of human origins.

COSMOLOGIES AND HUMAN ORIGINS

The most profound human questions are the ones that perplex us the most. Who are we? Where did we come from? Why are we here? What is our place in the universe? These questions have been shared by all peoples throughout history. Most cultures have developed sophisticated explanations that provide answers to these fundamental questions. **Cosmologies** are conceptual frameworks that present the universe (the *cosmos*) as an orderly system. They often include answers to these basic questions about the place of humankind in the universe and human origins, usually considered the most sacred of all cosmological conceptions.

Cosmologies account for the ways in which supernatural beings or forces formed the earth and people. They are transmitted from generation to generation through ritual, education, laws, art, and language. For example, the Navajo Indians believe that

This painting by Michelangelo in the Sistine Chapel represents the idea of spiritual creation, which was the dominant worldview in Western cosmology for centuries.

the Holy People, supernatural and sacred, lived below ground in twelve lower worlds. A massive underground flood forced the Holy People to crawl through a hollow reed to the surface of the earth, where they created the universe. A deity named Changing Woman gave birth to the Hero Twins, called Monster Slayer and Child of the Waters. Human mortals, called Earth Surface People, emerged, and First Man and First Woman were formed from the ears of white and yellow corn.

In the tradition of Taoism, male and female principles known as *yin* and *yang* are the spiritual and material sources for the origins of humans and other living forms. Yin is considered the passive, negative, feminine force or principle in the universe, the source of cold and darkness, whereas yang is the active, positive, masculine force or principle, the source of heat and light. Taoists believe that the interaction of these two opposite principles brought forth the universe and all living forms out of chaos.

WESTERN TRADITIONS OF ORIGIN

In the Western tradition, the ancient Greeks had various mythological explanations for the origin of humans. One early view was that Prometheus fashioned humans out of water and earth. Another had Zeus ordering Pyrrha, the inventor of fire, to throw stones behind his back; these stones became men and women. Later Greek cosmological views considered evolutionary ideas. Thales of Miletus (c. 636–546 B.C.) argued that life originated in the sea and that humans initially were fishlike, eventually moving onto dry land and evolving into mammals. A few hundred years later, Aristotle (384–322 B.C.) suggested another theory of creation through evolution. Based on comparative physiology and anatomy, his argument stated that life had evolved from simple lower forms to complex higher forms, such as humans.

The most important cosmological tradition affecting Western views of creation is recounted in the biblical Book of Genesis. This Judaic tradition describes how God created the cosmos. It begins with "In the beginning God created the heaven and the earth" and describes how Creation took six days, during which light, heaven, earth, vegetation, sun, moon, stars, birds, fish, animals, and humans originated. Yahweh, the creator, made man, Adam, from "dust" and placed him in the Garden of Eden. Woman, Eve, was created from Adam's rib. Later, as Christianity spread throughout Europe, this tradition became the dominant cosmological explanation of human origins.

THE SCIENTIFIC REVOLUTION

In the Western world following the medieval period (c. A.D. 1450), scientific discoveries began to influence conceptions about humanity's relationship to the rest of the universe. Copernicus and Galileo presented the novel idea that the earth is just one of many planets revolving around the sun. As this idea became accepted, humans could no longer view themselves and their planet as the center of the universe, as had been the traditional belief. This shift in cosmological thinking set the stage for entirely new views of humanity's links to the rest of the natural world. New developments in the geological sciences began to revise the estimates of the age of the earth radically. These and other scientific discoveries in astronomy, biology, chemistry, mathematics, and other disciplines dramatically transformed Western thought, including ideas about humankind.

The scientific theory of evolution, which sees plant and animal species originating through a gradual process of development from earlier forms, provides an explanation of human origin. Although it is not intended to contradict cosmologies, it is based on a different kind of knowledge. Cosmological explanations frequently involve divine or supernatural forces that are, by their nature, impossible for human beings to observe. We accept them, believe them, on the basis of faith. Scientific theories of evolution, in contrast, are derived from the belief that the universe operates according to regular processes that can be observed. The scientific method is not a rigid framework that provides indisputable answers. Instead, scientific theories are propositions that can be evaluated by future testing and observation. Acceptance of the theory of evolution is based on observations in many areas of geology, paleontology, and biology.

CATASTROPHISM VERSUS UNIFORMITARIANISM

In the Western world before the Renaissance, the Judeo-Christian view of Creation provided the only framework for understanding humanity's position in the universe. The versions of Creation discussed in the biblical text fostered a specific concept of time:

a linear, nonrepetitive, unique historical framework that began with divine Creation. These events were chronicled in the Bible—there was no concept of an ancient past stretching far back in time before human memory. In the seventeenth century, this view of Creation led Archbishop James Ussher of Ireland (1581–1665) to calculate the "precise" age of the earth. By calculating the number of generations mentioned in the Bible, Ussher dated the beginning of the universe to the year 4004 B.C. Thus, according to Bishop Ussher's estimate, the earth was approximately 6,000 years old.

The biblical account of Creation led to a particular view of the existence of plants and animals on earth. As the Bible recounted the creation of the world and everything on it in six days, medieval theologians reasoned that the various species of plants and animals must be fixed in nature. In other words, they had not changed since the time of divine Creation—God had created plants and animals to fit in perfectly with specific environments and did not intend for them to change. This idea regarding the permanence of species influenced the thinking of many early scientists.

CATASTROPHISM The view of a static universe with unchanging species posed problems for early geologists, who were beginning to study thick layers of stone and gravel deposits containing the fossilized remains of forms of life not represented in living species. Georges Cuvier (1769–1832), the father of zoology, found fossil bones from prehistoric elephantlike mammals called mammoths in the vicinity of Paris. To reconcile these fossils with prevailing theological views, Cuvier proposed the geological theory known as **catastrophism.** This concept suggested that many species had disappeared since Creation through catastrophes such as floods, earthquakes, and other major geological disasters of divine global proportions. Catastrophism became the best-known geological explanation consistent with the literal interpretation of the biblical account of Creation.

UNIFORMITARIANISM Other geologists challenged catastrophism and the rigidity of nature through scientific studies. One of the first critics was the French naturalist Comte Georges Louis Leclerc de Buffon, who in 1774 theorized that the earth changed through gradual, natural processes that were still observable. He proposed that rivers had created canyons, waves had changed shorelines, and other forces had transformed different features of the earth. After being criticized by theologians, Buffon attempted to coordinate his views with biblical beliefs. He suggested that the six days of Creation described in the Bible should not be interpreted literally. Buffon suggested that these passages actually refer to six epochs of gradual Creation rather than to 24-hour days. Each epoch consisted of thousands of years in which the earth and different species of life were transformed. Although Buffon's interpretation allowed more time for geological changes in the Earth's past, there was no evidence for the six epochs of gradual creation.

As information on the geological record accumulated, the uniformitarian view eventually became the mainstream position in geology. In 1795, James Hutton, in his landmark book *Theory of the Earth,* explained how natural processes of erosion and deposition of sediments had formed the various geological strata of the earth. Hutton indicated that these natural processes must have taken thousands of years. In his book, he estimated that the earth was at least several million years old. In 1833, Charles Lyell, the father of modern geology, reinforced the uniformitarian view. In *Principles of Geology,* he argued that scientists could deduce the age of the earth from the rate at which sediments are deposited and by measuring the thickness of rocks. Through these measurements, Lyell concluded that the earth was millions of years old. This view of gradual change, which provided the basis for later geological interpretations, was referred to as **uniformitarianism.**

Modern geologists have much more sophisticated means of dating the earth. As will be discussed later, the age of the earth is now estimated to be billions, rather than millions, of years, divided into five major ages and many other periods and epochs (see Table 3.1). In addition, recent evidence suggests that during some periods, major, violent changes affected the earth's geological conditions. Although the views of Hutton and Lyell have been superseded, they were historically important in challenging the traditional views of a static universe with fixed species. The uniformitarian view thus set the stage for an entirely new way of envisioning the universe, the earth, and the living forms on the planet.

THEORIES OF EVOLUTION

Evolution refers to the process of change in species over time. Evolutionary theory holds that existing species of plants and animals have emerged over millions of years from simple organisms. Before the mid–1800s, many thinkers had suggested evolutionary theories, but because they lacked an understanding of how old the earth really was and because no reasonable explanation for the evolutionary process had been formulated, most people could not accept these theories.

One such early theory of evolution was posited by the French chemist and biologist, Jean Baptiste de Lamarck (1744–1829). Lamarck proposed that species change and adapt to their environment through physical characteristics acquired in the course of their lifetime. He thought that when the environment changed, *besoin,* the will or desire for change within organisms, would enable them to adapt to their new circumstances. In other words, if a particular animal needed specialized organs to help in adaptation, these organs would evolve accordingly. In turn, the animals would pass on these new organs to their offspring.

The most famous example used by Lamarck was the long necks of giraffes. He suggested that the long neck of the giraffe evolved when a short-necked ancestor took to browsing on the leaves of trees instead of on grass. Lamarck speculated that the ancestral giraffe, in reaching up, stretched and elongated its neck. The offspring of this ancestral giraffe stretched still further. As this process repeated itself from generation to generation, the present long neck of the giraffe was eventually achieved.

Variations of Lamarck's view of inheritance are sometimes known as the *inheritance of acquired characteristics.* Many nineteenth-century scientists used this concept to explain how physical characteristics originated and were passed on to successive offspring. Today, however, this theory is rejected for several reasons. First, Lamarck overestimated the ability of a plant or an animal to "will" a trait or physical characteristic into being to adapt to an environment. In addition, we now know that physical traits acquired during an organism's lifetime cannot be inherited by the organism's offspring. For example, a weightlifter's musculature, an acquired characteristic, will not be passed on to his or her children.

TABLE 3.1 A Record of Geological Time

Era	Period	Epoch	Millions of Years Ago	Geological Conditions and Evolutionary Development
Cenozoic (Age of Mammals)	Quaternary	Recent	0.01	End of last Ice Age; warmer climate. Decline of woody plants; rise of herbaceous plants. Age of *Homo sapiens*.
		Pleistocene	2.0	Four Ice Ages; glaciers in Northern Hemisphere; uplift of Sierras. Extinction of many large mammals and other species.
	Tertiary	Pliocene	5	Uplift and mountain building; volcanoes; climate much cooler. Development of grasslands; flowering plants; decline of forests. Large carnivores; many grazing mammals; first humanlike primates.
		Miocene	25	Climate drier, cooler; mountain formation. Flowering plants continue to diversify. Many forms of mammals.
		Oligocene	38	Rise of Alps and Himalayas; most land low; volcanic activity in Rockies. Spread of forests; flowering plants; rise of monocotyledons. Apes evolve; all present mammal families represented.
		Eocene	55	Climate warmer. Gymnosperms and flowering plants dominant. Age of Mammals begins; modern birds.
		Paleocene	65	Climate mild to cool; continental seas disappear. Evolution of primate mammals.
Mesozoic (Age of Reptiles)	Cretaceous		144	Continents separated; formation of Rockies; other continents low; large inland seas and swamps. Rise of flowering plants; gymnosperms decline. Dinosaurs peak, then become extinct; toothed birds become extinct; first modern birds; primitive mammals.
	Jurassic		213	Climate mild; continents low; inland seas; mountains form; continental drift continues. Gymnosperms common. Large, specialized dinosaurs; first toothed birds; insectovorous marsupials.
	Triassic		248	Many mountains and deserts form; continental drift begins. Gymnosperms dominant. First dinosaurs; egg-laying mammals.
Paleozoic (Age of Ancient Life)	Permian		286	Continents merge as Pangaea; glaciers; formation of Appalachians. Conifers diversify; cycads evolve. Modern insects appear; mammal-like reptiles; extinction of many Paleozoic invertebrates.
	Carboniferous		360	Lands low; great coal swamps; climate warm and humid, then cooler. Forests of ferns, club mosses, horsetails, and gymnosperms. First reptiles; spread of ancient amphibians; many insect forms; ancient sharks abundant.

Devonian	408	Glaciers; inland seas. Terrestrial plants well established; first forests; gymnosperms and bryophytes appear. Age of Fishes; amphibians and wingless insects appear; many trilobites.
Silurian	438	Continents mainly flat; flooding. Vascular plants appear; algae dominant in aquatic environment. Fish evolve; terrestrial arthropods.
Ordovician	505	Sea covers continents; climate warm. Marine algae dominant; terrestrial plants appear. Invertebrates dominant; fish appear.
Cambrian	570	Climate mild; lands low; oldest rocks with abundant fossils. Algae dominant in aquatic environment. Age of marine inverterbrates; most modern phyla represented.
(Precambrian) Proterozoic	1,500	Planet cooled; glaciers; Earth's crust forms; mountains form. Primitive algae and fungi, marine protozoans. Toward end, marine invertebrates.
Archean	3.5 billion	Evidence of first prokaryotic cells.
Origin of the Earth	4.6 billion	
Origin of the Universe	15–20 billion	

DARWIN, WALLACE, AND NATURAL SELECTION

Two individuals affected strongly by the scientific revolution were Charles Darwin and Alfred Wallace, nineteenth-century British naturalists (a term used at that time to refer to biologists). Through their careful observations and their identification of a plausible mechanism for evolutionary change, they transformed perspectives of the origin of species. Impressed by the enormous variation present in living species, Darwin and Wallace independently developed an explanation of the basic mechanism of evolution. This mechanism is known as **natural selection.**

Beginning in 1831, Darwin traveled for five years on a British ship, the HMS *Beagle,* on a voyage around the world. During this journey, he collected numerous species of plants and animals from many different environments. Meanwhile, Wallace was observing different species of plants and animals on the islands off Malaysia. Although both Darwin and Wallace arrived at the theory of natural selection simultaneously and independently, Darwin went on to present a thorough and completely documented statement of the theory in his book *On the Origin of Species,* published in 1859.

In their theory of natural selection, Darwin and Wallace emphasized the enormous variation that exists in all plant and animal species. They combined this observation with those of Thomas Malthus, a nineteenth-century clergyman and political economist whose work focused on human populations. The relevance of Malthus's work was his observation that it is a basic principle of nature that living creatures produce more offspring than can generally be expected to survive and reproduce. For the thousands of tadpoles that hatch from eggs, few live to maturity. Similarly, only a small number of the seeds from a maple tree germinate and grow into trees. In recognizing the validity of this fact Darwin and Wallace realized that there would be *selection* in which organisms survived.

Variation within species and reproductive success are the basis of natural selection. Darwin and Wallace reasoned that certain individuals in a species may be born with particular characteristics or traits that make them better able to survive. For example, certain individuals in a plant species may naturally produce more seeds than others, or some frogs in a single population may have coloring that blends in with the environment better than others, making them less likely to be eaten by predators.

The photos of different dogs exhibit the wide variation in physical characteristics found within the same species.
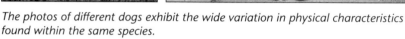
Source: Courtesy of (from left) C. R. DeCorse, Margaret Antonini, Joyce Perkins.

Individuals with these advantageous characteristics are more likely to reproduce and, subsequently, pass on these traits to their offspring. Darwin and Wallace called this process *natural selection* because nature, or the demands of the environment, actually determines which individuals (or which traits) survive. This process, repeated countless times over millions of years, is the means by which species change or evolve over time.

EXAMPLES OF NATURAL SELECTION

One problem Darwin faced in writing *The Origin of Species* was a lack of well-documented examples of natural selection at work. Most major changes in nature take place over thousands or millions of years; only when animals or plants are exposed to rapid, fundamental changes in their environment can we actually observe natural selection in action.

A classic case of natural selection was occurring in the industrial areas of England during Darwin's lifetime, though it was not documented until much later (Kettlewell, 1957). During the nineteenth century, populations of peppered moths *(Biston betularia)* in the British Isles showed variations in color—some were light, others were dark. When first noted in 1849, the dark variety was very rare. After this period, however, the darker-colored variety became more common.

What accounted for this sudden change? The reason was in the changing environment of industrializing Britain. Before the mid-nineteenth century, the light-colored moths blended with the color of the lichen-covered trees they inhabited and were, therefore, a less obvious target for birds. As pollution killed the lichen and soot darkened the trees' bark, this advantage was lost. The light-colored moths, being highly visible against the blackened trees, were

This photo illustrates the most widely documented case of natural selection. As industrial pollution caused the trees to darken, light-colored moths were more visible to predators than were dark-colored moths. Therefore, the dark-colored moths survived and reproduced more frequently, eventually becoming predominant in the population.

eaten by birds. In contrast, the dark-colored moths had an adaptive advantage because they were concealed against the background of the darkened trees; hence, this variety survived and reproduced in greater numbers. Referred to as *industrial melanism,* this example of natural selection has been confirmed experimentally with the release of light-colored and dark-colored moths in urban areas (Bishop, Cook, & Muggleton, 1978).

Natural selection is currently viewed as one of four major guiding forces in the evolution of species. It enabled Darwin to explain the mechanisms of biological evolution, and it remains a powerful explanation for the development of living species of plants and animals. Before turning to the other three processes that guide evolution, we will consider the way traits are passed on from one generation to the next.

PRINCIPLES OF INHERITANCE

Darwin contributed to the modern understanding of biological evolution by thoroughly documenting the *variation* of living forms and by identifying the key process of natural selection. Like most nineteenth-century scientists, however, he did not understand *heredity,* or how specific traits are passed on from one generation to the next. Darwin reasoned that during the reproductive process, the parental substances mix to produce new traits in the parents' offspring. These conclusions were based in part on his experiments with plants and animals, in which he had observed that the offspring often had characteristics from both parents. Darwin was unclear about how these traits were transmitted, but he thought that, as with an alloy such as bronze, which is a mixture of tin and copper, the traits of an offspring represented a blending of parental substances.

MENDEL AND MODERN GENETICS

Modern understanding of heredity emerged through the studies by an Austrian monk named Gregor Mendel (1822–1884). During the 1860s, Mendel began a series of breeding experiments with pea plants, experiments that revolutionized biological thought. Although his findings were not recognized until the twentieth century, Mendel laid the groundwork for what is today known as the science of *genetics,* the biological subfield that deals with heredity.

In compiling his rules of genetics, Mendel discredited earlier theories of inheritance. He demonstrated conclusively that traits are inherited in a particulate manner. In other words, individuals do not inherit traits through a blending of parental substances, such as fluids or blood, as Darwin had believed. Rather, traits are passed from parents to offspring in individual "particles," or packages.

MENDEL'S EXPERIMENTS   Some of Mendel's most important experiments involved the crossbreeding of pea plants that differed in certain key characteristics. For example, he carefully cross-pollinated purebred plants that produced only yellow peas with purebred plants that produced only green peas. By following the results of cross-pollination through several generations, Mendel discovered a distinct pattern of reproduction. The first generation of hybrid plants, that is, plants produced by the parents having different characteristics, were all yellow. However, when he crossbred these hybrid plants, the second generation contained both yellow and green plants. Thus, the green color that seemed to disappear in the first generation of hybrids reappeared in the second generation. Significantly, the ratio of yellow to green plants in the second generation was always approximately three-to-one (3:1).

DOMINANT AND RECESSIVE TRAITS   Mendel drew several important conclusions from these experiments. First, he rejected the earlier notions of inheritance, such as blending. None of the pea plants exhibited a mixed color; all were either yellow or green. In addition, he concluded that certain traits, like yellow color, prevailed over other traits, such as green color. The prevailing traits he termed **dominant.** In contrast, he labeled as **recessive** those traits that were unexpressed in one generation but expressed in subsequent generations. In pea plants, he found that yellow was dominant and green was recessive.

Mendel repeated these experiments focusing on size (tall versus dwarf plants), shape (round versus wrinkled peas), and other characteristics. In each case, he arrived at the same results. The crossbreeding of purebred plants with dominant and recessive traits produced only offspring that exhibited the dominant characteristics. However, the offspring of these hybrid plants exhibited dominant and recessive traits in the 3:1 ratio. The fact that this ratio reappeared consistently convinced Mendel that the key to heredity lay deep within the pea plant seeds. Mendel concluded that the particles responsible for passing traits from parents to offspring occurred in pairs, the individual receiving one from each parent. Purebred parents could pass on only the dominant or recessive trait, whereas hybrid parents could pass on either one. Mendel labeled the dominant traits *A* and the recessive ones *a*. Purebreds thus contain either *AA,* pure dominant, or *aa,* pure recessive; hybrids contain *Aa*. Individuals with *Aa,* of course, exhibit the dominant trait.

GENES AND HEREDITY   From Mendel's work and that of other biologists, we now know that traits are determined by genes. A **gene** is a discrete unit of hereditary information that determines specific physical characteristics of organisms. Most sexually reproducing plants and animals have two genes for every physical trait, one gene inherited from each parent. The alternate forms of the same genes, such as "tall" or "dwarf," are referred to as **alleles.** When an organism has two of the same kinds of alleles, it is **homozygous** for that gene. Thus, homozygous tall plants are *TT,* purebred dominant for tallness, whereas homozygous dwarf plants are *tt,* purebred recessive for shortness. In contrast, when an organism has two different alleles, it is **heterozygous** for that gene. Thus, the *Tt* hybrids are heterozygous plants.

When a heterozygous plant expresses only characteristics of one allele such as tallness, that allele is dominant. The allele whose expression is masked in a heterozygote (for example, shortness) is a recessive allele. Thus, two organisms with different allele combinations for a particular trait may have the same outward appearance: *TT* and *Tt* pea plants will appear the same.

Biologists distinguish between the genetic constitution and the outward appearance of an organism. The specific genetic constitution of an organism is referred to as its **genotype;** the external, observable characteristics of that organism, which are shaped in part by both the organism's genetic makeup and unique life history are called its **phenotype.** Genotype and phenotype are illustrated in Figure 3.1.

PRINCIPLE OF SEGREGATION   Mendel's theory explained how hybrid parents expressing only the dominant trait could produce offspring exhibiting the recessive trait. As Figure 3.2 illustrates, the mixing of two *Tt* configurations produces four combinations: one *TT,* two *Tt,* and one *tt*. This accounts for the 3:1 ratio that Mendel observed. From this calculation, Mendel concluded that the particle containing the recessive trait, which is masked by the dominant trait in one generation, can separate, or segregate, from that trait during reproduction. If this occurs in both parents, the offspring can inherit two recessive particles and thus exhibit the recessive trait. Mendel called this the *principle of segregation.*

PRINCIPLE OF INDEPENDENT ASSORTMENT   The experiments just discussed all focused on one trait. In subsequent studies, Mendel investigated the outcomes of fertilization between pea plants that differed in two ways, such as in both color and shape of the pea. As in the previous experiments, the offspring of purebred (homozygous) parents exhibited only the dominant characteristics. When Mendel cross-fertilized these hybrids, however, the offspring displayed the characteristics present in the first generation in a ratio of 9:3:3:1, as illustrated in Figure 3.3.

This experiment indicated that no two traits are always passed on together. Mendel concluded that during the reproductive process, the particles determining different traits separate from one another and then *recombine* to create variation in the next generation. Thus, in the experiment cited above, Mendel's plants produced peas that were yellow and round, yellow and wrinkled, green and wrinkled, and green and round. Mendel referred to this result as the *principle of independent assortment.*

Because Mendel did not have the advanced technology to investigate cellular biology, he did not know the inner workings of the units of inheritance. But his principles of segregation and independent assortment are still viewed as key operative mechanisms in the transmission of traits from one generation to another.

MOLECULAR GENETICS

Modern scientists have a better understanding than did Mendel of the dynamics of heredity at the cellular level. They have discovered that, like other animals, humans have two different forms of cells: **somatic cells** (body cells) and **gametes** (sex cells: eggs and sperm). Within the nucleus of the somatic

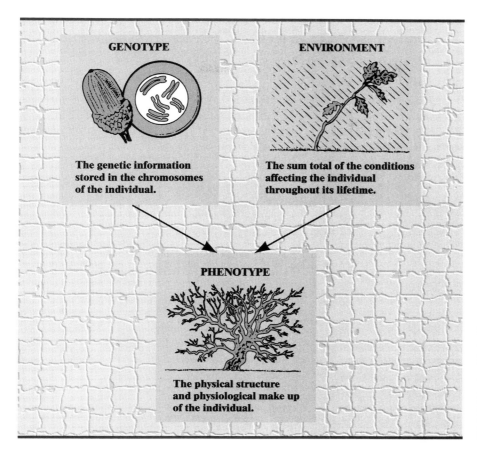

GENOTYPE

The genetic information
stored in the chromosomes
of the individual.

ENVIRONMENT

The sum total of the conditions
affecting the individual
throughout its lifetime.

PHENOTYPE

The physical structure
and physiological make up
of the individual.

FIGURE 3.1 The genotype interacts with the external environment to produce the phenotype.

Source: From *The Illustrated Origin of Species* by Charles Darwin, abridged by Richard Leakey (Rainbird/Faber & Faber, 1979). Copyright © The Rainbird Publishing Group Ltd, 1979. Reprinted by permission.

cells are pairs of **chromosomes,** which contain the hereditary units. Humans have twenty-three pairs of chromosomes, or forty-six chromosomes in all. When these somatic cells divide in the natural process of fissioning to produce new cells, a process biologists call **mitosis,** they replicate themselves to produce cells having forty-six chromosomes. Mitosis is simply a process for making identical cells within a single living individual.

In contrast, human sex cells, or gametes, are produced through the process of **meiosis**—two successive cell divisions that produce cells with only half the number of chromosomes (twenty-three). Meiosis reduces the amount of genetic material to half to prepare for sexual reproduction. During fertilization, when the two sex cells are joined together, they reproduce a new organism with forty-six chromosomes. It is during meiosis and sexual reproduction that Mendel's principles of segregation and independent assortment operate. This reshuffling, or *recombination,* of genetic material does not

change allele frequencies by itself. It ensures, however, that the entire range of traits present in a species is produced and can subsequently be acted on by evolutionary forces.

THE ROLE OF DNA How are hereditary units contained in chromosomes? Biologists have discovered that each chromosome contains the genetic material that determines the physical characteristics of an organism.

The secret of this genetic blueprint is a large molecule of **deoxyribonucleic,** or **DNA,** in each chromosome. The DNA molecule looks like a spiral ladder, or more poetically, like a *double helix* (Figure 3.4). The sides of the ladder consist of sugar (deoxyribose) and phosphate, and the rungs are made up of four nitrogen bases: adenine, thymine, guanine, and cytosine.

The DNA bases are arranged in sequences of three, called *codons.* These sequences determine the assembly of different amino acids. For example, the combination and arrangement of the bases guanine, thymine, and cytosine encode the amino acid gluta-mine. The amino acids join together in chains to pro-duce different proteins, chemical compounds that are fundamental to the makeup and running of the body's cells. There are twenty different kinds of amino acids that can, in differing combinations and amounts, produce millions of different proteins basic to life. The arrangement of bases in the DNA strands is copied by similar molecules, which enables the transfer of the pattern from the chromosomes to other parts of the cell where the production of pro-teins actually takes place. For the purposes of this discussion, a gene can be considered a DNA se-quence, divided into codons, that encodes the pro-duction of a particular protein chain. Different DNA sequences made up of codons encode the produc-tion of specific proteins.

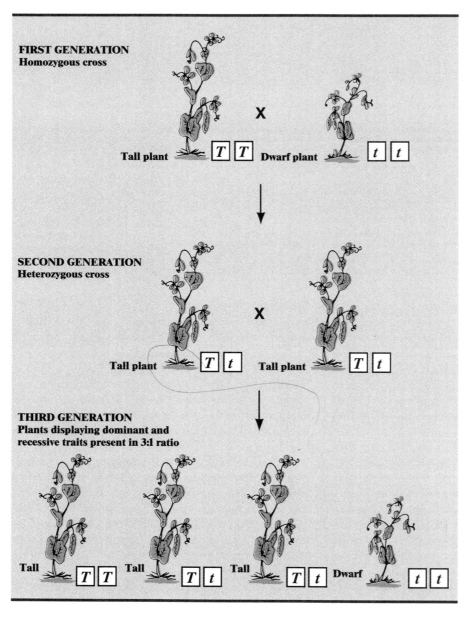

FIRST GENERATION
Homozygous cross

Tall plant — T T Dwarf plant — t t

SECOND GENERATION
Heterozygous cross

Tall plant — T t Tall plant — T t

THIRD GENERATION
Plants displaying dominant and recessive traits present in 3:1 ratio

Tall — T T Tall — T t Tall — T t Dwarf — t t

FIGURE 3.2 The Principle of Segregation In one of his experiments, Mendel crossbred plants that were purebred (homozygous) for particular traits, as illustrated here by the tall and dwarf pea plants. As tallness is a dominant trait, all the offspring of this cross were tall. In the third generation, however, the recessive traits reappear, or segregate.

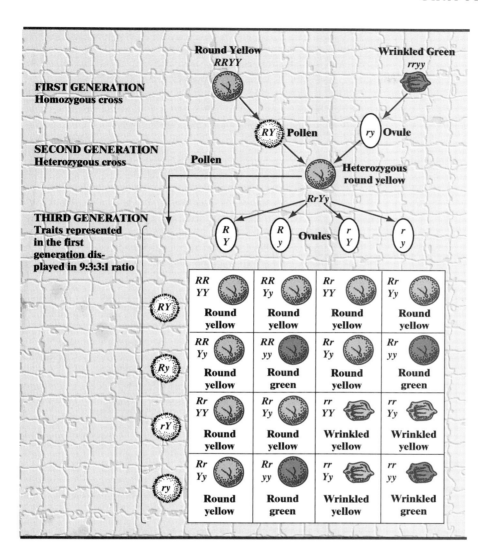

FIRST GENERATION
Homozygous cross

Round Yellow
RRYY

Wrinkled Green
rryy

RY Pollen

ry Ovule

SECOND GENERATION
Heterozygous cross

Pollen

Heterozygous
round yellow

RrYy

THIRD GENERATION
**Traits represented
in the first
generation dis-
played in 9:3:3:1 ratio**

R
Y

R
y

Ovules

r
Y

r
y

RY	*RR YY* Round yellow	*RR Yy* Round yellow	*Rr YY* Round yellow	*Rr Yy* Round yellow
Ry	*RR Yy* Round yellow	*RR yy* Round green	*Rr Yy* Round yellow	*Rr yy* Round green
rY	*Rr YY* Round yellow	*Rr Yy* Round yellow	*rr YY* Wrinkled yellow	*rr Yy* Wrinkled yellow
ry	*Rr Yy* Round yellow	*Rr yy* Round green	*rr Yy* Wrinkled yellow	*rr yy* Wrinkled green

**FIGURE 3.3   Traits Are
Passed on Independently**
This diagram, referred to as
an extended Punnett square,
demonstrates the principle of
independent assortment
discovered by Mendel. When
two heterozygous pea plants
are cross-fertilized, the alleles
that determine two different
traits sort independently of
each other, creating the
phenotypic ratio of 9:3:3:1.

POPULATION GENETICS
AND EVOLUTION

To understand the process of evolution fully, we cannot focus on individuals. A person's genetic makeup, fixed during conception, remains with that individual throughout his or her lifetime. Although people mature and may change in appearance, *they do not evolve.*

Evolution refers to change in the genetic makeup of a *population* of organisms. A **population** here refers to an interbreeding group of individuals. To understand evolution, a scientist must consider all the genes in a population. This assortment of genes is known as the **gene pool.** Any particular gene pool consists of different allele frequencies, the relative amounts of the alternate forms of genes that are present.

In terms of genetics, evolution can be defined as the process of change in allele frequencies between one generation and the next. Alternation of the gene pool of a population is influenced by four evolutionary processes, one of which, natural selection, has already been discussed in relation to the work of Charles Darwin and Alfred Wallace. The other three processes are mutation, gene flow, and genetic drift.

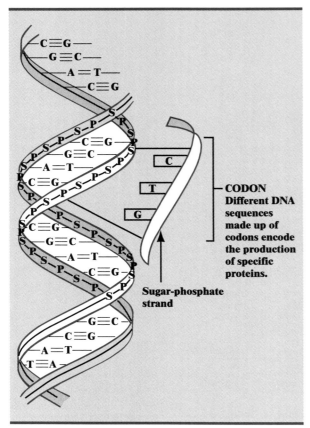

FIGURE 3.4 The Double Helix This illustration shows the chemical structure of DNA. The DNA molecule forms a spiral ladder of sugar (S) and phosphate (P) linked by four nitrogen bases: adenine (A), guanine (G), thymine (T), and cytosine (C).

MUTATIONS

Mutations are alterations of genetic material at the cellular level. They can occur spontaneously during the cell replication process, or they can be induced by environmental factors such as radiation. Although we frequently think of mutations as harmful, they introduce variation into the gene pool and may create new, advantageous characteristics. Mutation serves as the primary force behind evolution because it is the *only* source of new genetic variation. The other evolutionary processes act on the genetic variation introduced through mutation. The role of mutation was only recognized during this century with better understanding of molecular genetics.

Most mutations occur in the somatic cells of organisms. These types of mutations are not heritable. When the organism dies, the mutation dies with it. Some mutations, however, alter the DNA in reproductive cells. In this case, even change in a single DNA base, or a *point mutation*, may produce observable phenotypic change, for example, differences in blood chemistry. A mutation occurring in this instance will be passed on to the offspring.

Generally, the rates of mutations are relatively stable. But if we make the conservative estimate that humans have two copies of each of one hundred thousand genes and the average rate of mutations is 3×10^{-5}, then each of us on average carries six new mutant genes. When the size of the human population is considered, it is evident that the mutation process provides a large source of variability. It would, however, be unlikely for evolution to occur solely as a result of mutation. The rate of mutation of a particular trait within a specific population is likely to be relatively low—perhaps present only in one individual out of 10,000. Hence, mutation alone would be unlikely to affect great change in allele frequencies within the population. Yet if mutations are acted on by natural selection, they become a potentially important source of evolutionary change.

GENE FLOW

Gene flow is the exchange of alleles between populations as a result of interbreeding. When this exchange occurs, new genetic material may be introduced, changing the allele frequencies in a population. The process of gene flow has affected most human societies. Migrants from one society enter a new region and intermarry with the local population. Through reproduction, they transmit new genes into the population. In this way, new mutations arising in one population can be transmitted to other members of the species.

In addition to providing a mechanism for introducing new genetic material, gene flow can act to decrease variation between populations. If two distinct populations continue to interbreed, they will become progressively similar genetically. Migration and connections between different populations have long been a feature of human societies, and among early human ancestors. This genetic interconnectedness explains why new human species have not emerged over millions of years: There has been sufficient gene flow between populations to prevent the creation of substantial genetic distance.

With the development of modern transportation, gene flow occurs on a worldwide scale. In this context, however, it is useful to remember that many cultural or social factors play a role in gene flow in human populations. Religious practices, socioeconomic status, and ethnicity may all influence the selection of mates (see Chapter 6).

GENETIC DRIFT

Genetic drift is evolutionary change resulting from random sampling phenomena that eliminate or maintain certain alleles in a gene pool. It includes the influence of chance events that may affect evolutionary change that are in no way influenced by individuals' genetic makeup. For example, in any population, only a small sample of the potential array of genetic material is passed on from one generation to the next: Every human being produces hundreds of thousands of gametes, each representing a different genetic combination, yet people produce only a few offspring. The chance selection of genetic material that occurs during reproduction results in minor changes in allele frequencies from one generation to the next. Chance events, such as death by disease or accident, also effect allele frequencies. For example, if only ten individuals within a population carry a particular genetic trait and all of them die as a result of accident or disease, this genetic characteristic will not be passed on to the next generation.

Because evolution occurs in populations, change resulting from genetic drift is influenced by the size of the population as well as the relative allele frequencies represented. In larger populations, random events such as accidental deaths are unlikely to have as significant an effect as in smaller populations. A particular kind of genetic drift, known as the **founder effect,** results when only a small number of individuals in a population pass on their genes to the following generation. Such a situation might result when a famine decimates a large group, or when a small migrant population moves away and establishes a new settlement in an isolated area. In these instances, the founding members of the succeeding generation will have only a portion—a sample—of the full range of the genetic material that was present in the original population. Because early human ancestors and human populations lived in small bands

of people, perhaps consisting of family groups, genetic drift was likely an important evolutionary force.

NATURAL SELECTION

Natural selection provides the key to evolution. It can be defined as change resulting from differential reproductive success in the allele frequencies of a population. The other evolutionary forces already discussed are important in creating variation in allele frequencies within and between populations, but they provide no direction—no means for a population to adapt to changing conditions. This direction is provided by natural selection.

As illustrated in the case of the peppered moth, certain alleles (as expressed in particular physical traits) may be selected for by environmental factors. They may enable an organism to resist disease better, obtain food more efficiently, or avoid predators more effectively. Individuals with such advantages will, on average, be more successful in reproducing and thereby pass on their genes to the next generation at higher rates.

Evolutionary "success" can be evaluated only in relative terms: If the environment changes, selection pressures also change. In the case of the peppered moth, the lighter-colored variety was initially the more successful (or more "fit"), but as the trees became blackened by soot, the darker variety was favored. This shift in allele frequencies in response to changing environmental conditions is called **adaptation.** Through evolution, species develop characteristics that allow them to survive and reproduce successfully in particular environmental settings. The specific environmental conditions to which a species is adapted is referred to as its **ecological niche.**

HOW DOES EVOLUTION OCCUR?

Although it is useful to discuss mutation, gene flow, genetic drift, and natural selection as distinct processes, they all interact to affect evolutionary change. Mutation provides the ultimate source of new genetic variants, whereas gene flow, genetic drift, and natural selection alter the frequency of the new allele. The key consideration is change in the genetic characteristics of a population from one generation to the next. Over time, this change may produce major differences among populations that were originally very similar.

To measure evolutionary change, researchers find it useful to evaluate evolutionary processes operating on a population by comparing allele frequencies for a particular trait to an idealized, mathematical model known as the **Hardy-Weinberg theory of genetic equilibrium.** This model, developed independently by G. H. Hardy and W. Weinberg, sets hypothetical conditions under which none of the evolutionary processes is acting and no evolution is taking place. The model makes several important assumptions. It presumes that no mutation is taking place (there are no new alleles); there is no gene flow (no migration or movement in or out of the population); no genetic drift (a large enough population is represented that there is no variation in allele frequencies due to sampling); and that natural selection is not operating on any of the alleles represented. The model also assumes that mating is randomized within the population so that all individuals have equal potential of mating with all other individuals of the opposite sex.

Given these assumptions, there will be no change in allele frequencies from one generation to the next. If examination of genotype frequencies within a population matches the idealized model, no evolution is taking place and the population is said to be in *Hardy-Weinberg equilibrium.* If study suggests the

A characteristic of a species is that its members can successfully interbreed only with one another. The mule is the offspring of a female horse and a male donkey, two clearly distinct species. As mules are always sterile, however, the reproductive isolation of the two species is maintained.

genotype frequencies are not the same as the predicted model, then we know that at least one of the assumptions must be incorrect. Further research can then be undertaken to identify what the source of evolutionary change is. In practice, determining which assumptions are invalid is challenging. Different evolutionary processes may act against one another, giving the appearance that none is operating. Small amounts of change may also go unrecognized. Nevertheless, the Hardy-Weinberg theory provides a starting point for evaluating evolutionary change.

SPECIATION

One of the most interesting areas of research in evolutionary theory is how, why, and when new species arise. This is known as the study of **speciation.** Generally, biologists define a **species** as a group of organisms that have similar physical characteristics and can potentially interbreed with one another to produce fertile offspring, and who are reproductively isolated from other populations.

GRADUALISM   According to evolutionary theory, speciation occurs when there is an interruption in gene flow between populations that formerly were one species but became isolated by geographic barriers. In geographic isolation, these populations may reside in different types of environments, and natural selection, mutation, or genetic drift may lead to increasingly different allele frequencies. Eventually, through evolutionary change, these two populations become so different genetically that they are no longer the same species. Darwin hypothesized that speciation was a gradual process of evolution occurring very slowly as different populations became isolated. This view is called *gradualism,* or the **gradualistic theory of speciation.**

PUNCTUATED EQUILIBRIUM   Beginning in the early twentieth century, some scientists challenged the gradualistic interpretation of speciation, arguing that new species might appear rapidly. *Paleontologists* (fossil specialists) Stephen Jay Gould and Niles Eldredge (1972) proposed a theory known as **punctuated equilibrium.** When examining ancient fossil beds, paleontologists discovered that some plants or animals seemed to exhibit little change over millions of years. These creatures appeared to remain in a

The fossil record indicates that evolutionary change takes place at different rates.

Source: Reprinted with special permission of King Features Syndicate.

state of equilibrium with their habitats for long geological periods. However, the fossil record indicates that major changes, or *punctuations,* appear to have occurred every so often, leading to rapid speciation.

Punctuated equilibrium and gradualism (see Figure 3.5) present extreme perspectives of the rate at which evolution occurs, but the two views are not incompatible. The fossil record provides examples of both cases (Brown & Rose, 1987; Levinton, 1988). The particular rate of change in a particular species depends on its specific adaptive features and the surrounding environmental conditions. Most paleontologists, biologists, and anthropologists hypothesize that both types of evolution have occurred under different circumstances during different geological epochs. As our understanding of the fossil record in-

creases, we will be better able to specify when and where speciation and evolution occurred rapidly and when and where they occurred gradually.

ADAPTIVE RADIATION

Adaptive radiation provides a useful illustration of the evolutionary process and the factors that influence rates of evolutionary change. It can be defined as the rapid diversification and adaptation of an evolving population into new ecological niches.

As we have seen, organisms have tremendous reproductive potential, yet this potential is generally limited by the availability of food, shelter, and space in the environment. Competition with other organisms and natural selection limit the number of off-

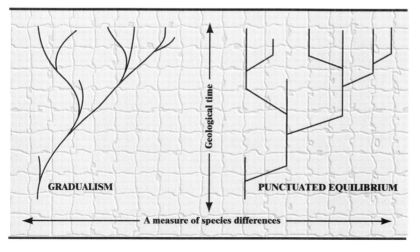

FIGURE 3.5   An illustration of two popular models of evolution: gradualism and punctuated equilibrium.

spring that live. Eventually, organisms may utilize all the resources available in a particular ecological niche. This compels some individuals in a population to exploit new niches where resources are more available and chances for survival greater. Evolutionary change may allow for rapid expansion into new environments. For example, when reptiles first adapted to land environments, they were able to expand into a vast array of ecological niches with little competition.

Sometimes environmental change creates conditions favorable for the adaptive radiation of some species. Natural disasters may lead to the extinction of many forms of life, as described in the box "The Nemesis Theory." Such conditions favor species that have the ability to exploit the changing conditions. Evolutionary processes acting on the expanding population may produce many new varieties and species adapted to ecological niches different from the parent population. The adaptive radiation of many species is recorded in the fossil record.

THE EVOLUTION OF LIFE

Modern scientific findings indicate that the universe as we know it began to develop between 15 billion and 20 billion years ago. At approximately 4.6 billion years ago, the sun and the earth formed, and about a billion years later, the first life appeared in the sea. Through evolution, living forms developed adaptive characteristics, survived, and reproduced. Geological forces and environmental changes, bringing about both gradual and rapid changes, led to new forms of life.

From studying the fossilized bones and teeth of different creatures, paleontologists have tracked the evolution of living forms throughout the world. They document the fossil record according to geological time, which is divided into *eras,* which are subdivided into *periods*, which in turn are composed of *epochs* (look back to Table 3.1 on pages 46–47).

ANALOGY AND HOMOLOGY

How do paleontologists determine evolutionary relationships? Two useful concepts in discussing the divergence and differentiation of living forms are analogy and homology. **Analogy** refers to similarities

in organisms that have no genetic relationship. Analogous forms result from convergent evolution, the process by which two unrelated types of organisms develop similar physical characteristics. These resemblances emerge when unrelated organisms adapt to similar environmental niches. For example, hummingbirds and hummingmoths resemble each other physically and have common behavioral characteristics. However, they share no direct evolutionary descent.

In contrast, **homology** refers to traits that have a common genetic origin but may differ greatly in form and function. For example, a human hand bears little resemblance to a whale's fin. Humans and whales live in very different environments, and the hand and fin perform in very different ways. Careful examination of human and whale skeletons, however, reveals many structural similarities (see Figure 3.6). These similarities indicate a common genetic ancestry. Thus, the hand and the fin are homologous.

FIGURE 3.6 The structural similarities between the human hand and the whale's fin are an example of homology.

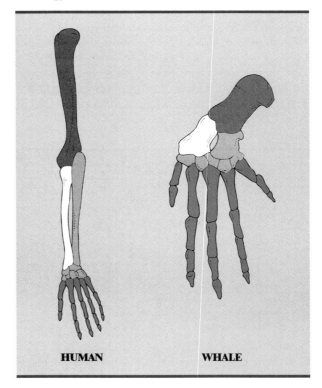

HUMAN   WHALE

CRITICAL PERSPECTIVES

THE NEMESIS THEORY

Most scientists today accept some form of punctuated equilibrium, the idea that dramatic changes, or punctuations, have occurred periodically in the evolution of life. One possible explanation for some of these punctuations is a hypothesis known as the *Nemesis theory.* This theory suggests that extraterrestrial forces are partly responsible for the pace and direction of the evolution of life. It argues that mass extinctions were caused by collisions of the earth and gigantic comets more than 6 miles in diameter and weighing a trillion tons.

The name *Nemesis* refers to a "death star," a hypothetical companion star to the sun, 2 light-years away and not yet seen by any astronomer. According to the Nemesis theory, this star has an eccentric orbit that periodically passes through the Oort Cloud, an envelope of billions of comets surrounding the solar system, deflecting comets randomly into space, some of which strike the earth. Paleontologist David Raup suggests that this Nemesis orbit occurs at a regular pattern every 26 million years (Raup, 1986).

It is hypothesized that the explosions produced by these collisions, millions of times greater than the largest nuclear bomb ever tested, threw dust and debris into the atmosphere, shutting out the sun's radiation and warmth for more than two years. These months of darkness, similar to a "nuclear winter," were followed by soaring temperatures, polluted air, and dead seas. With the blockage of sunlight, photosynthesis was suppressed, thereby choking off plant life, interrupting most of the earth's food chains, and ultimately causing mass extinctions of millions of plant and animal species (Alvarez et al., 1980). Some paleontologists think that these mass extinctions, occurring at intervals of 26 million years, help explain the patterns of evolution found on the earth. They believe, for example, that the extinction of the dinosaurs may have resulted from an extraterrestrial collision. Though many unsubstantiated hypotheses remain, this interpretation has been lent some support by increasing geological evidence for meteor strikes at critical periods in the Earth's evolutionary past. The stories of science fiction novels and film may have some basis in reality.

As Stephen Jay Gould (1985) has pointed out, however, the disastrous conditions caused by the death star would also have created conditions conducive to new forms of life. Thus, although the destruction caused by Nemesis may have resulted in the demise of the dinosaurs, it produced new environmental conditions that favored the adaptive radiation of creatures such as mammals. In other words, it set the stage for an explosion of new life forms.

The Nemesis theory is just that—a testable proposition that must be evaluated by many scientists from different disciplines. As with any theory, scientists have raised significant questions regarding this interpretation. While rapid extinctions of species are well documented in the fossil record, many scientists offer alternative explanations for these extinctions, for example, large-scale volcanic activity (Benton, 1986; Officer, 1990; Signor & Lipps, 1982). In any case, the Nemesis theory continues to be one contending explanation for the extinction and development of different forms of life.

PLATE TECTONICS AND CONTINENTAL DRIFT

In examining the evolution and distribution of living forms, it is important to consider the role of geological processes. The formation of natural features such as continents, mountains, and oceans provides an important mechanism for restricting or encouraging gene flow. **Plate tectonics** is the complex geological process that brings about the drift of continents. The outer shell of the earth is made up of plates that are in constant motion caused by the movement of molten

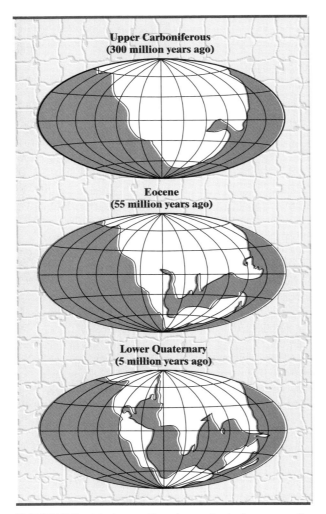

Upper Carboniferous
(300 million years ago)

Eocene
(55 million years ago)

Lower Quaternary
(5 million years ago)

FIGURE 3.7 Understanding the geological process of continental drift—the movement of continents as a result of plate tectonics—helps paleontologists understand the distribution of fossil species.

rocks deep within the earth. According to scientific investigation, the continents move a few centimeters a year (Tarling, 1985). Over millions of years, the continents have sometimes drifted together and then separated, a process known as **continental drift.**

Determining the precise location of different continents at specific geological time periods has helped scientists to understand evolutionary connections among different species of plants and animals. Scientists hypothesize that until about 200 million years ago, the earth's landmass was one gigantic, interconnected continent, which is referred to as *Pan-*

gaea. During the Mesozoic era, Pangaea began to break apart, forming two supercontinents. The southern supercontinent, known as *Gondwana,* consisted of what are now South America, Africa, Antarctica, Australia, and India. The northern continent, consisting of North America, Greenland, Europe, and Asia, is known as *Laurasia* (see Figure 3.7).

Throughout the Mesozoic and Cenozoic eras, the supercontinents continued to move. South America separated from Africa; North America, Greenland, and Europe divided; and Africa joined with Eurasia. Forty million years ago, North America separated from Europe. By 20 million years ago, the continued fracturing and movements of the geological plates resulted in the gradual migration of the continents to their present locations.

Examination of continental drift has helped paleontologists and other scientists understand the distribution of different plant and animal species. For example, the same types of fossil reptiles have been recovered from Mesozoic deposits in North America and the Gobi Desert in Asia, a good indication that these landmasses were connected at that time. In contrast, the separation of South America from other continents during the Cenozoic supports the fossil and biological evidence for the divergence of primates from Africa, Asia, and Europe and primates from the Americas (see Chapter 4).

BLOOD CHEMISTRY AND DNA

The majority of information on the evolution of life and human origin is provided by the fossil record. In recent years, however, studies of the genetic makeup of living organisms have received increasing attention. It is striking to note that despite the tremendous diversity of life, the DNA codes for the production of proteins—with few exceptions—dictate the joining of the same amino acids in all organisms, from the simplest one-celled plants and animals to humans. This semblance of genetic building blocks provides additional evidence for the common origin of all life.

Study of the differences and similarities in the arrangement of genetic material for living animals provides important insights into evolutionary relationships. Similarities in the DNA of different species indicate that they inherited genetic blueprints (with minor modifications) from a common ancestor. In most instances this information has provided independent confirmation of conclusions about evolutionary relationships based on the study of skeletal characteristics and fossil remains. In some instances, however, physical characteristics may be confused because of convergent evolution. Study of genetic information and blood chemistry helps to avoid this confusion.

Genetic material of living animals has also been used to estimate when different species diverged. A technique known as *molecular dating* was developed by Vincent Sarich and Allan Wilson of the University of California, Berkeley (1967). The technique involves comparing amino acid sequences or, more recently, using what is called *DNA hybridization* to compare DNA material itself. As a result, Sarich and Wilson provided useful insights into the genetic relationship of humans to other species and estimates regarding when species may have separated.

Molecular dating is based on two key assumptions: (1) that molecular evolution proceeds at a fairly constant rate over time, and (2) that the greater the similarity between animals in biochemical terms, the more likely it is that they share a close evolutionary relationship. Research based on these concepts has been applied to the interpretation of human evolution (see Chapters 4 and 5).

The reliability of this technique as a dating tool has been hotly debated. Many scientists challenge the assumption that molecular evolution is constant over time. Rather, they believe that variation in mutation rates and the disparate generation lengths of different species skew the measurements of the "molecular clock" (Goodman, Baba, and Darga, 1983; Li and Tanimura, 1987; Lovejoy and Meindl, 1972). Other researchers feel that the technique remains useful if the potential limitations are taken into consideration. Future work may help to resolve these issues.

THE PALEONTOLOGICAL RECORD

THE PRECAMBRIAN AND PALEOZOIC ERAS
The fossil evidence shows that during the Precambrian, simple forms of life resembling modern bacteria, including some species that may have been able to photosynthesize, had emerged. Apparently, the predominant organisms during this era were various kinds of algae. Beginning with the Paleozoic, which dates from 570 million to 248 million years ago, deposits of fossils become more abundant, enabling paleontologists to follow the adaptive radiation of jellyfish, worms, fish, amphibians, and early reptiles.

THE MESOZOIC ERA
The Mesozoic (248 million–65 million years ago) marks the adaptive radiation of reptiles. This era is divided into the Triassic, Jurassic, and Cretaceous periods. The Mesozoic is known as the Age of Reptiles. Unlike earlier forms of life, reptiles could exist entirely outside the water. They were the first successful land animals and reigned as the dominant species in this new environment. Many of the snakes, lizards, and turtles found in Mesozoic formations are similar to contemporary species. Of all the reptiles that lived during the Mesozoic, the dinosaurs are the most well-known today. They included the giant carnivore (meat eater) *Tyrannosaurus*; larger, plant-eating creatures such as the *Brachiosaurus*; and numerous other species, both large and small.

Although reptiles were the dominant animals of the Mesozoic, paleontologists have found fossils of many other organisms from this same era. For example, bird fossils, some even showing the outlines of feathers, have been preserved from the Jurassic period. The paleontological record demonstrates beyond a doubt that a direct evolutionary relationship exists between reptiles and birds. One classic fossil example is *Archaeopteryx*, an animal about the size of a crow, with small wings, teeth, and a long, reptilian tail.

Near the end of the Cretaceous period, many animals became extinct. Changing climatic conditions, competition from newly evolving mammals, and, possibly, extraterrestrial episodes led to the demise of many reptile species, including the dinosaurs, as well as many other organisms.

THE CENOZOIC ERA
The Cenozoic (65 million years ago–present), or Age of Mammals, was characterized by the dominance and adaptive radiation of mammals. This era is divided into two periods, the Tertiary, which encompassed 63 million years, and the Quaternary, which covers the last 2 million years. During the Cenozoic, various species of mammals came to occupy every environmental niche. Some, such as whales and dolphins, adapted to the sea. Others, such as bats, took to the air. Most mammals, however, are land animals, including dogs, horses, rats, bears, rabbits, apes, and humans. One of the major evolutionary advantages that enabled mammals to radiate so rapidly was their reproductive efficiency. In contrast to reptiles, which lay eggs that are vulnerable to predators, most mammals retain their eggs internally within the female. The eggs

are thus protected and nourished until they reach an advanced stage of growth. Consequently, a much higher percentage of the young survive into adulthood, when they can reproduce.

SCIENTIFIC CREATIONISM AND EVOLUTION

Not all segments of Western society have accepted evolutionary theory. Evolution offers a view of the origins and development of life that contradicts literal interpretations of the Bible. Although evolutionary theory does not implicitly preclude the possibility of a supernatural deity, it does not require divine intervention to explain the origins of life. For this reason, some people continue to reject evolutionary theory. Currently in U.S. society, some religious fundamentalists, who call themselves *scientific creationists,* are attempting to refute evolutionary explanations. Through groups such as the Institute of Creation Research in California, these people propose a biblically based explanation of the origins of the universe and of life.

Scientific creationists argue that the universe was created by divine fiat within a period of six days. They further argue that Creation occurred about 10,000 years ago, challenging modern scientific theories that indicate billions of years of geological history and evolution. According to the creationists, all of the evidence collected by modern evolutionists and geologists is biased in favor of an evolutionary position. They claim that the accepted chronological age of the earth—approximately 4.6 billion years—is based on inaccurate dating methods.

Creationists explain the existence of fossilized remains of ancient and prehistoric life by referring to a universal flood that covered the entire earth for forty days. Surviving creatures were saved by being taken aboard a wooden ship, or ark, constructed by Noah. Creatures that did not survive this flood, such as dinosaurs, became extinct.

THE SHORTCOMINGS OF CREATIONISM

Scientific creationists read the texts and theories presented by biologists, geologists, and paleontologists and then present their arguments against evolutionary views. They do very little, if any, direct biological or geological research to refute evolutionary hypotheses. Consequently, their arguments are derived from biblical sources, mixed with misunderstandings of evolutionary hypotheses (Futuyma, 1995; Kitcher, 1982).

The cosmological framework espoused by the scientific creationists is not buttressed by the systematic testing of hypotheses that is required by the scientific method. Creationist ideas regarding physical laws, geological findings, and biological processes are unsupported by any empirical observations. For example, scientists around the world find no physical evidence of a universal flood. Local floods did occur and these may be related to the story of Noah that appears in the Bible (and in earlier Babylonian texts). But to date, no evidence exists suggesting any type of global flood that had the potential to wipe out all human life and other creatures such as dinosaurs (Stiebing, 1984).

Thus, scientific creationists are not scientific at all. The basis for their model of Creation is a literal interpretation of biblical sources. They argue that no geological events or physical laws can take precedence over Scripture. This is an argument based on faith rather than on testable, verifiable hypotheses. Therefore, this creationist scenario is neither correctable nor falsifiable. The model proposed by the scientific creationists has not offered a challenge to evolutionary theory. Many theologians and major religions accept evolution as a valid scientific theory. Faith in a supreme deity is often reconciled with the compelling geological and scientific evidence for evolution in the belief that a divine plan guided the natural geological and evolutionary processes. Many theologians argue that while evolutionary theory may explain the biological origins of humankind, it provides no insight into the human soul. As in much of religious belief, the strength of this belief does not depend on observable or verifiable phenomena.

Theologian Langdon Gilkey, who has written extensively about the creationist–evolutionist controversy, states:

First, "creation science" is not a scientific model, and therefore it is not at all a direct alternative to the scientific theory of evolution. On the contrary, it represents a religious or theological model of the explanation of origins which neither conflicts with nor excludes scientific theories of origins. (Gilkey, 1986:174)

 SUMMARY

Following the scientific revolutions in the West, various developments in the natural sciences, including geology and biology, led to new perspectives on the age of the earth and humankind's origins. Geologists began to discover that the earth had an ancient history, much longer than the few thousand years allowed by a literal reading of biblical chronology. In the nineteenth century, these new ideas influenced biologists Charles Darwin and Alfred Wallace, who were documenting the tremendous variation of plants and animals around the world. From their observations, they developed the theory of natural selection to explain how organisms evolved over time, adapting and reproducing successfully in particular environments.

Although Darwin and Wallace recognized the vast amount of variation in organisms in different environments, they lacked an adequate understanding of how characteristics were passed on from one generation to the next or, in other words, the principles of heredity. Through his experiments on pea plants, an Austrian monk, Gregor Mendel, discovered the essential principles of heredity. Mendel's insights regarding the transmission of dominant and recessive traits, segregation, and independent assortment have remained basic to our understanding of heredity.

Modern scientists have refined Mendel's insights on the study of inheritance. Today, biologists have a better understanding of cell biology and what is known as molecular genetics. Through studies of cells, biologists have unraveled some of the processes of inheritance. The DNA molecule found in the cell is the key factor in determining the traits that organisms inherit from their parents. It contains the chemical information that provides the coding for specific proteins that determine the physical characteristics of organisms.

Modern biologists study evolution in populations of organisms. They identify changes that mutation, gene flow, genetic drift, and natural selection create in the allele frequencies of populations. Changes in allele frequencies may enhance or limit the abilities of a population to adjust successfully to a specific environment.

Recently, paleontologists have challenged the gradualist form of evolution as proposed by early thinkers such as Darwin. They have found that species of organisms sometimes remain unchanged for millions of years, and then new species suddenly develop successfully in varied environments. Today this punctuated-equilibrium model of evolution complements the earlier gradualist model in explaining the proliferation of different species of organisms on the earth.

Some people have not accepted the modern scientific theories regarding the development of the universe and living organisms. In particular, scientific creationists have rejected the evolutionary concepts of modern biology. They propose that the earth has a recent history dating back about 10,000 years, and they believe that different species of life were created spontaneously by divine fiat. The creationists do not conduct any scientific studies to refute the modern explanations of evolution. Instead, they criticize the findings based on a distorted understanding of evolutionary processes and of the scientific enterprise itself. Scientific creationists do not offer scientific explanations but rather base their views on religious faith. As emphasized in Chapter 1, scientific knowledge takes a distinctively different form from that of religious knowledge. Hypothesis testing and acts of faith are radically different means of acquiring knowledge of the universe. To mix one form of knowledge with another, as the creationist scientists do, is to misunderstand this fundamental difference.

 QUESTIONS TO THINK ABOUT

1. Compare the scientific contributions of Buffon, Cuvier, and Lyell. How did each attempt to provide a more scientific understanding of the past?

2. In your own words, explain what is meant by the terms *natural selection, adaptation,* and *evolution.* Create examples to illustrate the concepts.

3. What are the four mechanisms of evolution, and how do they operate to change gene frequencies in a population?

4. Is gene flow influenced by cultural factors such as religion, kinship, language, socioeconomic status, and ethnicity? If so, how? Give some examples.

5. Should contemporary models of human evolution be classified as "origin myths"? Why or why not?

6. What is scientific creationism? Is it science? Should scientific creationism be taught alongside the theory of evolution in public school science classrooms? Why or why not?

 KEY TERMS

adaptation
alleles
analogy
catastrophism
chromosomes
continental drift
cosmologies
deoxyribonucleic acid (DNA)
dominant trait
ecological niche
evolution
founder effect

gametes
gene
gene flow
gene pool
genetic drift
genotype
gradualistic theory
Hardy-Weinberg theory of genetic equilibrium
heterozygous
homology
homozygous
meiosis

mitosis
mutation
natural selection
phenotype
plate tectonics
population
punctuated equilibrium
recessive trait
somatic cells
speciation
species
uniformitarianism

 SUGGESTED READINGS

DARWIN, C. R. 1975. *On the Origin of Species by Means of Natural Selection or the Preservation of the Favored Races in the Struggle for Life.* New York: W. W. Norton. A readily obtainable reprint of one of the most important books of all time. This long essay by Darwin is still of great significance to modern readers.

FUTUYMA, DOUGLAS J. 1995. *Science on Trial: The Case for Evolution.* Fitchburg, MA: Sinauer. A concise summary of the evidence for biological evolution, it provides a point-by-point refutation of pseudo-scientific creationist claims and is an excellent survey of the major issues and sources of debate involved.

GOULD, STEPHEN JAY. 1989. *Wonderful Life: The Burgess Shale and the Nature of History.* New York: W. W. Norton. This fascinating work by North America's most famous biologist and paleontologist presents the most recent thinking on evolutionary processes, including punctuated equilibrium and related issues. Gould's earlier collection of essays from the magazine *Natural History,* published under such titles as *An Urchin in the Storm, Ever Since Darwin, The Panda's Thumb,* and *The Flamingo's Smile,* are also recommended for understanding the most important issues in modern evolutionary thought.

RAUP, DAVID M. 1986. *The Nemesis Affair: A Story of the Death of the Dinosaurs and the Ways of Science.* New York: W. W. Norton. The Nemesis theory proposes that "Nemesis," a companion star to our sun, circles in an orbit that brings it close to Earth every 26 million years. Theorists propose that comets generated by the passage of this "death star" caused dramatic changes in the direction of life on Earth. David Raup, a noted paleontologist, provides a stimulating and entertaining discussion of the origin of the Nemesis theory and the how and why of scientific inquiry.

SPROUL, BARBARA. 1979. *Primal Myths: Creating the World.* New York: Harper & Row. A superb collection of origin myths found throughout the world. It demonstrates the range of the human creative imagination in constructing cosmologies.

Chapter 4

THE PRIMATES

MEMBERS OF THE MAMMALIAN ORDER **primates** are a diverse group of animals, including humans, monkeys, prosimians, and apes. They share certain characteristics such as large brain size, keen vision, dexterous hands, and a generalized skeleton that allows for great physical agility. Primates also tend to have smaller litters than other animals, devoting more care and attention to the rearing of their offspring. Certain traits are prominent in some primates and hardly evident in others. Similar features can be found in many nonprimates. For example, the lion has very efficient eyesight, and the tree squirrel is exceedingly agile. However, the unique *combination* of traits found in the primates distinguishes them from other animals.

We can trace the striking similarities among primates to a series of shared evolutionary relationships. Many people hold a common misconception about human evolution—the mistaken belief that humans descended from modern apes, such as the gorilla and chimpanzee. This is a highly inaccurate interpretation of both Charles Darwin's thesis and contemporary scientific theories of human evolution, which suggest that millions of years ago some animals developed certain characteristics through evolutionary processes that made them precursors of later primates, including humans. Darwin posited that humans share a common ancestor (now extinct) with living apes but evolved along lines completely different from modern gorillas and chimpanzees.

Primate paleontologists and paleoanthropologists study fossil bones and teeth of early primates to trace lines of human evolution dating back millions of years. Meanwhile, *primatologists* concentrate on living, nonhuman primates, working to discover subtle similarities and differences among these creatures. As researchers weave together the fossil record and conclusions from primatological observations, they discern the outlines of a vivid tapestry of human evolution—how it occurred and what makes humans physically and behaviorally distinct from other primate species.

PRIMATE CHARACTERISTICS

Primate evolution has produced some key physical and anatomical traits that represent adaptations to **arboreal** conditions—that is, life in trees (LeGros Clark, 1962; Richard, 1985). Among the most important physical characteristics of primates is their generalized skeletal structure, which allows for a great deal of flexibility in movement. Consider, for example, the *clavicle,* or collarbone, a feature found in early mammals. This skeletal element has been lost

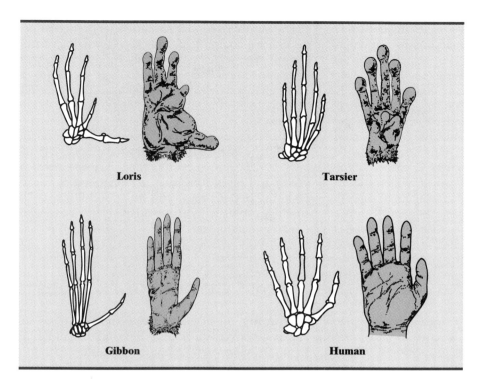

FIGURE 4.1 Examples of primate hands. Although they vary in form, they share a high degree of manual dexterity.

Source: Adapted from Fig. 2.20, p. 31 in *Primate Adaptation and Evolution* by John G. Fleagle. Copyright ©1988 by Academic Press, Inc. All rights reserved. Reprinted by permission of Academic Press and John G. Fleagle.

in faster, more specialized land animals, such as the dog and the horse, which have more rigid skeletons. In primates, the clavicle provides both support and flexibility, enabling them to rotate their shoulders and arms to perform a range of movements. In the wild, this skeletal structure gives primates the ability to reach for branches and food while roaming through treetops. Humans, of course, do not live in trees. However, their generalized primate skeleton greatly enhances their ability to drive cars, catch baseballs, and throw spears.

Dexterity in the digits (fingers and toes) of the feet and hands, another key primate trait, makes it easy for primates to grasp and manipulate objects. All primates (except for the callitrichids) have sensitive pads on their fingertips rather than claws, and many have five digits on their hands and feet, which can be used for grasping objects. Unlike cats or rodents, with claws, primates climb by wrapping their hands and feet around branches. An important distinguishing element of the primate hand is the **opposable thumb,** found in humans and many other primates (Figure 4.1). Humans can touch the tips of each of their fingers with the thumb, an ability that

makes them adept at manipulating small objects. Some primates do not have opposable thumbs, but all members of the Primate order share a high degree of digit mobility.

DENTITION, EYESIGHT, AND BRAIN SIZE

Dentition, the number, form, and arrangement of teeth, serves as a distinguishing characteristic of many types of animals. Because they are strong and are often better preserved in the fossil record than other parts of the skeleton, teeth are particularly valuable evidence for paleontologists, who use them to identify extinct primates. Compared to other mammals, primates have multipurpose teeth that can be used for either cutting or crushing foods. The dental structure of primates is consistent with an **omnivorous** diet, made up of a variety of foods, from plants, fruits, nuts, and seeds to insects and other animals.

Different teeth perform different functions. The anterior (front) teeth, including the incisors and canines, are used to transfer food into the mouth. They are designed for cutting, tearing, and biting. Their shape varies from one primate species to another as a result of evolutionary adaptations to particular food-processing situations. The posterior (back) teeth, or molars, are less specialized. They simply break down food by crushing or grinding.

Refined vision also sets primates apart. Whereas many animals are highly dependent on *olfaction,* the sense of smell, primates, having large eyes oriented to the front and protected by bony structures, rely heavily on vision. This visual orientation favors *binocular* and *stereoscopic vision,* in which the visual fields of the eyes overlap, transmitting images to both sides of the brain. Primates benefit from enhanced depth perception as a result. Evolution has also made the retina of primates' eyes sensitive to different wavelengths of light, producing color vision in most primates. Primates depend on their highly developed visual sense to identify food and coordinate grasping and leaping.

Distinguishing primates even more are the size and complexity of the brain. Relative to body size, primates have larger brains than any land animal; only the brains of marine mammals are of comparable relative size (Fleagle, 1988). In primates, the neocortex, the surface portion of the brain associated with sensory messages and voluntary control of movement, features a large number of *convolutions,* or folds, which maximize the surface area. As they evolved, these larger brains undoubtedly helped primates to locate and extract food and to avoid predators (Jolly, 1985).

REPRODUCTION AND MATURATION

In contrast to most other animals, primates reproduce few offspring, and these undergo long periods of growth and development. The **gestation period** for primates—that is, the length of time the young spend in the mother's womb—is longer than that of nonprimate animals of comparable size. Primate offspring are born helpless and unable to survive on their own. For this reason, they undergo long periods of maturation, during which they remain highly dependent on their parents and other adults. As an example, a kitten reaches full adulthood at one year, whereas a baboon takes seven or eight years to reach maturity. Humans, in contrast, have a period of infancy that lasts six years. Full adulthood, characterized by the appearance of the third molars, or "wisdom teeth," is reached at 20 years of age.

During this protracted maturation process, primates learn complex adaptive tasks from other members of their species—primarily by observing their mothers, fathers, and others in their social group. Through this *social learning,* primates gain the skills needed to locate food and shelter and to elude predators.

All these evolutionary trends can be traced through the lines of primates from the Eocene to the present day. Enhanced locomotion, refinements in vision and brain functions, an extended period of offspring dependency, and increased life span all combined to make primates adaptable to a variety of environmental niches.

CLASSIFICATION OF PRIMATES

Taxonomy, the science of classification, gives scientists a convenient way of referring to, and comparing, living and extinct organisms. Modern taxonomy is based on the work of the Swedish scientist Carl von Linnace, also known as Carolus Linnaeus (1707–1778). Linnaeus created a system of Latin names to categorize plants and animals based on their similarities. The Linnaean system follows a hierarchical pattern (see Table 4.1), ranging from large categories such as *kingdoms,* which encompass creatures sharing overarching characteristics, to small groups, or *species,* whose members can all potentially interbreed.

Human beings belong to the kingdom Animalia, one of the several major divisions of nature. Members of this kingdom are mobile, complex organisms that sustain themselves by eating plants and other animals. Other categories, or *taxa,* are based on more specific criteria (see Table 4.1). For example, humans have backbones, a feature that places them in the subphylum Vertebrata; presence of body hair and mammary glands further identifies them as members of the class Mammalia (mammals). Classes are subdivided into a number of orders. Humans belong to the order Primates. Like other mammals, primates are warm-blooded animals that possess hair for insulation and nourish their young with milk from mammary glands. However, primates' refined visual sense, manual dexterity, distinctive skeletal structure, and large brain size differentiate them from other mammals.

TABLE 4.1 Classification Relevant to Human Ancestry

The classification of living organisms is hierarchical. Membership in a kingdom is determined by very basic characteristics. Classification into other categories is based on increasingly specific criteria. The words *primate, anthropoid, hominoid,* and *hominid* are used to refer to members of the categories Primates, Anthropoidea, Hominoidea, and Hominidea. Superfamily names generally end in *oidea,* family names in *idae,* and subfamily names in *inae.*

Category	Taxon	Common Description
Kingdom	Animalia	Animals
Phylum	Chordata	Animals with notochords
Subphylum	Vertebrata	Animals with backbones
Superclass	Tetrapoda	Four-footed vertebrates
Class	Mammalia	Vertebrates with body hair and mammary glands
Order	Primates	All prosimians, monkeys, apes, and humans
Suborder	Anthropoidea	All monkeys, apes, and humans
Infraorder	Catarrhini	Old World anthropoids
Superfamily	Hominoidea	Apes and humans
Family	Hominidea	Bipedal apes
Genus	*Homo*	Humans and their immediate ancestors
Species	*Homo sapiens*	Modern human species

Source: From *Culture, People, Nature* 5th Ed., by Marvin Harris. Copyright ©1988 by Harper & Row, Publishers, Inc. Reprinted by permission of Addison-Wesley Educational Publishers, Inc.

Although all primates share certain basic characteristics, we do see a great deal of variation among species. There is some disagreement among primatologists about how particular species are related to one another and how the order should be divided. When Linnaeus developed his classification system to facilitate the comparison of organisms, Darwin had not yet introduced his theory of evolution. However, after Darwin's publication of *On The Origin of Species* in 1859, biologists increasingly applied theories of evolution to systems of classification, giving rise to a number of scientific disputes. Some scientists focus on physical similarities among species, traits that most likely emerged as evolutionary adaptations to specific environments. Other scientists stress actual genetic links.

How do scientists distinguish between genetic links and adaptive characteristics? Consider the classification of humans, chimpanzees, gorillas, and orangutans. On the one hand, African apes and orangutans, which have certain physical traits in common, have traditionally been placed in their own family, Pongidae. Humans, on the other hand, followed another evolutionary line entirely, making them distinct in appearance. In contrast to other primates, humans have developed a complex culture—material objects and nonmaterial concepts—that they use to interact with the environment. They have a much larger brain than the other primates and lack the others' thick covering of body hair. For this reason, scientists place humans and their immediate ancestors in the family Hominidea. However, careful study of ape and human anatomy and molecular studies of genetic material indicate that in actuality humans and the African apes are more closely related than either group is to the orangutans (Andrews & Martin, 1987). Today, in classifying species, many scientists try to bear in mind both the genetic relatedness of species and the characteristics produced by specialized adaptations.

PRIMATE SUBDIVISIONS

The order Primates is divided into two suborders: Prosimii, or prosimians, and Anthropoidea, or anthropoids. The prosimians include modern lemurs, lorises, and tarsiers, all of which are found exclusively in Asia and Africa. The modern anthropoids, comprising all monkeys, apes, and humans, can be separated into two smaller divisions, or infraorders: *Platyrrhini,* referring to all monkeys found in the Americas, and *Catarrhini,* or anthropoids found in Europe, Asia, and Africa. The catarrhines are subdivided into the superfamilies *Cercopithecoidea,* monkeys, and *Hominoidea,* including apes and humans. Paleoanthropologists take a particular interest in members of the superfamily *Hominoidea* because these are the primates most closely related to humans. The **hominoids** fall into three families: *Hylobatidae* (lesser apes), *Pongidae* (great apes), and *Hominidae* (hominids).

CLASSIFICATION OF FOSSIL PRIMATES

Biologists and primatologists sometimes disagree about the classification of living organisms. However, paleontologists and paleoanthropologists face even more daunting challenges in their attempts to classify extinct species because they must base their conclusions solely on characteristics preserved in the fossil record, such as skeletal structure and dentition. Natural variations within populations, the vagaries of climatic and geological changes, and the ravages of time all make identifying species on this basis exceedingly difficult.

In the past, scientists have placed great emphasis on the skull, or *cranium,* in distinguishing fossil primates from one another. The cranium provides clues to an extinct animal's vision, diet, cognitive abilities, and posture. However, as increasing numbers of fossils have been recovered, paleontologists are now looking more closely at the postcranial skeleton, all the bones of the body excluding the skull (Strasser & Dagosto, 1988). By examining the *postcranial skeleton,* we can determine a great deal about a primate's posture and locomotion, which in turn tell us a great deal about the animal's adaptations to disparate environments. As we bring together all this information, we can discern much more clearly how fossil primates looked and functioned in their environments.

The fossil record for certain periods is sketchy, at best. For example, we know that during the period spanning 4 million to 8 million years ago, there were important changes in primate adaptations. Unfortunately, fossils from that age are not well preserved. Because of the fragmentary nature of the fossil

record, we may never know precisely how primate evolution proceeded.

THE EVOLUTION OF THE PRIMATE ORDER

The 200 species of modern primates represent the products of millions of years of adaptation. Primates evolved during the Tertiary and Quaternary periods in the Cenozoic era (see Table 3.1, pages 46–47). The Tertiary period is subdivided into five successive epochs: Paleocene, Eocene, Oligocene, Miocene, and Pliocene. The Quaternary period comprises the Pleistocene and Holocene epochs.

Scientists speculate that the first mammals related to the primates appeared during the Paleocene, approximately 65 million years ago (Figure 4.2). Although not true primates, these early mammals resembled primates in some respects, particularly in the shape and arrangement of their teeth. The oldest recognizable primates emerged in the Eocene, bearing some resemblance to modern primates, yet identical to no species alive today.

Because many living primates inhabit environments similar to those occupied by extinct species, the study of these modern-day primates casts light on the adaptive characteristics of earlier primate species. In some ways, modern nonhuman primates embody the evolutionary phases through which the primates passed over millions of years. According to the fossil record, the prosimians developed as the earliest forms of primates, followed by the anthropoids.

PRIMATE ORIGINS

The ancestors of modern prosimians and anthropoids evolved during the late Cretaceous and early Tertiary epochs (70 million to 55 million years ago). These creatures resembled modern squirrels or tree shrews in size and appearance and were well adapted to an arboreal lifestyle. Fossil remains of these ancestral primates have been unearthed in Asia, Europe, Africa, and America. One of the earliest primate ancestors, a tiny animal, *Purgatorius,* has been described on the basis of finds at a number of sites in North America (Clemens, 1974). Paleontologists who have found *Purgatorius* jaw fragments and many isolated teeth conjecture that this creature subsisted primarily on a

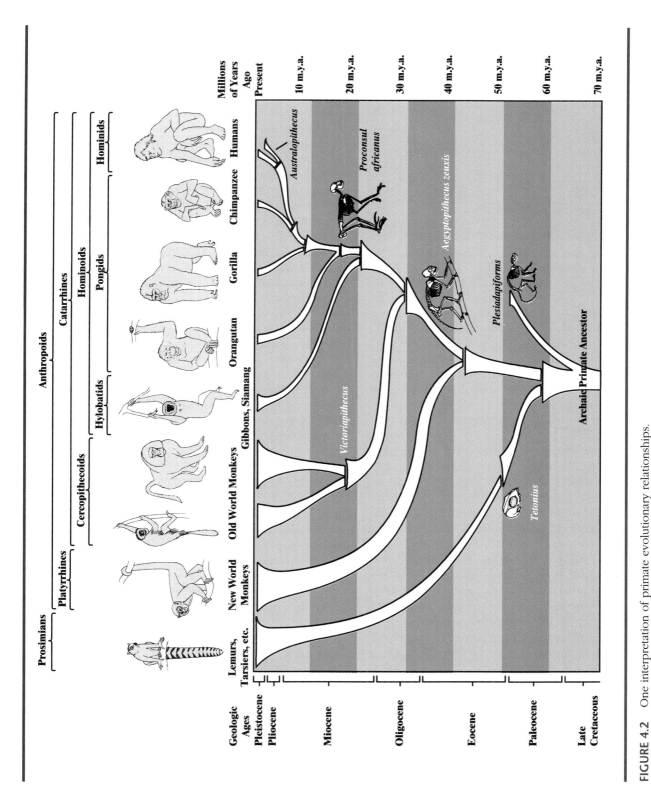

FIGURE 4.2 One interpretation of primate evolutionary relationships.

Source: From *Biology* by Curtis and Barnes, ©1968, 1975, 1979, 1983, 1989 by Worth Publishers. Used with permission.

Primates most likely evolved from creatures similar in appearance to modern tree shrews.

diet of insects. The dentition of *Purgatorius* barely resembles that of later primates, but its primatelike molars indicate that this diminutive creature had a more omnivorous diet than that of other insectivores. These molars distinguish *Purgatorius* from other early mammals and suggest an ancestral relationship between this creature and more recent primates.

Although the remains of *Purgatorius* have been uncovered only in North America, we cannot assume that primates first evolved there. It's important to note that North America was still attached to Europe when *Purgatorius* thrived, and primates could have moved easily between the continents. The discovery of *Purgatorius* on this continent may reflect the limited number of finds from this period in other areas. The earliest primates may have evolved in other parts of the world, including Africa and Asia. The discovery of 60-million-year-old fossil teeth near the Atlas mountains of Morocco lends support to the theory that early primates emerged in Africa (Gingerich, 1990). These teeth define a previously unknown species, *Altialasius koulchii*, tiny creatures probably weighing no more than 3.5 ounces. Given the fragmentary nature of the finds, the precise relationship of this species to primates is uncertain. These creatures may be precursors of anthropoid primates, or they may represent a side branch, such as the plesiadapiforms, which we examine next.

PLESIADAPIFORMS Many primatelike animals appeared during the middle Paleocene, and a number of different genera—collectively referred to as *plesiadapiforms*—were among the most successful. Sometimes classified as a distinct suborder of Primates, plesiadapiforms have no modern representatives. Plesiadapiform fossils have been unearthed in Europe and North America; the genus *Plesiadapis*, perhaps the best known, was described on the basis of fossil finds in the western United States and France (Gingerich, 1986). Plesiadapis had a long snout and a squat, stocky body, similar to the modern groundhog. A delicate impression in limestone indicates that these creatures had long, bushy tails.

Certain aspects of plesiadapiform dentition distinguish these creatures from earlier animals. These basic features, which are retained as *primitive characteristics* in later primates, suggest that the plesiadapiforms are related to later primates. However, the plesiadapiforms also exhibit selected adaptations to particular environmental niches. These distinctive adaptations, or *derived characteristics,* include specialized teeth, cranial structure, and limb bones unlike those of any later primates. In contrast to *Purgatorius,* the plesiadapiforms' specialized, derived nature indicates that they are not precursors of any later primate groups. Of the early primatelike creatures, only *Purgatorius* has enough of an unspecialized nature to be considered ancestral to modern primates.

For 30 million to 35 million years, these early creatures thrived in the forest areas of the world. Presumably, they fed on insects, seeds, bird eggs, and other small life forms. At the end of the Eocene, many of these early animals became extinct for reasons that are unclear. Some scientists speculate that changing climatic conditions modified the animals' environments, making survival impossible. Others conjecture that these creatures could not compete effectively with more successful animals such as rodents and prosimian primates.

FOSSIL PROSIMIANS

The prosimians, which appeared at the beginning of the Eocene, were the first true primates, and, like the earlier plesiadapiforms, early prosimian primates flourished. Researchers have located the most complete prosimian fossil finds in North America and Europe, but more fragmentary examples of prosimian fossils have been discovered in Asia and Africa (Fleagle, 1988).

In examining the cranial structure of the Eocene prosimians, scientists have found striking evidence to indicate that these animals relied much more on vision than on their sense of smell. Consider the *endocast*—a cast of the brain—of *Tetonius,* a 50-million-year-old prosimian found in Wyoming (Radinsky, 1967). Although the olfactory portions of the brain are small compared to earlier creatures, the occipital and temporal lobes—the sections of the cerebrum associated with vision—are relatively large. From its large eyes, researchers deduce that *Tetonius* was **nocturnal**—that is, it searched for food at night when other animals were sleeping. Present-day (extant) nocturnal animals have large eyes to take in the greater amounts of light needed for night vision. A nocturnal orientation can be inferred in extinct animals whose crania reveal large orbits (spaces in the skull for eyes).

Scientists have unearthed few postcranial skeletons of the early prosimians, but the available finds suggest that these creatures were evolving toward the more generalized skeletal structure that characterizes later primates. With slender limbs and modified hands and feet, these primates most likely had some of the locomotor and grasping abilities of modern species. And in some species, nails were replacing claws (Dagosto, 1988). These characteristics prefigure modern families of prosimians.

Interestingly, some fossil prosimians also exhibit features that resemble the most primitive characteristics in the anthropoid primates, such as the orbits of the eye and the structure of foot and leg bones. Paleoanthropologist Elwyn Simons (1972) has described these early prosimians as "the first primates of modern aspect," signaling their important place in primate evolution.

MODERN PROSIMIANS

Modern prosimians have changed relatively little from their ancestors and reside in small, isolated populations. They include the lemurs and indris of Madagascar (an island off the east coast of Africa), the lorises of tropical Southeast Asia and Africa, and the tarsiers of Southeast Asia.

A heightened visual sense makes most of the living prosimians nocturnal, although they do not have color vision. Like their fellow primates, however, they possess stereoscopic vision and enlarged brains relative to other animals, which help them to coordinate leaping and food gathering in their arboreal environment. A keen sense of smell, more highly developed than that of the anthropoid primates, helps them to seek out food and shelter at night.

Like other primates, the prosimians use the five dexterous digits on their hands and feet to grasp objects and to move easily through trees. Some prosimii, such as the indris, move by vertical clinging and leaping from branch to branch in an upright position, springing with their hind legs and landing with their arms and legs. All of the living prosimians have nails instead of claws on their digits, but some have retained grooming claws on their hind feet, which are used to clean their fur. Because of the extensive destruction of the tropical rain forests in which the prosimians live, all are endangered animals.

EVOLUTION OF THE ANTHROPOIDS

As many of the early prosimians gradually became extinct at the end of the Eocene epoch, new types of primates emerged in tropical forest environments. Among these new primates were creatures ancestral to modern anthropoids, including humans. Scientists divide the anthropoids into three groups based on their distinct ecological niches (Fleagle, 1988). The three categories are the platyrrhines (all monkeys from the Americas, or New World monkeys) and the two superfamilies of catarrhines (monkeys, or cercopithecoids, from Europe, Asia, and Africa, also called Old World monkeys, and the hominoids, including apes and humans).

The divergent evolution of the higher primates is closely tied to plate tectonics and continental drift, examined in Chapter 3. During the late Cretaceous period, approximately 65 million years ago, North America and Europe were joined, allowing free migration of primates between what later became the European, African, and Asian continents and the Americas. About 40 million years ago, the Americas separated from Europe, Asia, and Africa, ending contact between primates from the Eastern and Western Hemispheres. This geological development resulted in disparate lines of primate evolution. Platyrrhines evolved in southern Mexico and Central and South America, and cercopithecoids developed in Africa and Asia. We shall first examine the monkeys in the Americas and then delve into the evolution of monkeys and hominoids in Europe, Asia, and Africa.

EVOLUTION OF THE PLATYRRHINES The fossil record for the evolution of the monkeys in the Americas (platyrrhines) is sparse and consists only of fragmentary fossil evidence from Bolivia, Colombia, Argentina, and Jamaica. Indeed, all of the primate fossils found in South America would fill no more than a shoe box (Fleagle, 1988). Paleontologists, therefore, have not been able to reconstruct a detailed account of how anthropoid evolution proceeded in the Americas. Current interpretations favor the theory that the platyrrhines evolved from African anthropoid primates, a theory consistent with the limited fossil evidence and shared similarities in dentition and biochemical characteristics.

Although African origins for the platyrrhines present the most likely interpretation of the available fossil evidence, scientists remain puzzled about how ancestral anthropoids may have arrived in South America from Africa. Uncertainty surrounds much of our understanding of continental drift during this period, making it difficult to date precisely when various landmasses separated. South America is believed to have been an island continent during most of the early Cenozoic era.

Although periods of low sea levels during the Oligocene (38 million to 25 million years ago) may have exposed landmasses and created areas of relatively shallow water in the South Atlantic, movement between the continents would still have involved crossing open water. Transportation over large expanses of water on floating masses of vegetation has been used to explain the movement of some species; and, in fact, ocean currents would have been favorable to an African–South American crossing. Yet, it has been questioned whether this method of dispersal is consistent with primate dietary and climatic requirements.

North and Central America have also been suggested as the home of the ancestral platyrrhines, but there is no fossil evidence to bolster such an interpretation (Fleagle, 1988). Paleogeographical reconstructions would also seem to confound this theory, as they suggest that during the Oligocene the distance between South America and landmasses to the north was even greater than that between South America and Africa. However, the geological history of the Caribbean is poorly known. Another candi-

A spider monkey hanging from its prehensile tail. Primates with prehensile tails are found only in the Americas.

date for the platyrrhine origins is Antarctica. Though it was joined to South America during the appropriate time period, the absence of any information on the evolution of mammalian species on Antarctica makes this scenario impossible to evaluate.

MODERN MONKEYS OF THE AMERICAS The monkeys of the Americas encompass fifty-two different species—including the marmosets, the tamarins, the sakis, and the squirrel, howler, spider, and wooly monkeys—having a tremendous range in physical appearance and adaptations (Jolly, 1985). One feature that distinguishes anthropoids from the Americas from those from Europe, Asia, or Africa is the shape of the nose. The former monkeys have broad, widely flaring noses with nostrils facing outward. The latter anthropoids have narrow noses with the nostrils facing downward.

Monkeys from the Americas spend their days in trees, coming to the ground only to move from one tree to another. Their elongated limbs are ideal for grasping tree branches. As **quadrupeds,** they use all four limbs for locomotion. In addition, many of these monkeys have developed a unique grasping, or *prehensile,* tail. Prehensile tails serve as a fifth limb, enabling some platyrrhines to hang from branches and feed with their other limbs. This unusual tail also gives the monkeys greater coordination and balance as they move through trees. Most monkeys from the Americas eat a varied, omnivorous diet that includes fruits, insects, and small animals.

EVOLUTION OF THE CATARRHINES Compared to the meager fossil findings of monkeys from the Americ-

as, paleontologists have unearthed extensive anthropoids from Europe, Asia, and Africa. Specimens have been recovered from all over these continents, including many regions that primates no longer inhabit. Among the most significant fossil localities are those in the Fayum Depression in Egypt, an extremely arid region rich in fossil-bearing strata dating from 45 million to 31 million years ago (see Figure 4.3). Anthropoid fossils abound in the upper part of the formation, dating back about 37 million years ago. During this period, the Fayum was a lush tropical forest, ideal for primates; and an amazing variety of plant and animal fossils have been unearthed in this once-fertile region (Simons & Rasmussen, 1990).

Parapithecids Among the most interesting of the early anthropoid primates are representatives of the

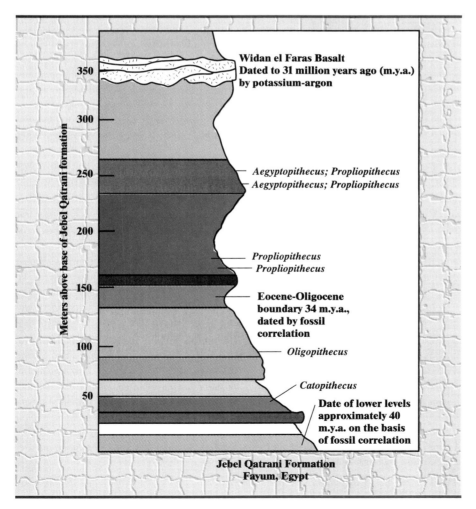

FIGURE 4.3 A stratigraphic profile of the fossil-bearing deposits in the Fayum Depression in Egypt.

family Parapithecidae. However, scientists have not determined the precise relationship of these animals to later primates. Skeletal and dental structures of certain parapithecids resemble those of platyrrhines. Discovering these features in African fossils, researchers conjectured that parapithecids may have emerged sometime near the divergence of the anthropoids from the Americas and those from Europe, Asia, and Africa. However, the parapithecids exhibit many specialized, derived dental traits that suggest evolutionary adaptations leading away from the lines of later anthropoids. One theory holds that these early anthropoid primates may represent a distinct branch early in the anthropoid line (Fleagle, 1988; Fleagle & Kay, 1987).

Apidium, one particularly well-known parapithecid identified in hundreds of fossil finds, was a small, quadrupedal primate with a short snout. It had thirty-six teeth, the same number found in modern monkeys from the Americas. Its comparatively small *orbits* indicate that *Apidium* was **diurnal,** that is, active during the day. Postcranial bones, such as the long hind legs and flexible ankles, lead researchers to believe that *Apidium* was an effective leaper.

CERCOPITHECOIDS   In tracing the earliest potential ancestors of monkeys (cercopithecoids) of Europe, Asia, and Africa, paleontologists cite the Miocene deposits in northern and eastern Africa, pegged at 26 million to 5 million years old. *Victoriapithecus,* one of these fossil monkeys, thrived near Lake Victoria in Kenya (Fleagle, 1988). A small to medium-sized monkey, weighing between 10 and 55 pounds, Victoriapithecus was quadrupedal and appears to have been equally well suited for an arboreal or **terrestrial** (ground-dwelling) life. Such adaptable capabilities would have been useful in the middle Miocene environment of Kenya, which scientists believe was an open woodland.

Victoriapithecus and related genera had a number of characteristics of modern cercopithecoid species. For example, the chewing surfaces of *Victoriapithecus* molars had distinctive patterns of cusps and ridges that are found in extant species. In addition, males and females had different-sized canine teeth, a feature *Victoriapithecus* had in common with all monkeys of Europe, Asia, and Africa. However, some of the more derived features found in later monkeys were absent from *Victoriapithecus.* Thus, these primates cannot be placed in any modern subfamily.

Rather, because they represent a transitional form between earlier catarrhines and modern monkeys, they make up their own subfamily of extinct primitive monkeys, the Victoriapithecinae.

Remains of many other species of monkeys have been found in fossil beds of Pliocene and Pleistocene age (6 million years of age up to the recent past). The majority of these extinct species can be conveniently organized into the same subfamilies as living monkeys, providing a neat link between *Victoriapithecus* and the present.

MODERN MONKEYS OF EUROPE, ASIA, AND AFRICA
Modern-day cercopithecoids encompass an extremely diverse group, consisting of seventy-eight different species scattered throughout sub-Saharan Africa, southern Asia, and northern Japan. Next to humans, they are the most widely distributed primate group. These monkeys include different species of macaques, langurs, savanna and hamadryas baboons, geladas, colobus monkeys, and proboscis monkeys. Like the anthropoids of the Americas, these monkeys are quadrupedal. Although primarily arboreal, some, such as the savanna baboons, are terrestrial. Unlike their American counterparts, these monkeys do not have prehensile tails; however, some do use their tails as aids in communication and balance.

These monkeys break down into two subfamilies: Cercopithecinea and Colobinea. Most of the cercopithecines live in Africa, but some, such as the macaques, are also found in Asia. The colobines include the langurs of Asia and the African colobus monkeys. The two subfamilies differ most with respect to nutrition. The cercopithecines have a slightly more varied diet, feeding on many types of vegetation and fruit. Their cheek pouches act like storage bins: These pouches can be filled with food intended to be eaten in another place. In contrast, the colobines subsist primarily on leaves; often referred to as "leaf-eating monkeys," they have specialized teeth, stomachs, and intestines that enable them to digest the woody, cellulose parts of plants. Compared to the cercopithecines, the colobines are less mobile, often residing in the trees where they feed.

Pronounced **sexual dimorphism,** in which males and females of the same species exhibit separate characteristics, distinguishes many cercopithecine monkeys. The male baboon, for example, has a much larger body and larger canine teeth than the female.

These differences arise as a result of rivalry among males for females. Species in which competition for sexual mates is limited exhibit little sexual dimorphism, whereas species characterized by intense competition display the most dramatic dimorphism (Kay et al., 1988). Other features, including relative differences in the size of the testes and sexual swelling in females during ovulation, also correlate with mating rivalry and patterns of social organization.

EMERGENCE OF THE HOMINOIDS

The Fayum region in Egypt has revealed the most intriguing and significant information on the early evolution of the apes (see the box, "Interpreting the Fayum Fossils"). Consider representatives of the genus *Aegyptopithecus*, discovered there and dating from approximately 33 million years ago. The fossil evidence for *Aegyptopithecus* includes a skull, numerous jaw fragments, and several limb bones. *Aegyptopithecus* is believed to have been comparable in size to a modern howler monkey, which weighs only 13 to 18 pounds (Fleagle, 1983). The postcranial bones reflect the skeletal structure of an arboreal quadruped.

The dentition of *Aegyptopithecus* offers several useful insights. Dental structure, for example, indicates that this early ape subsisted on a diet of fruit and leaves. Researchers have also noted a great deal of variation in the size of canine teeth, which may indicate that the species was sexually dimorphic. Building on this hypothesis, scientists make the theoretical leap that *Aegyptopithecus* lived in social groups in which competition over females was intense.

Aegyptopithecus resembles primitive monkeys and prosimians in several key respects, including its small brain and diminutive skeletal structure. However, other features, such as its thirty-two teeth—the same number found in humans and apes—suggest that *Aegyptopithecus* may represent an ancestor of later hominoids.

HOMINOID EVOLUTION The fossil record sheds little light on the period spanning the time when *Aegyptopithecus* flourished through the early Miocene, some 10 million years later. However, we can state with certainty that the Miocene (24 million to 5 million years ago) brought apes to the fore. According to the fossil evidence, the earliest forms of protoapes evolved in Africa before 18 million years ago. After

Skull of Aegyptopithecus zeuxis, *an Oligocene anthropoid, excavated from the Fayum Depression in Egypt.*
Source: Courtesy of D. Tab Rasmussen.

that, Africa was connected to Europe and Asia through the Arabian peninsula, enabling hominoids to migrate to these other continents. The fossil evidence indicates that the late Miocene apes made remarkable adaptations to all sorts of geographic and climatic conditions in Europe, the Middle East, Asia, and Africa.

A study of various Miocene fossil species spotlights the intermediate stages through which modern hominoids passed. However, determining the exact lineages leading to specific living species is complicated.

Scientists attempting to classify Miocene apes are hampered by two problems: the vast number of species and fragmentary fossil evidence. They categorize these apes primarily on the basis of their teeth, which prompted some imaginative researchers to dub them the "dental apes." This method of identification has one major weakness, however: Convergent evolution in similar environmental settings (say, the tropical forests of present-day Africa and the tropical

CRITICAL PERSPECTIVES

INTERPRETING THE FAYUM FOSSILS

Popular magazines and museum displays tend to arrange information into neat packages. Sophisticated illustrations present the viewer with a concise graphic summary of millions of years of human or primate evolution. Yet these sharply drawn illustrations and explanations belie the tenuous nature of many interpretations of the age, environment, and relationship among different fossil species. Consider the challenges faced by primate paleontologists conducting research in the Fayum Depression, located 100 miles southwest of Cairo, Egypt, along the northwestern margin of the Sahara Desert. Here researchers have made some of the most exciting discoveries of early primate fossils, gaining insights into the ancient environment of the region and piecing together pictures of extinct primates that were ancestral to modern monkeys, apes, and humans.

Paleontologists confront an apparent contradiction in tracing the roots of modern primates: The living catarrhines thrive in the tropical forests of Africa and Southeast Asia, yet fossil specimens come primarily from desert regions like the Fayum. This disparity stems from the incomplete nature of the fossil record. Fossil strata are not exposed in tropical forest regions, and the lush vegetation makes the identification of sites difficult. In contrast, the Jebel Qatrani Forma-

tion at the Fayum constitutes a geological "layer cake," hundreds of feet thick, that spans the Eocene and Oligocene epochs (see Figure 4.3). Over millions of years of wind and water erosion, these early deposits have been exposed for scientists to examine. Larger fossils and resistant pieces of stone lie in scattered gravel patches called "desert pavement"; lighter fossils have disappeared, carried away by desert winds.

Uniquely accessible and fossil-rich, the Fayum has attracted researchers for more than a century, most prominently Elwyn Simons of Duke University, who has spent thirty years piecing together the region's past (Simons & Rasmussen, 1990). Like most paleontologists, scientists working in the Fayum sometimes expose fossils by carefully digging away hard, overlying sediments with hammers, chisels, and brushes. However, nature has been enlisted as an ally in the Fayum excavations as well. To expose new fossil-bearing deposits, researchers sweep away hardened layers of desert pavement, revealing unweathered layers of sediment. Within a year the harsh winds may scour away as much as six inches of loose, underlying sand, uncovering new fossils. This technique works especially well when it comes to more delicate fossil remains.

In dating the Fayum fossils, various methods have come into play.

Through faunal correlation with dated fossils from other areas, the earliest marine sediments at the Fayum can be traced to the middle Eocene (less than 45 million years ago), whereas the earliest primates in the region most likely emerged around 37 million years ago in the late Eocene. The fossil-bearing deposits are capped by layers of basalt, a rock of volcanic origin that dates from 31 million years ago, as determined by potassium-argon dating (Fleagle et al., 1986). Scientists have assessed the relative ages of fossils in the deposit by carefully plotting their stratigraphic position in the formation.

Dry, bleak, and inhospitable today, the Fayum underwent major environmental changes over millions of years. Originally covered by the ancient Mediterranean Sea—the earliest layers in the Fayum are marine sediments—the Fayum experienced a gradual geological shift to a terrestrial environment. Fossils recovered from the terrestrial deposits of the Eocene and Oligocene number in the tens of thousands, including a wide array of extinct rodents, water birds, bats, snakes, and turtles (Simons & Rasmussen, 1990). All of the Fayum finds clearly indicate that the late Eocene and Oligocene environment was wet and tropical. The bones of paleomastodons, the oldest known elephants, have also been unearthed at the Fayum, as well as insects, tropical plants, and fossilized trees of various kinds. In fact, new species are constantly being discovered. Here we can see the importance of long-term research at individual fossil sites.

The ample supplies of fresh water and food in the Oligocene tropical forest of the Fayum provided an ideal environment for early primates. Almost a thousand primate fossils have been found there, representing eleven species (Rasmussen & Simons, 1988; Simons, 1989a, 1990; Simons & Rasmussen, 1990). Among the most interesting of these finds are the remains of *Aegyptopithecus* (Simons, 1984). A number of other early hominoids have also been identified, including *Catopithecus, Oligopithecus,* and *Propliopithecus.* How are all these different species related? Undoubtedly, some of these creatures contributed to the evolution of modern anthropoids. Yet determining how evolution actually proceeded—which forms of life preceded others and how and when successful adaptations emerged—remains a daunting task.

To establish evolutionary relationships, scientists scrutinize fossil remains for key identifying features. Basic characteristics shared by all members of a family, such as similarities in dentition and the structure of the skull, are called *primitive characteristics,* and they are helpful in grouping the animals together as early anthropoids. However, because the same

Egypt's Fayum Depression is desert today, but fossils of fruit, nuts, and vines indicate that a tropical forest covered that region 33 million years ago.

primitive characters are found in all the early anthropoids, they are of no help in determining the relationships within the group. To track down these relationships, scientists examine *derived characteristics,* the more distinct or specialized traits in individual fossil species that evolved as adaptations to particular environments. More closely related animals will have more of these characteristics in common.

Primate paleontologists cite the physical features of some of the Fayum anthropoids as evidence that these creatures are part of the same lineage—that is, a group of animals sharing an evolutionary

relationship that makes up a branch in the order Primates. Animals with a common lineage can be arranged in a ladderlike progression. For example, *Catopithecus* may have given rise to *Oligopithecus,* which eventually evolved into the characteristics seen in *Aegyptopithecus.* Some species of *Propliopithecus* may represent either intermediary stages in this progression or members of a related lineage.

Even though work at a fossil locality as fruitful as the Fayum Depression may raise more questions for paleontologists than it answers, little by little the fossil finds help fill in gaps about primate and human evolution. And as these finds flesh out the picture of primate evolution, they shed light on how primates developed, how they negotiated their early environments, and perhaps how thinking, sentient beings emerged on the planet.

Points to Ponder

1. How does the fragmentary nature of the fossil evidence affect the validity of anthropological interpretations and hypotheses?

2. What types of evidence are missing from the fossil record that would contribute to our understanding of extinct species?

3. When, if ever, do you think our understanding of the evolutionary process will be "complete"?

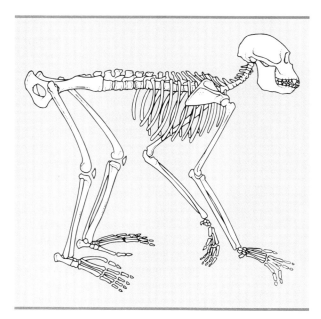

FIGURE 4.4   A reconstructed skeleton of *Proconsul africanus* based on finds from East Africa.

Source: Adapted from Roger Lewin, *Human Evolution: An Illustrated Introduction,* Blackwell Scientific Publications, 1989, p. 58. Used by permission of Blackwell Science Ltd.

forests of contemporary South Asia) may have produced nearly identical dentition in primates of markedly different ancestry. The postcranial bones of Miocene apes and protoapes reveal a number of apelike and monkeylike trait combinations dissimilar from any found in living hominoids. This leads researchers to speculate that these fossil species engaged in locomotive and behavior patterns unlike any exhibited by modern species (Fleagle, 1983).

Proconsul   The best-known early Miocene protoape, *Proconsul africanus* (Figure 4.4), has been reconstructed, offering a good illustration of the mix of features found in some of the Miocene primates. This protoape's pronounced snout, or muzzle, contrasted significantly with the more diminutive snout in the majority of later monkeys and apes. Yet the auditory region of the brain in this protoape was indistinguishable from that of living apes and monkeys from Eurasia and Africa. Other regions of the brain also mirrored those of living monkeys, but much of the sensory and mental development seen in living apes was absent (Falk, 1983). Examining the postcranial skeleton of *P. africanus,* researchers have

noted similarities to both apes and monkeys (Beard et al., 1986; Walker & Pickford, 1983). Moreover, the size and structure of the leg bones (especially the fibula) and the configuration of the foot resembled those of apes. Like living apes, *P. africanus* lacked a tail. However, the arrangements of the arm and hand bones had more in common with some monkeys than with apes. Despite these similarities, *P. africanus* lacked many of the more specialized features of modern apes or monkeys. A quadrupedal creature, adapted to an arboreal environment, *P. africanus* nonetheless lacked the swinging and leaping capabilities of apes and monkeys. Although *Proconsul* stands as an ancestor of later hominoids, the precise relationship between these primates has yet to be specified.

ANCESTORS OF MODERN HOMINOIDS   Early researchers studying the middle and late Miocene apes attempted to trace clear ancestral lines back in time to specific species. But the evolutionary status of these various species has become increasingly complicated as new finds are made. Current research casts doubt on the hope that any such neat and tidy picture will ever emerge.

Hylobatids   Modern representatives of the family Hylobatidae, including the siamang and the gibbon, are the most specialized and the smallest of the living apes. Various fossil specimens dating back as far as the Oligocene and Miocene have occasionally been classified as ancestral gibbons (Fleagle, 1988). However, these identifications are based on very superficial similarities between fossil and extant species. Looking closely at specific features, we see that these extinct creatures were far more primitive, lacking a number of refined cranial and postcranial skeletal developments found in all the living catarrhines. These discrepancies, combined with the fragmentary nature of the fossil finds, make interpretations of early hylobatid ancestry extremely tenuous (Fleagle, 1988). The earliest definitive examples of gibbon species date from the Pleistocene (less than 2 million years ago) in China and Indonesia.

The Orangutan   The only existing hominoid whose evolutionary history is comparatively well understood is the Asian member of the family Pongidae, the orangutan (genus *Pongo*). Fossil ancestors of the orangutan fall into two subgroups of Miocene hominoids, *Ramapithecus* and *Sivapithecus,* dating to about 16 million to 8 million years ago. These hominoids thrived in a variety of environments, including tropical forests as well as drier, open bushlands. Although *Sivapithecus* and *Ramapithecus* were long regarded as distinct groups, they are closely related, and today's researchers tend to combine them into the same genus.

The dentition and facial anatomy of the sivapithecines resemble those of modern apes and humans more than those of earlier hominoids. For this reason, some paleoanthropologists previously viewed these creatures as ancestors of human beings. However, because the more specific features of the lower face of some sivapithecines bear remarkable similarity to those of the modern orangutan, paleoanthropologists now agree that the sivapithecines were ancient relatives of modern orangutans, although they were not actual orangutans. Recent finds in Turkey, India, and Pakistan suggest that sivapithecines may have moved on all fours, like modern chimpanzees and gorillas, rather than spending a great deal of time climbing trees, as do orangutans.

Gigantopithecus   Some Miocene apes were enormous, including an extinct group of hominoids related to the Asian sivapithecines (Ciochon et al., 1990). Classified as *Gigantopithecus* (Figure 4.5),

these massive creatures were first identified on the basis of their teeth, which were being sold in Chinese shops as "dragon bones." Very fragmentary gigantopithecine fossils have been found in China, Vietnam, Pakistan, and northern India. The earliest gigantopithecine are from late Miocene fossil localities (dating from approximately 5 million years ago), and the most recent are from Pleistocene caves dating from less than 1 million years ago. Judging from the fossil remains, the gigantopithecines towered over other primates; no doubt, their huge teeth and jaws presented an intimidating sight. The larger of the two species, *G. blacki,* may have reached a height of 6 to 9 feet and weighed up to 600 pounds.

The dentition of *Gigantopithecus* resembles that of *Sivapithecus;* both share such features as thick enamel and low, flat cusps. These similarities suggest that the gigantopithecines descended from the earlier apes. Judging from the large teeth and thick, heavy mandibles, paleontologists theorize that *Gigantopithecus* subsisted on a diet of hard, fibrous plants such as bamboo. This hypothesis gained support recently when scientists discovered microscopic traces of plant residue called "phytoliths" on fossil teeth. These deposits may be consistent with a diet of bamboo.

The oversized features of *Gigantopithecus* have prompted speculation that this creature may be an

FIGURE 4.5 Drawing on fossils of extinct species and associated remains of other plants and animals, researchers are able to reconstruct the appearance and behavior of ancient primates, as in this illustration of *Gigantopithecus.*

Source: Adapted from Fig. 13.20, p. 386 in *Primate Adaptation and Evolution* by John G. Fleagle. Copyright ©1988 by Academic Press, Inc. All rights reserved. Reprinted with permission of Academic Press and John G. Fleagle.

ancestor of an as-yet-undiscovered Abominable Snowman residing in the frozen heights of the Himalayas. Most primatologists believe, however, that *Gigantopithecus* represents a highly specialized form of hominoid that diverged from the other lines and became extinct about 1 million years ago.

AFRICAN PONGIDS AND HOMINIDS   Like detectives who follow all sorts of trails—some fruitful, some dead ends—to reconstruct the sequence of events leading to a particular incident, paleontologists work doggedly to trace ape and human origins. Unfortunately, the trail has gone cold in the search for the Miocene and Pliocene ancestors of the African pongids and hominids. One line of Miocene apes that probably prefigures modern pongids and hominids are members of the genus *Proconsul.* For several decades two *Proconsul* species (*africanus* and *major*) were thought to be ancestral to modern chimps and gorillas (see Figure 4.4). This relationship to a modern species, however, was based on size and geographic distribution, not shared morphological characteristics (Fleagle, 1988: 391). Further studies demonstrate that *Proconsul* and other Miocene apes were far more primitive in structure than living species. Furthermore, relatively recent molecular dating studies lead experts to believe that *Proconsul* lived long before the divergence of the pongid and hominid lines (Ciochon & Corruccini, 1983). Moreover, scientists have come upon almost no hominoid fossils from the period between 8 million and 4 million years ago, which appears to have been a key time in the emergence of the hominids (see Chapter 5).

Recent discoveries indicate that the radiation of the Miocene hominoids was far more complex than previously thought, including as many as twelve genera—providing many more clues but no solutions to the question of hominoid origins. Other representatives of the genus *Proconsul*—mostly teeth but a few fragments of skulls, limb bones, and vertebra—show considerable variation. Weight estimates range from 10 to 150 pounds. Available evidence also suggests varying environmental adaptations and locomotive patterns. Although none of these species can be pointed to as the specific ancestor of modern hominoids, Proconsul species likely represent hominoid ancestors near the divergence of hylobatids, pongids, and hominids (see Figure 4.2).

THE EXTINCTION OF MOST MIOCENE APES   During the early Miocene, ape species proliferated, but most became extinct for reasons that still elude scientists (Conroy, 1990). By the middle Miocene (approximately 16 million to 10 million years ago), ape species became dramatically less common compared to monkeys, reversing the earlier trend. Apes all but disappeared from the fossil assemblages of the late Miocene (10 million to 5 million years ago), and, as yet, there have not been any fossil finds of apes in Eurasia and Africa after the Miocene. In contrast, the middle Miocene fossil record provides evidence of abundant and diverse species of monkeys and the radiation of African monkeys into Eurasia. At the end of the Miocene, the specialized bipedal apes, the hominids, made their appearance (see Chapter 5).

Global climatic and ecological changes undoubtedly played a role in the extinction of the Miocene apes. Although it is difficult to generalize about a time period spanning 20 million years, the trend in continental climates was toward drier and cooler conditions (Conroy, 1990). Sixteen million years ago, the tropical rain forests of Africa were replaced by more open woodlands and savannahs. In the circum-Mediterranean region, the climate became more temperate and seasonal.

Many of the Miocene apes probably had difficulty adjusting to the cooler, drier climates. It is only in the less temperate, more tropical regions of Asia and Africa that apes continue to survive. However, some of the hominoids became more terrestrial, successfully adjusting to the new, open environments. It is these species that became the precursors of modern apes and humans.

MODERN APES

The modern apes, descendants of the Miocene hominoids, are found only in Asia and Africa (Richard, 1985). In Asia, the surviving species include the gibbon, the siamang, and the orangutan. The African hominoids are the chimpanzee and the gorilla.

THE GIBBON AND SIAMANG   The gibbon and siamang are the most numerous of the living apes, inhabiting evergreen forests throughout Southeast Asia. There are several species of gibbon, which weigh between 11 and 14 pounds. The siamang is larger, weighing up to 25 pounds. The hylobatids savor a diverse diet, ranging from fruits and leaves to spiders and termites, although the choice of food depends on local environments and the time of year.

All the hylobatids have relatively short trunks; extremely long, slender arms; curved fingers (see Figure 4.1); and powerful shoulder muscles. These characteristics enable them to negotiate their arboreal environment through brachiation. **Brachiation** refers to arm-over-arm suspensory locomotion used to move from branch to branch. Many primates can easily hang from tree branches with one hand while feeding or wrestling with a playmate with the other. However, brachiating primates such as the gibbon swing through the trees like acrobats, covering up to 30 feet in a single motion.

The gibbons and siamangs live in monogamous family groups consisting of male-female pairs and as many as four immature offspring. These young may stay in the family group for up to ten years. Foraging for food together or individually, families sometimes range over large areas and fiercely defend their territories. The hylobatids are noisy, often calling and vocalizing to signal their presence to other groups.

THE ORANGUTAN The orangutan, the only Asian pongid alive today, lives exclusively in the heavily forested regions of the Indonesian islands of Borneo and Sumatra. Orangs are large, sexually dimorphic apes. Males weigh about 150 pounds, whereas females reach about half that size. Because of their large size, orangs are not as agile as the lesser apes. Nevertheless, their long arms and fingers allow them to move quite efficiently through the trees. When traveling long distances, they occasionally drop to the ground and move in a quadrupedal fashion. The orangutan has a distinctive *noyau* social organization, in which adult males and females do not live in large social groups or pairs. Instead, adult females, together with their immature offspring, range over comparatively small areas searching for leaves, fruits, and seeds. Adult males, in contrast, cover larger areas, often encountering several females with whom they may mate.

Increasingly, these shy, mostly solitary creatures are facing extinction in many areas as development eliminates their ranges and the depletion of the tropical rain forests continues apace. At one time, more than 500,000 orangutans inhabited the tropics of Asia, but today the orangutan has become an en-

dangered species. A total of perhaps 30,000 survive in Borneo and Sumatra.

THE GORILLA The best-known apes still roaming the earth are the gorilla and the chimpanzee. As with the orangutan, the habitats of these great apes are being threatened by humans. Today these apes—all confined to restricted areas in Africa—are listed as endangered species. About 40,000 lowland gorillas (the type usually seen in zoos) live in the forests of western and central Africa. In the lake areas of East Africa, only 400 mountain gorillas remain.

The gorilla is the largest living primate (Richards, 1985; Schaller, 1976). The adult male weighs up to 400 pounds; the female grows to about 200 pounds. Although they can climb, gorillas are the most terrestrial of all primates, besides humans, because of their great size. On the ground, they use an unusual quadrupedal form of locomotion called **knuckle walking.** Rather than supporting their forelimbs with the palms of their hands (as do most other primates), knuckle walkers rest their hands on the ground in a curled position, with the tops of their curled middle and index fingers bearing their weight. Big-boned creatures, gorillas also have large, powerful jaws and chewing muscles for eating a wide variety of terrestrial vegetation such as roots, shoots, and bark. Yet despite their tremendous size and strength, they are shy, gentle creatures.

Gorillas thrive in social groups of about twelve animals, although the group's size may range from two to twenty, and lone males are occasionally seen. Groups are dominated by an older male, or silverback. Observing mountain gorillas over a long period, Dian Fossey discovered that their groups consist of unrelated females and immature males (see the box "Primatologists in the Field: Jane Goodall and Dian Fossey) . Female gorillas may transfer from one group to another once or many times during their lives. This pattern differs markedly from other primate groups, which are made up of related individuals. Males appear to transfer less frequently than females, and when leaving a group they generally do not join another. New groups may form when females join a lone male.

THE CHIMPANZEE Chimpanzees inhabit a broad belt across equatorial Africa from the west coast to Lake Tanganyika in the east. Two species of

ANTHROPOLOGISTS AT WORK

PRIMATOLOGISTS IN THE FIELD: JANE GOODALL AND DIAN FOSSEY

For many years, our ideas about the behaviors of nonhuman primates came from observations of animals in captivity. Removed from their natural environments and often secluded from other animals, these primates did not exhibit normal behavior patterns. By observing primates in the wild, we can sketch a much more accurate picture of their activities and responses. Such studies, though, pose special problems. Animals like chimpanzees and gorillas live in areas that are heavily forested and difficult to reach. They keep their distance from humans, making close observation challenging. In addition, primates are long-lived (great apes may reach an age of 50 years), which makes it difficult for researchers to draw conclusions about maturation and long-term social relations. Today we know a great deal about the behavior of chimpanzees, gorillas, and other primates in their natural habitats. This information was gathered through the persistence of many primatologists, but the pioneering work of two researchers—Jane van Lawick Goodall and Dian Fossey—deserves special note.

Jane Goodall was born in Bournemouth, England, in 1934.

As a child, she was always fascinated by living creatures. In one of her books, *In the Shadow of Man,* she recounts how at age 4 she crept into a henhouse and patiently waited five hours to see how a chicken laid an egg. At age 8, Goodall decided that when she grew up she would go to Africa and live with wild animals. This long-standing desire took her to Kenya in the late 1950s, where she met Louis Leakey, noted for his research on human origins (see Chapter 5). Leakey encouraged Goodall's interest in African wildlife, suggesting that she study chimpanzees. At that time, almost nothing was known about the behavior of wild chimpanzees. Only one study had been carried out, and that had lasted less than three months, far too short a period to gather much meaningful information. Leakey secured funding for Goodall to study chimpanzees in an isolated area near Lake Tanganyika in East Africa.

Arriving in Gombe National Park in 1960, Goodall found the chimpanzees frustratingly uncooperative. It took her more than six months to overcome their deeply rooted fear of humans. Goodall's (1971, 1990) work on the Gombe chimps has now spanned more

than thirty years, providing invaluable insight into chimpanzee behavior. She was the first to observe chimpanzees manufacturing simple tools to extract termites—a favorite food—from their nests. She has documented the long-term bonds between mother and child and the complexity of social relations. Goodall was also the first to spot the darker side of chimpanzee behavior, including occasional instances of cannibalism, murder, and warfare. We owe these important observations about the social behavior of chimps to Goodall's extraordinary commitment. Hers is the longest ongoing study of any primate group. She continues her work today at the Gombe Stream Research Center, which she established.

Another key contributor to primate studies was Dian Fossey, who spent eighteen years documenting the life of the mountain gorillas of Parc des Virungas, Zaire, and Parc des Volcans, Rwanda (Fossey, 1983). Dian Fossey was born in 1936 in San Francisco. She had always dreamed of visiting Africa to study the wildlife, a dream she fulfilled in 1963 when she borrowed enough money for a seven-week safari to

chimpanzees have been identified in Africa: the "common" chimpanzee (*Pan troglodytes*) and the "pygmy" chimpanzee (*Pan paniscus*). At present, pygmy chimpanzees are found only in a small forested area of central Africa, whereas the common chimp inhabits both rain forests and mountain forests as well as dry woodland regions. Like gorillas, chimpanzees are knuckle walkers whose anato-

observe the mountain gorillas of Mount Mikeno, Zaire. While in East Africa, Fossey sought out Louis Leakey, who was then supporting Jane Goodall's work. Convinced of Fossey's commitment, Leakey was instrumental in obtaining funding for her to initiate research for a long-term study of the mountain gorilla.

Fossey faced particularly difficult problems in her studies. The habitat of the mountain gorilla is threatened by the extension of cultivation and herd lands. In addition, the area is hunted by people from nearby African settlements who often use snares to capture animals like the antelope; tragically, gorillas are sometimes caught in these traps. In other cases, gorillas themselves have been the target of poachers who have cultivated a thriving market in gorilla body parts. Some tribes use gorilla ears, tongues, testicles, and small fingers to make a virility potion. Gorilla heads and hands have sometimes been sold as gruesome souvenirs to European tourists for about twenty dollars apiece. Fossey estimated that two-thirds of gorilla deaths were the result of poaching.

Fossey adopted a strong stand against poachers, practicing what she called "active conservation." She destroyed traps, raided temporary shelters used by poachers,

Dian Fossey's patience eventually allowed her to make careful observations of mountain gorillas. Her tragic death in the field has been recounted in the film Gorillas in the Mist.

and herded gorillas away from areas where hunters proliferated. In 1985, she was brutally murdered in her cabin at the Karisoke Research Center in Rwanda. Some of her experiences are recounted in the book *Gorillas in the Mist,* published shortly before her death and subsequently made into a movie.

Significantly, Goodall and Fossey had received no formal training in primatology before beginning their fieldwork. Although later granted a doctorate from Cambridge University, Goodall had only a secondary school education and experience in business when she first went to Kenya. Fossey had been an occu-

pational therapist before going to Africa, and she garnered official recognition for her work only after years of field experience. With their extraordinary courage and commitment, these women overcame tremendous obstacles to undertake their research.

The work of researchers like Goodall and Fossey has greatly broadened our knowledge of the behavior of living primates, but this knowledge is accumulating far too slowly. The fossil record indicates that primates once ranged over a much wider portion of the globe than they do today. Many species that are now extinct suffered from environmental changes over the past 30 million years; however, the perilous position of some modern primates can be laid directly at the doorstep of humans. In many areas, primates are still hunted for food. Expanding human settlements and the destruction of forests directly endanger nonhuman primates, threatening the very survival of animals like the mountain gorilla of central Africa, the lion-tail macaque of India, the golden-lion tamarin of Brazil, and many others. The work of researchers like Goodall and Fossey highlights the plight of humankind's closest living relatives, and may awaken people to the myriad preventable threats these creatures face.

my suits this form of locomotion. However, chimps also spend a good deal of time swinging in the trees and feeding on all sorts of fruit and vegetation. In addition, primatologists have discovered that chimps occasionally hunt, eating birds and small mammals (Goodall, 1986). Chimps and gorillas are recognized as the most intelligent of all the apes. Recent studies indicate that from a genetic standpoint,

An infant chimpanzee (Pan troglodytes) *in northeastern Sierra Leone. Genetically, the chimpanzee and the gorilla are almost identical to humans.*

chimpanzees and humans are almost 99 percent identical (Goodman & Lasker, 1975).

Chimpanzees band together in less structured social organizations than those of other anthropoid primates. A chimpanzee community may number from fifteen individuals to several dozen, but adults of both sexes sometimes forage alone and sometimes band together into groups of between four and ten. There is no overall leader, and the makeup of the smaller feeding groups is constantly changing. Chimpanzee communities are defined by groups of males that generally maintain friendly relations and utilize the same range.

Primatologist Jane Goodall (1986) has observed that male-female sexual bonding among chim-

panzees is extremely fluid. One female may mate with a number of males in the group. In other cases, a male and a receptive female may form a temporary sexual bond, or *consortship,* and travel together for several days. As Goodall has also noted (1986), the fluid nature of chimpanzee social life makes the day-to-day experiences of a chimpanzee far more varied than those of most other primates.

PRIMATE BEHAVIOR

Modern primates have complex learning abilities and engage in multifaceted social activities. Some primates of the same species exhibit disparate forms of social organization and behavior because of differing ecological circumstances. The behavior of other primates varies within the species, even when they live in similar environmental conditions. Despite this diversity, primatologists studying these creatures in their natural habitats have identified some common primate behaviors.

A cornerstone of primate life is the relationship between mothers and infants. Many animals virtually ignore their offspring after birth. However, as mentioned earlier, primates tend to have longer periods of maturation than other animals. During infancy, primates are cared for by their mothers, often forming a lasting bond (Harlow & Harlow, 1961). This *mother-infant attachment* is particularly strong among the anthropoid apes, and it may continue throughout a primate's lifetime. We find poignant examples of chimps tenderly caring for an injured infant or mother in Jane Goodall's studies of chimpanzees in the wild. These close attachments undergird primate social organization.

PRIMATE SOCIAL GROUPS

Most primates congregate in social groups, or communities, sometimes known as *troops.* Living in groups confers many advantages, including a more effective defense against predators, enhanced food gathering, increased reproductive opportunities, more intensive social learning, and greater assistance in rearing offspring. By ensuring that infants receive adequate care and nourishment, primates bolster their reproductive success.

As part of group living, primates engage in various kinds of affiliative behavior that help resolve conflicts and promote group cohesion. This conduct, which may involve interpersonal behavior such as hugging, kissing, and sex, can be illustrated by **social grooming.** Whereas other animals groom themselves, social grooming is unique to primates. Chimpanzees will sit quietly as other members of the troop carefully pick through their hair, removing ticks, fleas, dirt, and other debris. Social grooming not only promotes hygiene but also reduces conflict and friction between males and females and between young and old in the troop. Grooming works like a "social cement," bonding together members of the troop by maintaining social interaction, organization, and order among them all (Jolly, 1985).

Social grooming is an important aspect of life in living primate species and it was likely also part of early hominid life.

THE FAR SIDE By GARY LARSON

"So then Sheila says to Betty that Arnold told her what Harry was up to, but Betty told me she already heard it from Blanche, don't you know . . ."

Source: "The Far Side" ©Farworks, Inc. Used by permission. All rights reserved.

Primates also use a wide variety of *displays*—including body movements, vocalizations, olfactory (odor-related) signals, and facial gestures—to communicate with one another. In making these displays, primates express basic emotions such as fear and affection, as well as convey threats, greetings, courtship signals, and warnings of impending danger. The primates communicate through grunts, hoots, ground slapping, head bobbing, screams, scent marking (among prosimians), and facial gestures (among anthropoids of Eurasia and Africa). The more intelligent apes, especially the gorillas and chimpanzees, draw on a more highly developed repertoire of communication tools.

DOMINANCE HIERARCHY A social order known as **dominance hierarchy** characterizes the group living arrangements among primates (Fedigan, 1983). Dominance refers to the relative social status or rank of a primate, which is determined by its ability to compete successfully with its peers for objects of value such as food and sexual partners. In most primate groups, a specific dominance or rank hierarchy is based on size, strength, and age. In general, males dominate females of the same age, and older, stronger individuals acquire a higher rank than younger and weaker ones. Certain females may dominate others because they are better able to compete for food, thus enhancing their reproductive success. Each member of the group must learn its rank in the hierarchy and act accordingly. Even species like the chimpanzee, which do not have single, dominant group leaders, are organized into a dominance hierarchy. Once the order of dominance is established in the group, this hierarchical structure serves to head off conflict and socially disruptive activities. In some ecological circumstances, such as a harsh savannah, the dominance hierarchy is extremely rigid. Under more forgiving conditions, such as a forested region, the hierarchy is more loosely structured. In either case, the dominance hierarchy reduces chaotic behaviors and promotes orderly, adaptive conduct.

PRIMATE AGGRESSION Primatologists have observed that dominance hierarchies inhibit outright aggression and conflict (Goodall, 1986; Richards, 1985). Goodall describes some circumstances in which violence and aggression erupt in chimpanzee communities. Adult males patrol the perimeters of their home range, looking for trespassers. Though chimpanzee males rarely cross into the home ranges of other communities, when trespassing does occur, the patrol party viciously attacks the outsider.

In one incident witnessed by Goodall, a number of adult males split off from their community and moved from one area of their home range in Gombe National Park to another area. During a three-year period, the original group attacked and killed most of the members of this renegade community. Goodall concludes that when a dominance hierarchy is disrupted or when dominance hierarchies are not well developed (as, for instance, when chimps break away from one community to form another), violence and warfare may result.

PRIMATE SEXUAL BEHAVIOR   Male-female sexual behavior among primates does not follow a single pattern; rather, it varies according to environmental circumstances and the particular species of primate. As in the case of the gibbon, there may be monogamous sexual bonds between one male and one female. However, most of the higher primates do not form a close, singular sexual bond. Primatologists usually find in a group a single dominant adult male that has exclusive access to females. Sexual relations in all primate groups depend on many complex social factors. Friendships and alliances with other individuals underlie access to mates. Sometimes female gorillas side with a young dominant male against an older one (Campbell, 1987; Fossey, 1983). Such social subtleties and nuances, especially pronounced among the anthropoid primates, distinguish primate behavior from that of other species.

SOCIOBIOLOGY

Sociobiology provides an important theoretical perspective on primate behavior. Sociobiologists begin with the assumption that natural selection has acted on behavior in the same way it has acted on physical characteristics (Wilson, 1975, 1978). Thus some behaviors survive over others on the basis of how well they contribute to survival and reproductive success. Sociobiologists are not concerned with how specific genes may lead to specific forms of behavior; rather, they are interested in general strategies of behavior and how these might be adaptive in contributing to the reproductive success of a species or

individual. According to this approach, successful reproductive strategies have led to innate predispositions that influence behavior. Sociobiology has provided an important explanatory mechanism for many primatologists.

It has been demonstrated that some of the behavior seen in social insects, invertebrates, and some lower vertebrates is genetically controlled. These organisms will exhibit certain patterns of behavior, even if they have never been exposed to any other members of their species and, thus, never had the opportunity to learn their behavior. Such behavior patterns are termed *innate*. On the other hand, *learned* behavior is clearly of great importance in other vertebrates, and including primates. The challenge is to decipher which aspects of the behavior in a particular animal or individual are genetically influenced and which are learned.

In light of this, consider, for example, how sociobiologists view sexual reproductive strategies in primates. Sociobiologists observe that nature has assigned females and males very different parts to play in the reproductive process. Whereas males release millions of sperm in a single ejaculation, females produce only a relatively few eggs, perhaps hundreds over a lifetime. Furthermore, females are pregnant for months and cannot become pregnant again for a period of time. Thus, in principle at least, males are biologically capable of fathering thousands of offspring, whereas females are able to bear a much smaller number of children.

From these biological givens, sociobiologists argue that, from an adaptive viewpoint, males may reproduce their genes most efficiently through a strategy of sexual promiscuity that maximizes their number of offspring. This strategy does not, however, serve the reproductive interests of females. Each pregnancy demands great expenditures of energy. Thus, the efficient female strategy is to choose carefully a male whose qualities will contribute to her child's ability to survive and reproduce most successfully (Symons, 1979). Sociobiologists do not argue that such male and female reproductive strategies are conscious processes, just that in the course of evolution people who did not practice these behaviors have been gradually eliminated. Thus, over a long period of time certain innate predispositions survived.

Some sociobiologists have suggested that such innate predispositions influence a wide variety of human behaviors, including economic practices, kinship relations, aggression, the presence or absence of warfare in small-scale societies, and male–female relationships (Barash, 1987) (see the box on "Human Aggression," in Chapter 6). Many anthropologists have found such suggestions problematic and controversial. They instead underscore the role of learned behavior and human culture. Anthropologist Marshall Sahlins (1976), for example, views this approach as an abuse of evolutionary theory in explaining cultural phenomena. Focusing on kin selection, he cites cases in which human kinship systems are organized by cultural rather than by biological categories. For example, he notes that in some societies there are kinship categories for "brothers," "aunts," "uncles," and so on, that have little relationship to actual genealogical relationships. Sahlins concludes that human kinship is not always organized according to degree of genetic relatedness, as sociobiology predicts, and he dismisses sociobiology as a valid theory. In a different vein, biologists Stephen Jay Gould and Richard Lewontin (1979) have critiqued sociobiology as a circular reasoned argument that presents simplistic, idealized situations rarely found in nature.

Sociobiologists have responded to such critiques generally by emphasizing that their hypotheses are tentative and need to be evaluated and tested through empirical research. They argue that to dismiss the entire field of sociobiology through examples of ethnographic data that do not appear to support the theory is extreme. Furthermore, sociobiologists doubt that biological forces will ever be shown to "determine" human behavior and that human behavior is always a result of both biological and cultural factors. Still, they underscore that some biological tendencies may make some cultural patterns easier to learn than others.

THE HUMAN PRIMATE

As members of the order Primates, humans share many physical and anatomical similarities, as well as some behavioral characteristics, with other species of primates. Like other primates, modern humans can rotate their arms fully, grasp branches, and stand erect. The striking resemblance between the skeletons of a chimpanzee and a human being clearly identifies humans as primates. Yet humans possess certain novel capacities and abilities that make them unique.

For example, humans alone walk upright on two legs. Chimps, gorillas, and orangutans may stand upright for short periods, but only humans maintain a completely erect posture and consistently walk upright on two legs. The human pelvis, legs, and feet provide the balance and coordination that make this type of movement possible. Because human hands are not needed for locomotion, they have evolved into highly precise organs for the manipulation of objects. Human hands have short finger bones (or phalanges) compared to other primates (see Figure 4.1). This trait further enhances humans' manual dexterity. We will examine in Chapter 5 the adaptive aspects of *bipedalism* (walking on two legs) and the consequences of this evolutionary advance for early hominid behavior.

Humans' sensory organs bear a striking similarity to those of some of the other primates. For example, humans, as well as apes and monkeys, have keen visual acuity, including stereoscopic, binocular, and color vision. They also have diminished olfactory abilities compared to other animals. Thus, humans, apes, and monkeys all appear to perceive the external world in much the same way.

Although all primates have large brains relative to body size, the human brain is three times as large as we would expect for a primate of this size and build (Passingham, 1982). The human cerebrum, referred to in common usage as the "gray matter," and its outer covering, the neocortex (the section that controls higher brain functions), are far more highly developed than those of other primates, allowing humans to engage in complex learning, abstract thought, and the storing and processing of vast amounts of information. The size and complexity of the human brain, together with the protracted period of dependence and maturation characteristic of young humans, stand as the most significant differences between humans and other primates and give rise to the former's extraordinary capacity to learn; to their imaginative social interactions; and to their facility—unique among all life forms—to use and produce symbols, language, and culture.

 SUMMARY

Members of the order Primates are a diverse group of animals, classified together on the basis of certain shared characteristics. These include a high degree of manual dexterity, keen eyesight, complexity of the brain, prolonged periods of maturation, sophisticated social learning, and a generalized skeleton that allows a great deal of versatility in movement.

Although all primates share certain fundamental characteristics, they also exhibit a great deal of variation. The order Primates is divided into smaller categories—families, genera, and species—based on more specific criteria. This classification scheme reflects the evolutionary relationship of primates to one another. Although scientists essentially agree on these divisions, there are differences of opinion concerning which criteria should be used to define how one species relates to another.

Problems faced in classifying fossil organisms are especially complicated. Primate paleontologists excavate fossil localities to uncover the remains of early primates, and thousands of fossil primates have been unearthed. However, only a small percentage of the organisms that have lived are preserved in the fossil record. Remains are often very fragmentary, making the identification of fossil species difficult. As researchers accumulate more information, they revise or modify earlier interpretations.

The evolution of modern primates, including humans, can be traced back in time through the fossil record. Fossil primates resemble modern species, but none was identical to any creature living today. Modern primates represent the culmination of millions of years of adaptation to particular environmental niches. Nevertheless, modern primates embody some attributes found in ancestral forms. Many living primates occupy environments similar to those inhabited by extinct species. Thus, by studying modern primates, paleontologists gain insight into the abilities and adaptive qualities of species that are no longer

extant. Studies of primate fossils, combined with observations of the behavior of living primates, suggest that primates evolved in a forest environment.

Modern primates display a wide variety of behavioral patterns. Nevertheless, some generalized observations can be made. Many primate species congregate in social groups, sometimes referred to as troops, which afford increased protection from predators, more reproductive opportunities, and more opportunities for social learning. Within groups, members are often organized into a social order, or dominance hierarchy. Often this order is based on age, size, and strength. In general, males dominate females of the same, and larger, stronger individuals acquire more status than younger and weaker individuals. Primates often engage in various kinds of affiliative behavior, such as kissing, hugging, and social grooming. This friendly behavior is coupled with a variety of displays of emotions, from greetings to warnings.

An important theoretical perspective of primate behavior is presented by sociobiology. This interpretation is based on the premise that natural selection has acted upon behavior patterns in the same way that it has acted on physical characteristics. This does not presume a genetic basis for certain behaviors, but rather that certain behaviors might lead to reproductive success. Some researchers have challenged this theory, yet many primatologists, as well as some anthropologists, see it as a potential explanatory mechanism for some aspects of both primate and human behavior.

Even though humans have many unique abilities, they are still a primate species. Their ability to perform physical tasks, ranging from grasping a doorknob to driving a car, depends on physical abilities that evolved in earlier primates. Insight into the origins of human social behavior, too, can be gleaned from observations of nonhuman primates.

QUESTIONS TO THINK ABOUT

1. Primate evolution has produced some physical and anatomical traits that are related to an arboreal adaptation. What are some of these characteristics? Do all primates share these features?

2. How are fossil primates classified? Is it easier to classify a living species or one that is extinct? What is the definition of a species?

3. What is the difference between derived and primitive traits? How are these characteristics used to classify fossil primates?

4. Discuss the characteristics unique to the hominoids. When did the hominoid radiation take place? How is hominoid evolution relevant to human evolution?

5. Consider the distinctive aspects of primate behavior. How do you think this might provide clues to the behavior of early human ancestors?

6. How does the human primate differ from other primates?

 KEY TERMS

arboreal

brachiation

dentition

diurnal

dominance hierarchy

gestation period

hominoids

knuckle walking

nocturnal

omnivorous

opposable thumb

primates

quadrupeds

sexual dimorphism

social grooming

taxonomy

terrestrial

 SUGGESTED READINGS

CONROY, GLEN C. 1990. *Primate Evolution.* New York: W. W. Norton. A detailed but readable synthesis of current scholarship on primate evolution. Contains excellent introductory sections that examine evolutionary change, paleontological methods, and the ways scientists grapple with disparate interpretations of the fossil record.

FLEAGLE, JOHN G. 1988. *Primate Adaptation and Evolution.* New York: Academic Press. A survey of the fossil evidence on primate evolution, interspersed with information on the adaptive strategies of modern primates. A richly detailed and comprehensive work.

FOSSEY, DIAN. 1983. *Gorillas in the Mist.* Boston: Houghton Mifflin. A personal account of Fossey's research on the mountain gorilla in Zaire and Rwanda, and her struggle to preserve the gorilla against human encroachments.

GOODALL, JANE. 1990. *Through a Window: Thirty Years Observing the Chimpanzees of Gombe.* Boston: Houghton Mifflin. An extremely readable account by a leading primatologist of fieldwork among chimpanzees in their natural environment in Africa.

JOLLY, ALISON. 1985. *The Evolution of Primate Behavior.* New York: Macmillan. A comprehensive summary of existing knowledge on primate behavior. This revised edition incorporates much new information culled from contemporary primate studies.

Chapter 5

HOMINID EVOLUTION

CHAPTER OUTLINE

THE EVOLUTION OF THE PRIMATES IN THE Eocene, Oligocene, and Miocene epochs serves as a backdrop for the emergence of early human ancestors. By the Miocene epoch (25 million to 5 million years ago), primates in various forms—the precursors of modern prosimians, monkeys, and apes—proliferated in many geographic regions. But sometime in the late Miocene or early Pliocene, new and distinct forms of primates emerged. Classified as the family Hominidae, or **hominids,** these primates have subtle similarities in their teeth, jaws, and brains. However, the primary characteristic that identifies them as a distinct genus is the structural anatomy needed for **bipedalism,** the ability to walk upright on two legs.

Paleoanthropologists have advanced and discounted hypotheses about hominid evolution for more than a century. Although opinions diverge on the proper naming and classification of individual fossil specimens, paleoanthropologists are in broad agreement about the more general relationships among the different finds. Hominids are generally divided into two overarching genuses: *Australopithecus,* which emerged first, and *Homo.* Going back

some 4 million years, the earliest known hominid, *Australopithecus afarensis,* is known from African fossil finds. Scientists believe that *A. afarensis* stands near the base of the hominid family tree, with two branches leading to later australopithecines on one side and one branch giving rise to the genus *Homo* on the other.

Australopithecine species evolved from roughly 4 million to 1 million years ago. After that, there is no trace of this genus in the fossil record, leading paleoanthropologists to conclude that the australopithecines became extinct at about that time.

The genus *Homo* includes the direct ancestors of modern humans. Interestingly, early protohuman hominids, which bear some resemblance to the australopithecines, coexisted with the later australopithecines between 2 million and 1 million years ago. What distinguishes the first *Homo* specimens from the australopithecine line is a trend toward larger brain capacity. The earliest member of the *Homo* line to be identified in the fossil record is *H. habilis,* dating between 2.2 million and 1.6 million years ago, followed by *H. erectus,* whose oldest specimens are pegged at 1.6 million years ago.

93

Homo erectus, in turn, evolved into *H. sapiens,* the species that encompasses modern humans, during the past 400,000 years.

TRENDS IN HOMINID EVOLUTION

The hominids are members of the order Primates. As such, they share the basic primate characteristics discussed in Chapter 4, including a generalized skeleton, manual dexterity, and prolonged infant dependency. But the hominids have evolved with several distinctive characteristics, including three critical ones: (1) bipedalism, (2) reduction of face, jaw, and anterior teeth, and (3) enlargement of cranial capacity in the genus *Homo.* Paleoanthropologists have discovered evidence of these features in the skeletal structures of early hominids, and their evolution can be traced through fossil remains. These characteristics are least pronounced in the earliest hominids; modern humans exemplify their full development. Other trends, such as degrees of social complexity and the origins of human culture, are also of great importance. They are inferred on the basis of early hominid tools, food remains, and living sites, items examined in Chapter 7.

BIPEDALISM

Hominids are the only primates that are fully bipedal. Although gorillas and chimpanzees can stand upright, they are primarily knuckle walkers that spend most of their time on all fours. As with other types of locomotion, bipedalism is reflected in skeletal structure. For example, the hips and knees of humans differ markedly from those of knuckle walkers like the chimpanzee (Lovejoy, 1988). Paleoanthropologists focus a great deal of attention on the **foramen magnum,** the opening in the base of the skull through which the spinal cord passes. In quadrupedal animals, this aperture is located at the back of the skull, which causes the head to extend out in front of the body. In contrast, the foramen magnum in bipedal creatures is located on the bottom of the skull, sitting squarely above the body. Structures of the skull associated with bipedalism are especially telling because the postcranial bones of many fossil hominids have not been preserved.

Bipedalism stands as the earliest and the most important trend in hominid evolution. Initially, many paleoanthropologists believed that the first known hominids were not proficient at bipedalism, perhaps moving with a swinging, slouched gait like that of chimpanzees or gorillas. These interpretations were based on limited fossil finds and have not been supported by recent studies. Fossil remains of the oldest known hominids indicate that these creatures walked as well as modern humans (Figure 5.1). Our best scientific guess places the appearance of bipedalism in hominids sometime between 4 million and 10 million years ago, a period of time poorly represented in the fossil record.

ADAPTIVE ASPECTS OF BIPEDALISM Scientists have advanced a number of intriguing but unsubstantiated theories to explain the evolution of bipedalism in early hominids. It is important to recognize that whereas bipedalism requires a minimal expenditure of energy, it serves as a relatively slow means of locomotion—clearly, a disadvantage in avoiding predators. Some researchers have speculated that bipedalism developed in early hominids because of changes in the African environment during the late Miocene and early Pliocene. The climate during this period became drier, and the dense Miocene forests gave way to more open savanna grasslands. In this type of environment, upright posture may have allowed ancestral hominids to see over the high grass to identify both predators and food sources. Such behavior has been observed among modern ground-dwelling primates like baboons and chimpanzees, which adopt upright postures to survey their surroundings (Oakley, 1964). Although this is a tantalizing theory, it is impossible to validate on the basis of the fossil evidence. In addition, although the Miocene environment did become increasingly dry, it produced a wider variety of habitats than the widespread grasslands initially envisioned by some researchers (Lovejoy, 1981). In fact, some areas retained canopy forests. Whether upright posture had clear advantages over quadrupedality in this sort of environment remains to be determined.

TOOL USE Other theories on the origins of bipedalism hold that this type of locomotion served early hominids by freeing the hands to perform such tasks

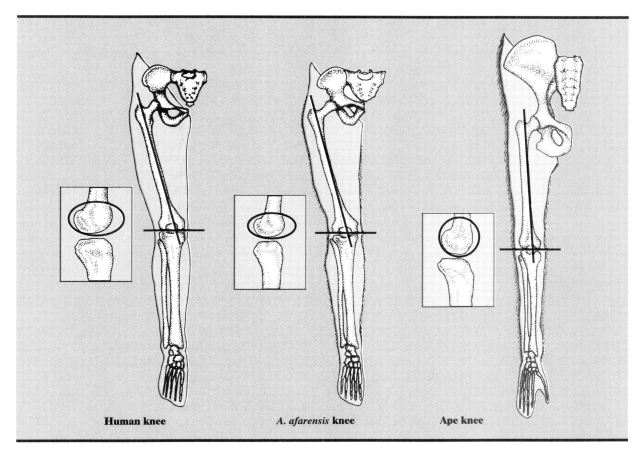

Human knee   *A. afarensis* **knee**   **Ape knee**

FIGURE 5.1   Drawing of Lucy's knee bones and hips compared with those of humans and apes.

Source: Adapted from Donald Johanson and Maitland Edey, *Lucy: The Beginnings of Humankind,* Simon & Schuster, 1981, p. 157. Copyright © 1981 by Donald C. Johanson and Maitland A. Edey. Drawings by Luba Dmytryk Gudz/Brill Atlanta.

as transporting food, carrying infants, and using tools. Because early hominids lacked sharp teeth and strong jaws, the ability to use tools would have given them the option of choosing from a greater variety of food sources (Pilbeam, 1972; Shipman, 1984; Washburn, 1960). In fact, some anthropologists attribute the emergence of bipedalism in hominids exclusively to the need to use tools. The existing evidence, however, does not support this hypothesis: The earliest stone tools date only to 2.5 million years ago, whereas the origin of bipedalism can be traced to more than 4 million years ago.

Nevertheless, advocates of the argument that the need to use tools gave rise to bipedalism discount this evidence as inconclusive. They assert that tools made of other materials, such as wood, plants, or bone, may have been used much earlier. They point out that modern nonhuman primates such as chimpanzees make simple tools of twigs and grass to extract food from tight spots, and tools of this type are unlikely to be preserved or even recognized as tools if they do appear in the archaeological record (Wolpoff, 1983).

TRANSPORT OF FOOD AND OFFSPRING   Consider, too, the ability to transport food and offspring— important activities among the hominids. Yet how these abilities relate to the selective pressures that initially prompted the evolution of bipedalism is unclear (Lovejoy, 1981, 1984). Paleoanthropologists conjecture that, as with other primates, the offspring of early hominids clung firmly to the mother. This allowed the female to move freely in search of food or to find safety. Although bipedalism allows for infants to be carried, that advantage would seem to be greatly outweighed by the inability to move quickly through the trees to elude predators. Likewise, food carrying would seem to have limited adaptive significance because animals tend to consume food where it is found.

Paleoanthropologist Owen Lovejoy has asserted that the evolution of bipedalism turned on much more than merely the ability to carry objects. Because it involved the modification of a wide range of biological and behavioral traits, it must have conferred some adaptive advantage on early hominids, even before they had fully developed the physical capabilities for bipedalism. Lovejoy posits that the crucial advantage may have been the ability to transport food back to a mate by walking upright and using simple implements such as broad leaves to maximize the amount of food that could be carried. Provisioning by the male would have allowed the female to increase the quality and quantity of time devoted to infant care. This intensification of parental attention, in turn, would have promoted the survival of infants and, therefore, the species. Taking the theory a step further, Lovejoy asserts that food sharing and the cooperation that underlies this behavior may have produced a reproductive strategy that favored sexual fidelity and close, long-term relations between a male and a female.

Although future fossil finds may shed more light on the origins of bipedalism, this trait probably evolved as a result of a confluence of factors. Sorting out which ones played the most prominent roles in the development of this adaptation continues to challenge anthropologists.

REDUCTION OF THE FACE, TEETH, AND JAWS

We also see in hominid evolution a series of interrelated changes primarily associated with diet and food-processing requirements. The oldest fossil hominids have a protruding, or *prognathic,* face. In addition, their incisor and canine teeth are large compared to those of modern humans. To accommodate the larger canines, which extend beyond the other teeth, there are gaps between the teeth of the opposing jaw. Finally, the teeth of early hominids are arranged in a U-shaped pattern, and the teeth on opposite sides of the mouth are parallel. All these characteristics are similar to features found in modern pongids.

Approximately 2 million years ago, these characteristics started to become less pronounced in hominids. Early representatives of the genus *Homo* have smaller canines, and the gaps associated with larger teeth disappear. In humans, the canine teeth retain a distinctive shape, but they are almost the same size as the other teeth. Of all the hominids, the faces of modern humans are the least protruding.

As noted in Chapter 4, primate teeth can handle an omnivorous diet with ease. However, hominid teeth, with flat molar crowns and thick tooth enamel, are highly specialized for grinding. Early primates as well as living prosimians and anthropoids had large canine and incisor teeth, which are well suited to cutting and slicing. In contrast, the size of these teeth is greatly reduced in later hominids.

Some australopithecines developed massive chewing muscles and extremely large molars compared to those of modern humans. This strong dentition earned one species, *Australopithecus boisei,* the nickname "nutcracker man." Scientists believe that these features most likely evolved in response to a diet of coarse, fibrous vegetation. Paleoanthropologists cite several key skeletal structures in the jaw and the cranium as evidence of this creature's powerful chewing capacity. Thick, enlarged jaws and cheekbones provided attachments for these huge muscles. Some australopithecine fossil specimens have a **sagittal crest,** a bony ridge along the top of the skull, which grows larger as more chewing muscles reach up along the midline of the cranium.

In contrast to the australopithecines, evolving *Homo* species may have consumed a more varied diet based on gathering vegetation, hunting animals, and scavenging. This theory corresponds with the size and contour of their molars—similar to those of modern humans—and the absence of such features as sagittal crests, which accompany specialized chewing muscles.

INCREASE IN CRANIAL CAPACITY

In the genus *Homo* a distinctive characteristic is a tendency toward increased cranial capacity and the complexity of the brain. Like the changes in dentition, growth in cranial capacity first appears in hominids dating from less than 2 million years ago. Before that, the size of the hominid brain underwent comparatively little change. *Australopithecus afarensis,* the earliest known hominid (living some 3 million to 4 million years ago), had a cranium about the size of a softball, barely surpassing that of a modern chimpanzee. Hominid cranial capacity remained fairly constant at this size for 2 million years, averaging just over 400 cubic centimeters (cc). Then, sometime after 2 million years ago, members of the genus *Homo* began to show a steady increase in cranial size. The brain in *Homo erectus* averaged 1,000 cc, and the modern human brain measures, on the average, 1,350 cc, a threefold increase from the australopithecines. Significantly, this constitutes an increase in both *relative* and *absolute* size. Even taking into account that modern humans are substantially larger than australopithecines, the relative size of the hominid brain more than doubled in the last 2 million years (McHenry, 1982).

Changes in the cranial capacity of early hominids undoubtedly influenced physical and social developments less easily studied through fossil remains. For instance, increasing brain size almost certainly prompted numerous modifications in hominid diet, the use of tools, the evolution of language, and the intricacies of social organization. Greater sophistication in any of these areas may have improved early hominids' chances of survival.

OTHER PHYSICAL CHANGES

The evolution of certain physical characteristics of contemporary hominids is difficult to trace because the characteristics are not preserved in the fossil record. For example, unlike other surviving primates, modern humans are not completely covered with hair. Presumably, there was a tendency toward less body hair during hominid evolution, but we can find no indication of this in fossilized bones. Loss of body hair might be a relatively recent phenomenon.

FOSSIL EVIDENCE FOR HOMINID EVOLUTION

In *The Origin of Species,* Charles Darwin devoted relatively little attention to human evolution, noting simply, "Much light will be thrown on the origin of

"IT'S A WONDERFUL FIND, AND YET THERE'S SOMETHING SUSPICIOUS ABOUT IT."

One of the problems that paleoanthropologists must face as they attempt to interpret the fossil record is the fragmentary and incomplete nature of the discoveries. Many finds are only represented by fragmentary skeletal remains.

Source: Courtesy of Sidney Harris.

man and his history" (Darwin, 1979: 222). In the mid-nineteenth century, when Darwin was writing his treatise, scientists had scant fossil evidence for hominid origins. Since Darwin's time, however, thousands of hominid fossils have been recovered, most of them in Africa. The Hominid Vault of the Kenya National Museum alone contains almost 300 hominid specimen finds from Kenya and Tanzania, and more than 1,500 specimens have been recovered from South African sites. Specimens range from isolated teeth to nearly complete skeletons. Although paleoanthropologists have uncovered many spectacular finds, some discoveries merit special attention because they prompted anthropologists to modify theories of human evolution. In this section, we will examine several of the most important fossil finds, beginning with the earliest. The locations of some of these key discoveries are illustrated in Figure 5.2.

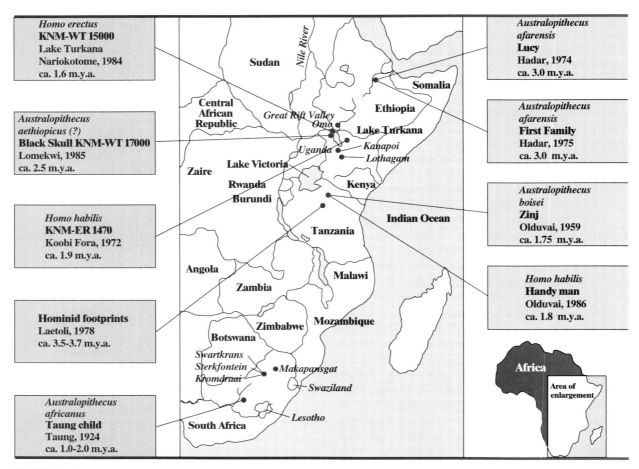

FIGURE 5.2 Map of African fossil finds.

Source: From Roger Lewin, *In the Age of Mankind: A Smithsonian Book of Human Evolution,* Smithsonian Books, 1988, p. 71. (Art by Phil Jordan and Julie Scheiber)

THE OLDEST HOMINIDS

Fossil evidence of the early evolution of the hominids remains very incomplete. Evidence predating 4 million years ago is limited and, for the most part, fragmentary. The earliest probable hominid remains come from Lothagam in northern Kenya. The find, dated to between 5 million and 6 million years ago, consists of a lower jaw fragment. Based on similarities to other finds, some researchers have suggested that the specimen is hominid, but the lack of associated postcranial bones makes it impossible to be certain. Other early finds from Kenya and Ethiopia are similarly fragmentary.

More information is provided by fossils found at the site of Aramis in northeastern Ethiopia. This re-

gion, located at the intersection of the Rift Valley, the Red Sea, and the Gulf of Aden, has produced some of the most spectacular and earliest hominid fossil finds, including Lucy and other examples of *Australopithecus afarensis* (discussed in the following sections). Aramis was explored in 1992 and 1993 by Timothy White of the University of California, Gen Suwa of the University of Tokyo, and Berhane Asfaw of the Ethiopian Ministry of Culture and Sports Affairs (White et al., 1995). This research, and additional discoveries in 1994 and 1995, produced the remains of several dozen individuals dated to approximately 4.4 million years ago. Although the discoveries are likely hominid, they are very primitive compared to later australopithecines such as *Australopithecus afarensis* and *Australop-*

ithecus africanus (discussed subsequently). The cranial capacity is quite small—at least as small as that of other early hominids—but the form of the cranium is more apelike, and the canine teeth are larger. Despite these distinctive features, the placement of the cranium over the spinal column, the shape of the pelvis, and the limb bones are consistent with bipedal locomotion—the hallmark of hominids.

Because of the distinctive, primitive aspects of the Aramis finds, Tim White and his colleagues have argued that the Aramis discoveries represent the earliest and most primitive hominids yet found and that they should be designated by new genus and species names: *Ardipithecus ramidus.*

THE FIRST AUSTRALOPITHECINES: KANAPOI AND EAST TURKANA

The region around Lake Turkana in northern Kenya has also yielded a host of important fossil finds, including the discoveries of *Australopithecus aethiopicus, Homo habilis,* and *Homo erectus* (discussed subsequently). The earliest hominid remains are represented by a number of finds made over the past 30 years at Kanapoi, south of Lake Turkana, and Allia Bay, East Turkana. Some of the most recent discoveries and assessments of the finds have been made by Meave Leakey (Leakey et al., 1995).

The fossils are fragmentary, including teeth and jaw fragments and some postcranial bones. The age of the finds is placed between 3.9 million and 4.2 million years ago. The leg bones are consistent with bipedal—hominid—posture, but the finds also present some distinctive attributes. Like *Ardipithecus,* the skull and the teeth are quite primitive. The external ear openings are also distinct from more recent hominids. However, in contrast to the *Ardipithecus* remains, the molar enamel on these specimens is thick, more analogous to more recent hominids. Hence, the finds may represent a transitional link between species such as *Ardipithecus* and the australopithecines. Because of their similarity to later finds, Meave Leakey and her colleagues have placed the discoveries in genus *Australopithecus* but assigned them a new species designation, *anamensis,* in recognition of their distinctive attributes. The relationship of these finds to *Australopithecus afarensis* is still being evaluated.

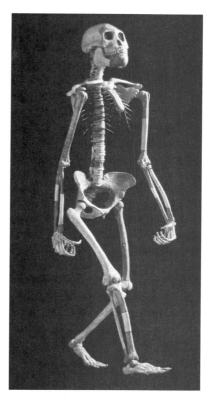

Reconstruction of skeletal remains of Australopithecus afarensis, *recovered by Donald Johanson at Hadar, Ethiopia, in 1974.*

AUSTRALOPITHECUS AFARENSIS

During the 1970s, a joint American–French team of paleoanthropologists led by Donald Johanson and Maurice Taieb made several exciting finds in the well-preserved geological beds near the Great Rift Valley in the Afar region of Ethiopia (Johanson et al., 1982; see the box "Donald Johanson: Paleoanthropologist"). This valley had experienced mountain building and volcanic activity over the last several million years, which had brought many fossils to the surface, where they could be spotted by researchers.

Expeditions to Hadar have produced thousands of fossils. These include many hominid remains, providing scientists with a more complete understanding of early hominid variation. Because many subtle differences distinguish the dentition of the various Hadar hominids, Johanson initially believed that more than one species was represented. After subsequent study of the remains, however, Johanson

ANTHROPOLOGISTS AT WORK

DONALD JOHANSON: PALEOANTHROPOLOGIST

Born in Illinois in 1943, Donald Johanson is one of the world's most influential and best-known paleoanthropologists. His firsthand accounts of research in Ethiopia (Johanson & Edey, 1981) and Olduvai Gorge (Johanson & Shreeve, 1989) have earned popular acclaim for their readable and thought-provoking insights into the field of paleoanthropology.

In 1973, Johanson, in conjunction with French geologist Maurice Taieb, began research in an area known as Hadar in the Afar triangle of northeastern Ethiopia. At that time the region had been largely unexplored by paleoanthropologists. Although the present climate is arid and inhospitable, the fossil record indicates that the region supported a variety of life forms 4 million to 3 million years ago. In Hadar, Johanson and his fellow researchers uncovered many finds that cast light on early hominids and their environment. Two spectacular finds in the Hadar have received particular attention. The first, uncovered in 1974, was a strikingly complete (40 percent) skeleton of an early hominid affectionately referred to as "Lucy." The second find, unearthed in 1975 at a site designated AL 333, consisted of a remarkable collection of hun-

dreds of hominid bones, representing at least thirteen individuals. Johanson believes that all the creatures at AL 333 may have died at the same time in a sudden catastrophic event like a flash flood. Both discoveries were representative of a previously undescribed species, which Johanson named *Australopithecus.*

In 1978, Johanson, in conjunction with paleoanthropologist Timothy White, reinterpreted the prevailing notions about hominid ancestry. They surveyed existing information and integrated it with Johanson's finds from Hadar. Then they restructured the hominid family tree, placing *A. afarensis* at the base, with two branches, one sprouting toward the genus *Australopithecus* and the other giving rise to the genus *Homo.* Although more recent discoveries lead to the conclusion that the hominids may be divided into more than two branches, the majority of paleoanthropologists accept *A. afarensis* as the earliest known hominid.

Some of Johanson's interpretations have been called into question. Other researchers have challenged his classification of *A. afarensis,* arguing that the fossils represent more than one species. His major critics include paleoan-

Donald Johanson (far left) with other paleoanthropologists analyzing hominid fossils from the Afar region of Ethiopia.

thropologists Mary and Richard Leakey, known for their own research at Olduvai Gorge, Laetoli, Koobi Fora, and Lake Turkana. Johanson has observed: "Frustrating as it is, the distantly tantalizing truths about our origins will probably not be revealed before we ourselves are buried under the earth. But that will not stop me from testing and retesting new hypotheses, exploring further possibilities" (Johanson & Shreeve, 1989: 133).

and his colleagues concluded that the hominids all belong to a single species, which they designated *Australopithecus afarensis* (Bower, 1991; Johanson, White, & Coppens, 1978). The researchers argue that the differences they discerned are the result of natural variation and sexual dimorphism within the species. Others scholars, however, still maintain that more than one species is represented.

LUCY   Among the most spectacular finds made by Johanson's team at Hadar was a fossilized skeleton of an ancient hominid that was almost 40 percent intact. This find, scientifically designated *Australopithecus afarensis,* became popularly known as "Lucy" (named after a Beatles song, "Lucy in the Sky with Diamonds"). Initial discussions of the discovery concluded that Lucy was a female, but some more recent analyses have suggested the remains are actually those of a male. Such interpretive disagreements underscore the challenges researchers face in analyzing fragmentary remains of nonliving species. Lucy had a small cranium (440 cc) and large canine teeth. In fact, Lucy's skull resembles that of a modern chimpanzee. However, below the neck the anatomy of the spine, pelvis, hips, thigh bones, and feet clearly shows that Lucy walked on two feet (Lovejoy, 1988). Lucy was a fairly small creature, weighing approximately 75 pounds, and she stood about 3.5 to 4 feet tall.

THE FIRST FAMILY   Another important discovery at Hadar came in 1975 at a fossil locality known as site 333. Johanson and his crew found many hominid bones scattered along a hillside. Painstakingly piecing them together, the researchers reconstructed thirteen individuals, including both adults and infants, with anatomical characteristics similar to those of Lucy. Experts hypothesize that these finds may represent one social group that died at the same time, perhaps during a flash flood. The group has, in fact, been referred to as the "First Family of Humankind" (Johanson & Edey, 1981). However, as the precise conditions that produced the site remain uncertain, this interpretation must be regarded as tentative.

The *Australopithecus afarensis* fossils discovered at Hadar have been dated between 3 million and 4 million years ago, making these some of the earliest well-described hominid remains. The fossils are remarkably primitive in comparison to other australopithecines; from the neck up, including the teeth, cranium, and jaw, *A. afarensis* is definitely apelike. However, the abundant limb bones and anatomical structures indicating pelvic orientation, as well as the position of the hips and knees, indicate that *A. afarensis* was a fully erect, bipedal creature (Lovejoy, 1988). Other finds, comparable to those from Hadar, have been reported from Laetoli and also from Chad in Central Africa. The latter find is especially notable, as it was recovered some 1,500 miles farther west than any other early hominid find.

THE LAETOLI FOOTPRINTS

The site of Laetoli, some 30 miles south of Olduvai Gorge in northern Tanzania, has produced a number of fossil finds, including possible examples of *Australopithecus afarensis,* the fossil species described at Hadar. However, the site is best known for the remarkable discovery of fossilized footprints. Thousands of footprints of various species of ancient animals are preserved in an ancient layer of mud covered with volcanic ash. A finding by Mary Leakey in 1978 confirms that fully bipedal creatures roamed the earth approximately 3.5 million years ago. The evidence consists of a trail more than 75 feet long made by at least two hominids, maybe three. Intensive studies of these footprints revealed that the mode of

A hominid footprint fossilized in volcanic ash at Laetoli, Tanzania is shown next to a modern foot. Dated at over 3.5 million years, the trail of footprints at the site provides dramatic evidence that early hominids such as Australopithecus afarensis *were fully bipedal.*

CRITICAL PERSPECTIVES
NEW PERSPECTIVES ON THE TAUNG CHILD

Since the days of Raymond Dart, paleoanthropologists have been using advanced scientific techniques to analyze ancient hominids. In one of the newest technological advances, scientists Glenn C. Conroy and Michael Vanier (1990) of Washington University in St. Louis peered inside the fossilized skull of the Taung child by means of computer image-enhancing techniques and sophisticated electronic scanning equipment. Scientists had previously been unable to examine the inside of the skull because it was packed tightly with ancient sediment. Conventional X-rays could not penetrate the dense mineral deposits, and attempts to remove the sediment risked damaging the fossil itself.

Conroy and Vanier combined more than forty computerized tomography (CT) pictures into a single three-dimensional image, modeling the inside and outside of

the Taung child's skull. Tomography is a medical procedure that doctors normally use to probe the inside of a solid object, such as an arm or a leg, by sweeping an X-ray over an electronically predetermined cross section, yielding a two-dimensional likeness. Once Conroy and Vanier produced the image of the Taung fossil, they electronically sliced, dissected, and scanned the skull without damag-

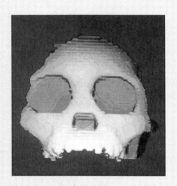

A CT scan of the Taung child.
Source: Courtesy of Glenn C. Conroy, Washington University.

ing it. They detected thin bones and patterns of dentition no paleoanthropologist had seen before.

These tomographic pictures have led to the conclusion that *Australopithecus africanus* is much more primitive than had previously been thought. Looking at the outer portions of the fossil, some researchers, including Dart, had concluded that *A. africanus* was ancestral to the genus *Homo.* The new information may spur a reevaluation of this interpretation.

Imaging procedures and computer modeling give paleoanthropologists important new tools in their search into human origins. Using these techniques, scientists are now able to scan dense material without damaging it. These approaches should help paleoanthropologists refine their hypotheses, leading to a clearer picture of humankind's roots.

locomotion for these early hominids was fully bipedal and was similar to that of modern humans.

TAUNG CHILD: A SOUTH AFRICAN AUSTRALOPITHECINE

The Taung child is memorable as the first example of the genus *Australopithecus* described. The find was named *Australopithecus africanus,* the "southern ape of Africa," by its discoverer (see the box "New Perspectives on the Taung Child"). Dated be-

tween 2 million and 3 million years old, it represents a species of hominid that lived after the species discussed in the preceding sections. However, it possessed smaller cranial capacity than more recent examples of genus *Homo. A. africanus* is primarily known from sites in South Africa.

The discovery of the Taung child stemmed from a remarkable combination of coincidence and luck. The man responsible for the find was Raymond Dart, an Australian-born anatomist living in South Africa. In 1924, in a box of fossils from the rubble of a limestone quarry near the town of Taung, South Africa, Dart found the front of a skull, a jaw, and an endocranial cast of a previously unknown hominid. On the basis of the teeth, Dart judged the creature to have been quite young at death, and he called his discovery the "Taung child." Today the individual is estimated to have been between 3 and 4 years old at the time of death.

Although the Taung child had certain apelike features, it also exhibited a number of unique characteristics. For example, the foramen magnum was further forward in the Taung child than in modern apes, indicating that this creature's head was balanced above the spine. In other words, it moved with the upright posture characteristic of a biped. The brain of the Taung child was very small, hardly larger than that of a chimpanzee. Its structure, however, differed from that of apes and was more highly developed in some regions. The teeth were much closer in size to a human child's than to an infant ape's. Dart astounded the scientific world by announcing that the Taung child was a hominid, an intermediate link between humans and earlier primates.

Many paleoanthropologists challenged Dart's conclusion, arguing that the Taung child was really an ape. Popular evolutionary theories suggested that large cranial capacity was the critical characteristic of hominid evolution, and critics pointed out that the cranial capacity of Dart's find was too small for it to have been ancestral to humans (see the box on Charles Dawson's "dawn man," "The Piltdown Fraud"). But Dart's critics were proved wrong. In the decades following the discovery of the Taung child, a number of similar finds were made in South Africa. During the 1940s, Dart excavated additional fossils from Makapansgat Cave. Scottish paleontologist Robert Broom also came upon a number of similar fossils at Sterkfontein. Some of these new finds were adult specimens of creatures like the Taung child. With their humanlike dentition, bipedal capabilities, and small cranial capacity, they were unquestionably hominids. These discoveries clearly established the Taung child as a hominid and *Australopithecus* as a valid genus.

Adult specimens of *A. africanus* probably weighed between 40 and 60 pounds and were between 3.5 and 4.75 feet tall; their cranial capacity averaged around 450 cc. Although the age of the South African

Australopithecine skulls from the limestone deposits of South Africa. The example at lower left is a robust form; a gracile form is at top right.

finds is still uncertain, the gracile australopithecines seem to date to between 3 million and 2 million years ago, a conclusion based on study of the fossils of extinct animals of known age found in the same deposits.

AUSTRALOPITHECUS AETHIOPICUS: THE "BLACK SKULL"

The incomplete puzzle of hominid ancestry was filled in with one more piece in 1985, this one dug out of the fossil beds west of Lake Turkana, at a fossil locality known as Lomekwi I. Discovered by English paleoanthropologist Alan Walker, the find consists of the fragments of an australopithecine but of a type far more robust than *A. africanus*. Because the fossil had been stained blue-black by manganese in the soil, it became known as the "Black Skull," or, by its Kenya National Museum catalogue number, KNM-WT 17000 (Walker et al., 1986). Possibly another member of the same species is represented by more incomplete remains found earlier in the Omo River valley of Ethiopia. These have been tentative-

CRITICAL PERSPECTIVES
THE PILTDOWN FRAUD

One of the most bizarre stories in the history of paleoanthropology involves the fossil known as "Piltdown man." Widely discussed and debated for several decades, this discovery was eventually exposed as an elaborate fraud. Although it does not figure in current theories of hominid evolution, we examine the Piltdown man controversy because the alleged specimen was accepted as a legitimate human ancestor during the early decades of the twentieth century and influenced interpretations of human evolution (Blinderman, 1986; Weiner, 1955). This cautionary tale illustrates the efficiency of modern scientific techniques but also serves as a warning about how scientists can be swayed by their own preconceived ideas.

Piltdown man was "discovered" in 1912 in a gravel quarry near Sussex, England, by a lawyer and amateur geologist named Charles Dawson. The quarry had previously produced the bones of extinct animals dating to the early Pleistocene (approximately 1.8 million years ago). The supposed hominid remains uncovered there consisted of the upper portion of the cranium and jaw. The skull was very large, with a cranial capacity of about 1,400 cc, which placed it within the range of modern humans. However, the lower jaw was apelike, the canine teeth large and pointed. This picture of early hominids mirrored popular early twentieth-century notions of the unique intellectual capabilities of humanity. Humans,

so the interpretation went, evolved their large brains first, with other characteristics appearing later. In fact, a great deal of evidence points to just the opposite evolutionary pattern.

Piltdown man was officially classified as *Eoanthropus dawsoni* ("Dawson's dawn man") and accepted by the scientific community as the earliest known representative of humans found in western Europe. A number of paleoanthropologists in France, Germany, and the United States remained skeptical about the findings, but they were unable to disprove the consensus of the English scientific community. As time went by, however, more hominid fossils were discovered, and none of them exhibited the combina-

ly designated *Australopithecus aethiopicus* by some researchers.

The Black Skull allows us to sketch a more complex, intricate picture of the hominid family tree. Found in sediments dating to approximately 2.5 million years ago, the fossil's cranium is small, comparable in size and shape to that of the *A. afarensis* fossils from Hadar. The movement of the *A. aethiopicus* jaw is also similar to that of *A. afarensis*. Yet the face is large, prognathic, and very robust, boasting massive teeth and a pronounced sagittal crest. In these respects the Black Skull resembles finds such as *Australopithecus boisei*, which dates to 750,000 years later. We can, therefore, venture a reasonable guess that the structure and age of *Australopithecus aethiopicus* place it between *A. afarensis* and *A. boisei*.

AUSTRALOPITHECUS BOISEI: THE "NUTCRACKER MAN"

Following the initial discoveries in South Africa, many additional australopithecine fossils have come to light. One of the most exciting finds, called *Australopithecus boisei,* was the first of many discoveries made in eastern Africa by paleoanthropologists Louis and Mary Leakey.

Australopithecus boisei was found in the Olduvai Gorge, a 30-mile canyon stretching across the Serengeti Plain of Tanzania. In 1959, Mary Leakey recovered an almost complete fossil skull from the gorge. The find was a robust australopithecine but was even more robust than the examples known from South Africa. The teeth of *A. boisei* were distinctly hominid in form but were much larger than

tion of an apelike jaw and a large, humanlike cranium.

With contradictory evidence mounting, skepticism grew in the paleoanthropological community concerning the legitimacy of the Piltdown fossils. Finally, in the early 1950s, scientists completed a detailed reexamination of the Piltdown material. Using fluorine analysis (see Chapter 2) they discovered that the skull and jaw were of relatively recent vintage; the jaw, in fact, was younger than the skull. In reality, the Piltdown fossil consisted of a human skull from a grave a few thousand years old attached to the jaw of a recently deceased adolescent orangutan. The apelike teeth embedded in the jaw had been filed down to resemble human teeth, and the place where the jaw joined the skull had been broken away. The jaw was stained with a chemical to match the color of the skull.

Clearly, whoever perpetrated the Piltdown hoax had some knowledge of paleoanthropology. By the time the hoax was unmasked, most of the people who could have been implicated had died (Blinderman, 1986). Putting aside the question of who was responsible for the hoax, we now recognize that paleoanthropological research between 1912 and the 1950s was definitely harmed by the Piltdown find, because the scientific community pursued a false path in hominid research. The initial acceptance of the Piltdown fraud as legitimate may partially explain why one of the most startling hominid fossil discoveries of the early twentieth century was relegated to relative obscurity for so many years. This discovery was the Taung child.

Points to Ponder

1. What lessons does the Piltdown fraud provide for the way paleoanthropological research should proceed and how findings should be validated?

2. The recovery methods and the limited information on the context of the find clearly contributed to the success of the Piltdown fraud. Contrast the details of the Piltdown discovery with more recent finds at Olduvai Gorge, Tanzania, or Hadar, Ethiopia.

3. Can you think of other cases in which researchers' theoretical perspectives have affected their interpretation of the evidence?

those of any other australopithecines, a feature that earned *A. boisei* the nickname "Nutcracker Man." Louis Leakey (1959) formally dubbed it *Zinjanthropus boisei,* but similarities with the robust forms from South Africa led to its eventual inclusion in the genus *Australopithecus.* Today it is formally referred to as *Australopithecus boisei* and less formally as "Zinj." What makes the find particularly notable is that it was the first early hominid find to be dated using a numerical dating technique.

DATING *AUSTRALOPITHECUS BOISEI* Because of the vagaries of nature, the australopithecine finds of South Africa have been difficult to date. Scientists cannot precisely determine the conditions that formed the fossil localities where they were found. However, researchers know with certainty that the deposits have been eroded and disturbed by nature and that fossils of varying ages have been mixed together. In contrast, the fossil deposits at Olduvai Gorge lie in undisturbed strata, occupying the same relative positions in which they were originally deposited. In addition, the area around Olduvai Gorge was volcanically active in the past. As a result, deposits of *tuff,* a porous rock formed from volcanic ash, created distinct layers within the Olduvai deposits, and these layers can be dated by using the potassium-argon method (see Chapter 2). Potassium-argon dates on tuffs above and below Zinj placed the fossil's age at approximately 1.75 million years old. This date, and later estimates on other fossil finds, revolutionized paleoanthropology by finally providing numerical ages for specific fossil specimens.

ROBUST AUSTRALOPITHECINES FROM SOUTH AFRICA

In addition to remains of *Australopithecus africanus,* South African sites have produced remains of other fossil hominids more recent in age than *A. africanus.* Because of this variation, Dart, Broom, and other researchers gave these discoveries a number of new genus and species designations. Although differences of opinion still exist about their exact relationship to other species, for convenience they are generally designated *Australopithecus robustus* to distinguish them from the more delicate, or gracile, *A. africanus.*

South African examples of *A. robustus* are poorly dated, but available evidence suggests that they are more recent than *A. africanus,* perhaps dating between 2 million and 1 million years ago. Scientists have posited that *A. robustus* was larger than *A. africanus,* weighing more than 200 pounds. We can only venture educated guesses about the height and weight of these creatures, however, because of the relatively small pieces of postcranial bone that have been recovered. In fact, the body sizes of the gracile and robust forms may have been comparable (McHenry, 1988), although the skull and dentition of *A. robustus* clearly distinguish it. Specimens have a large, broad face and enormous teeth and jaws. Another feature found in *A. robustus* but absent in *A. africanus* is a sagittal crest. All of these features indicate that *A. robustus* could chew tough, fibrous foods.

HOMO HABILIS: THE "HANDYMAN"

The discovery of Zinj in 1959 sparked a flurry of activity at Olduvai Gorge. Between 1960 and 1964, the Leakeys and their colleagues excavated the fragmentary remains of approximately twenty fossil hominids (Leakey, 1961). Some of these were clearly *Homo erectus;* others appeared comparable to the Zinj find. However, still other fossils pointed to the existence of a creature unlike any of the known australopithecines or *H. erectus.* The distinguishing characteristic of the new species was its cranial capacity, which Louis Leakey estimated at close to 640 cc, significantly larger than that of any australopithecine but still substantially smaller than that of *Homo erectus.* The Leakeys named the creature *Homo habilis,* or "handyman."

For years, many critics challenged this conclusion, maintaining that the fossils fell within the normal cranial range of australopithecines. Eventually, the Leakeys' son, Richard, confirmed his father's interpretations through discoveries from the fossil deposits of Koobi Fora on the eastern shores of Lake Turkana, Kenya. Excavations by Richard Leakey and his coworkers produced several specimens that have been classified as *Homo habilis.* The most complete of these, found in 1972, is known by its Kenya National Museum catalogue number, KNM-ER 1470. Unlike the fragmentary remains of the Olduvai specimens, the 1470 skull is relatively complete and has a cranial capacity of 775 cc. The various *H. habilis* remains from Olduvai and Koobi Fora date from between 2.2 million and 1.6 million years ago (Simons, 1989b). Hence, the range of these hominids overlaps those of the robust australopithecines and, perhaps, the later range of *Australopithecus aethiopicus* and *africanus.*

Recently, the dating of finds from Longgupo Cave in China has suggested that early *Homo* species lived outside Africa. The age of the remains has been placed between 1.9 million and 1.7 million years ago. These finds are quite fragmentary, however, and their relationship to other hominids can be more fully assessed only in light of more recent work.

JAVA MAN: THE "FIRST" *HOMO ERECTUS*

The first bona fide discovery of a pre-*Homo sapiens* hominid was made in 1891 by the Dutch doctor Eugene Dubois. Digging near Trinil on the Solo River in northern Java (an Indonesian island), Dubois found a leg bone, two molars, and the top of a cranium of a hominid. The leg was indistinguishable from that of a modern human, but the cranium, small and flat compared to those of modern humans, had heavy browridges. Dubois named his find *Pithecanthropus erectus* ("erect ape-man"). Today the species is classified as *Homo erectus.*

Dubois (1894) considered his find a missing link between humans and apes, but this view betrayed faulty understanding of Darwin's theory of evolution. Darwin's "missing link" referred to a common ancestor of the human and ape lineages; he never proposed a direct link between modern humans and apes, which represent the end points of distinct evolutionary lines. Other scientists correctly placed *Pithecanthropus* as an intermediary form on the evolutionary track between *Homo sapiens* and an earlier hominid ancestor.

PEKING MAN
AND OTHER *HOMO ERECTUS*

Following the discovery of "Java man," scientists started gathering information about similar creatures at a rapid pace in the first decades of the twentieth century. Many of the most important finds came from Zhoukoudian, about 30 miles southwest of Beijing (then spelled Peking in English transliteration), China. In 1929, a team of researchers led by Chinese geologist W. C. Pei found a skull embedded in limestone during an excavation. Pei showed the skull to Davidson Black, a Canadian anatomist who was teaching at Peking Union Medical College. Concluding from his analysis that this skull represented a form of early human, Black labeled the creature *Sinanthropus pekinensis,* commonly known as "Peking man."

His curiosity piqued, Black undertook additional work at the site, which eventually produced 6 skulls, 12 skull fragments, 15 pieces of lower jaw, 157 teeth, and miscellaneous pieces of postcranial skeleton (Rukang Wu & Lin, 1983). Also unearthed at the site were traces of charcoal (possibly the remains of cooking fires) and stone tools. Anatomist Franz Weidenreich succeeded Black at the medical college and prepared casts, photos, and drawings of the Zhoukoudian fossils. Before the Japanese invasion during World War II, Weidenreich fled China with these reproductions—a fortuitous move for science, because the actual fossils were lost during the war and have never been recovered. Recent dating of the stratum where the Zhoukoudian fossils were recovered suggests that Peking man lived between 460,000 and 230,000 years ago.

In addition to more finds in China, discoveries were made in other areas. Forty years after Dubois's excavations in Java, anthropologist G. H. R. Koenigswald uncovered the remains of comparable early hominids in the same area. Initially, scientists, working with few finds and lacking comparative specimens, speculated that each of these discoveries constituted a new evolutionary branch. We now know that, despite their disparate locations, these early discoveries are all representatives of a single genus and species, today classified as *Homo erectus*. In many respects, *Homo erectus* is identical to modern humans, although the postcranial skeleton is generally heavier and more robust. What most sets this species apart from

Homo sapiens is the cranium, which lacks the high, vaulted appearance of that of modern humans and has a smaller average brain capacity.

Homo erectus was a highly successful and widely dispersed species. Well-dated fossil finds date between 1.6 million and 400,000 years ago. Recent dating of animal bones associated with *Homo erectus* fossils from the sites of Ngandong and Sambungmacan, Java, suggest that pockets of *Homo erectus* populations may have existed as recently as 50,000 years ago. With the exception of *Homo sapiens,* this species had the widest distribution of any hominid. Some of the more recent examples of *Homo erectus* share many similarities with modern humans, perhaps illustrating both the interrelatedness of the species with *H. sapiens* and the arbitrary nature of classification.

Fossil finds bearing *H. erectus* features have been recovered from Kenya, Tanzania, Zambia, Algeria, Morocco, China, and Java (Day, 1986). One of the oldest and most complete finds, known as "Turkana boy," was recovered at the Nariokotome site near Lake Turkana in Kenya. Recently, finds that may date from the same period as Nariokotome have been found outside Africa. Though not fully studied, a possible *H. erectus* mandible from the Republic of Georgia in eastern Europe has been provisionally dated to 1.5 million years ago. The recent redating of finds from fossil localities in Sangriran and Mojokerto in Java to 1.6 million years ago has also raised questions about both the distribution and the evolution of *H. erectus* (Swisher et al., 1994).

INTERPRETING THE FOSSIL RECORD

As paleoanthropologists unearth more and more early hominid fossils, interpretations of hominid evolution become increasingly complex. Initially, scientists drew a straight evolutionary line from *Australopithecus africanus* to *Homo erectus* and on to *Homo sapiens.* But recent finds clearly demonstrate that in several instances more than one species of hominid roamed the earth at the same time. How were these different species related, and how do they relate to the evolution of *Homo sapiens?*

Fundamental to the study of hominid evolution is the question of which features should be used to classify genuses and species (Tattersall, 1986). Because the size and complexity of the brain are the most distinctive physical characteristics of modern humans, increasing cranial capacity is clearly an important feature of the *Homo* line. Yet the range of cranial capacities overlaps among hominids, making it more difficult to identify characteristics that can be used to distinguish individual species. There is also a great deal of variation within species, in features such as body size and skeletal structure. This can be illustrated by studies of living primate species. For example, chimps from Tanzania's Gombe National Park display an astonishing degree of variation in size and skeletal structure (Bower, 1990). In interpreting fragmentary hominid fossils from widely separated localities, we must take into account such natural variations.

In the preceding discussion of fossil evidence for hominid evolution, the names designating specific genuses and species are the most widely accepted appellations used by paleoanthropologists. They are not, however, universally accepted. Perspectives of hominid classification lie between two extremes. Some scientists, known as *splitters,* argue that the current categories do not reflect all the species represented. For instance, some researchers have argued that the *A. afarensis* finds from Hadar do not constitute a single, sexually dimorphic species but rather at least two distinct species. Others in this camp contend that further divisions are called for within the gracile and robust australopithecines (Tattersall, 1986). At the opposite extreme from the splitters are the *lumpers,* who argue that current taxonomic designations place too much emphasis on differences among individuals and not enough on variations within species. This position is best advocated by C. Loring Brace and his colleagues at the University of Michigan (Brace, 1967, 1989; Brace & Montagu, 1965). Brace asserts that the information available on *Homo habilis, A. afarensis,* and *A. aethiopicus* is insufficient to categorize each as a distinct species.

Many different interpretations of hominid evolution have been advanced through the years. Some of these, explored in the following section, are illustrated in Figure 5.3. When they were proposed, they represented valid attempts to explain the fossil evidence available. Like all sciences, paleoanthropology proceeds by formulating hypotheses and then testing them against empirical data. In contrast to most sciences, however, the

data from the fossil record cannot be obtained by laboratory experiments. Rather, paleoanthropologists must await the next unpredictable fossil find. As new evidence is uncovered, new hypotheses are developed, and old ones are modified or discarded.

AUSTRALOPITHECUS AFRICANUS AS ANCESTOR

A number of theories propounded in the 1960s and 1970s placed *A. africanus* at the base of the hominid family tree, as illustrated in Figure 5.3(a). These interpretations of evolution basically held that hominids developed along two main branches. As the most sophisticated of the australopithecines, *A. africanus* was considered the most likely to have given rise to the genus *Homo* and was therefore placed at the bottom of the branch leading to *Homo habilis, H. erectus,* and ultimately *H. sapiens.* The robust australopithecines occupied their own branch, eventually becoming extinct around 1 million years ago. Because of their large teeth and specialized chewing apparatus, the robust australopithecines were not viewed as directly ancestral to *Homo.* In some interpretations, *A. africanus* was located at the base of the hominid tree, suspected of being ancestral to both the australopithecines and *Homo.* Other variations saw *A. boisei* and *A. robustus* on separate branches entirely.

AUSTRALOPITHECUS AFARENSIS AS ANCESTOR

Following the discovery of Lucy and the First Family at Hadar in the 1970s, Donald Johanson and Timothy White proposed a new interpretation of hominid evolution, which is illustrated in Figure 5.3(b). They hypothesized that the genus *Australopithecus* began with *A. afarensis,* dated at about 4 million to 3 million years ago. They contended that *A. afarensis* was the common ancestor of all subsequent hominids. In their scheme, one of the branches from *A. afarensis* leads to *A. africanus* and *A. robustus.* The other major branch leads toward the evolution of *Homo habilis* and succeeding species, culminating in modern *Homo sapiens* (Johanson & White, 1979). The majority of paleoanthropologists concurred with this model of hominid evolution until the mid-1980s.

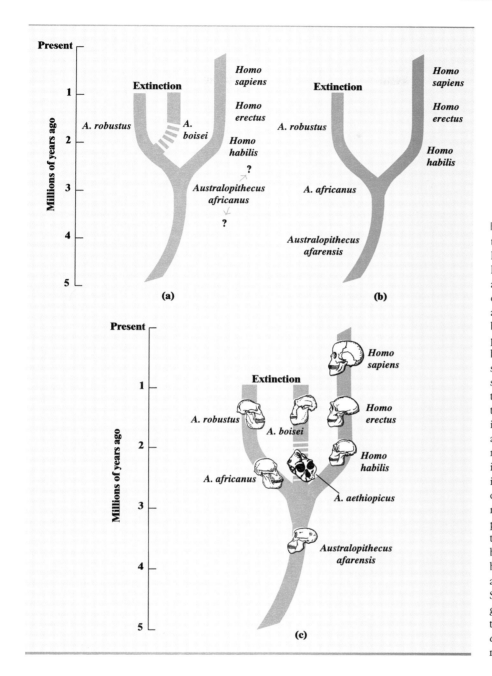

FIGURE 5.3 Different interpretations of hominid evolution: (a) Various theories have placed *A. africanus* in a position ancestral to *Homo* or to both *Homo* and later australopithecines. The robust australopithecines are placed on their own side branch. (b) In 1979, Johanson and White named a new species, *A. afarensis,* which they placed at the base of the hominid family tree leading to both *Australopithecus* and *Homo. A. africanus* was moved to a side branch leading to the robust australopithecines. (c) In 1985, the discovery of *A. aethiopicus* made the picture more complex, suggesting that all of the australopithecines cannot be located on a single side branch. Other interpretations are also being evaluated. Some researchers have suggested that the split between the *Homo* and *Australopithecus* lines occurred prior to 4 million years ago.

REVISED MODELS

More recent discoveries have extended the hominid family tree further back in time. Models must take into account older and more primitive species like *Ardipithecus ramidus* and *Australopithecus ana-*

mensis. Interpretations have also attempted to provide better interpretation of the relationship of more recent hominids. With the discovery of the Black Skull in 1985, a comparatively neat picture of human evolution suddenly grew clouded and more complex. Johanson and White had placed *A. boisei* at the end of the extinct line of australopithecines, a sort of hyper-robust form of *A. robustus*. This was a logical conclusion, given the available fossil evidence at the time. Unfortunately, *A. aethiopicus* doesn't fit into this crisp picture. It has certain characteristics found in *A. boisei* but not in *A. africanus* and *A. robustus,* yet *A. aethiopicus* is as old as *A. africanus.* Scientists wishing to insert the Black Skull into the Johanson and White evolutionary tree would be hard pressed to explain how certain features appeared in *A. aethiopicus,* disappeared in *A. africanus* and *A. robustus,* and then reappeared in *A. boisei.* A more logical and workable interpretation places the Black Skull and *A. boisei* on a branch of their own (Johanson & Shreeve, 1989). Of the two other limbs, one would lead to *A. africanus* and *A. robustus,* the other to the genus *Homo,* as pictured in Figure 5.3(c). Other interpretations are also being explored.

MISSING PIECES IN THE FOSSIL RECORD?

All these views of hominid evolution are based on excavated fossils. Another hypothesis denies that any known australopithecine was ancestral to the genus *Homo.* Rather, according to this view, an as yet unidentified common ancestor of both genera thrived in the remote past, perhaps 6 million years ago. Some scholars contend that australopithecines like Lucy emerged after the split between the *Australopithecus* and *Homo* lineages. In other words, the ancestral form for the *Homo* line is still not known (Shipman, 1986a).

This interpretation recognizes the inadequacy of the fossil record and the incomplete nature of the available data. However, this theory is problematic because it is based on hypothetical evidence—a common ancestor that has not yet been discovered. Fossils of hominids dating earlier than 5 million years ago are fragmentary and their genera remain uncertain. Despite the limited information, the majority of paleoanthropologists accept interpretations that place *A. afarensis* at the base of the hominid family tree because these interpretations stem from known fossil evidence. It is possible that future discoveries will extend the human lineage further back in time and produce an increasingly "bushy" hominid family tree.

GENETIC DIFFERENCES AND HOMINID EVOLUTION

All of the previous interpretations are rooted in studies of the fossil record. During the past several decades, some researchers have approached the study of human evolution from a completely different direction. As noted in Chapter 3, scientists have studied the genetic differences of living animals and attempted to time the divergence of different species.

Comparing modern prosimians, monkeys, and apes, researchers have demonstrated that chimpanzees and humans are identical in many respects and evince differences of less than 2 percent on the biochemical-genetic level. There is slightly more distance between humans and gorillas, but genetically the similarities approach 97 percent. Estimating the time it took for evolution to produce this amount of genetic distance, some researchers have interpreted these results to mean that the separation of the pongids and hominids occurred between 10 million and 4 million years ago (Brown, 1990; Silby & Aquist, 1984). The genetic information, therefore, would appear to complement the australopithecine fossil evidence, which suggests that the divergence had occurred by 4 million years ago.

When cautiously applied, genetic information may facilitate classification and aid in formulating workable hypotheses concerning human evolution. However, the process is not universally accepted. Furthermore, molecular dating at best provides only a rough approximation of the relative genetic distance between different species and the possible time of divergence. It provides no clues to how ancestral hominids adapted to different environments, to their feeding habits, to their geographic range, to their lifeways, or to any of the myriad other questions that concern paleoanthropology.

FROM *HOMO ERECTUS* TO *HOMO SAPIENS*

Several sources of evidence indicate that the earliest human ancestors evolved in Africa. Australopithecines—the oldest hominids—and the first stone tools are from Africa, and our closest genetic relatives, the chimpanzee and gorilla, also emerged there. Climatic conditions on the continent during the Pliocene and Pleistocene were warm and well suited to evolving hominids. The consensus among scientists has been that *Homo erectus* also evolved in Africa. Until recently, although remains of this species had been identified in Asia, the earliest finds, before 1 million years ago, were from Africa. This view has become less viable in light of the recent redating of *H. erectus* finds from Indonesia to dates comparable in age to the oldest African material.

Fossils of *H. erectus* range in age from 1.6 million to 400,000 years old. The longevity of the species is a testament to how well *H. erectus* adapted to different environmental conditions, having ranged across the diverse climates from Africa and southern Europe to Asia.

Scientists cannot pinpoint which selective pressures prompted *H. erectus* to evolve into *H. sapiens.* Presumably, *H. sapiens* must have had some adaptive advantage over earlier hominids, but no consensus has emerged about what specific selective pressures were involved. Among the physical changes found in *H. sapiens* are a larger brain and full speech capabilities, which undoubtedly sparked concomitant behavioral consequences. Yet many of the distinctive characteristics seen in *H. sapiens* stem from cultural factors as well. As will be seen in Chapter 7, *H. erectus* made increasing use of socially learned technology to interact with and control the environment. This trend intensifies in later human populations.

Hominid remains from the period between 400,000 and 200,000 years ago are difficult to classify because they exhibit physical traits characteristic of both *H. erectus* and *H. sapiens*. These hominids, which can be alternately viewed as either advanced *H. erectus* or early *H. sapiens,* are referred to as **transitional forms.** The discovery of finds that do not fit neatly into taxonomic categories is not surprising. As we saw in Chapter 4, related species have many similar characteristics that reflect their evolutionary relationships. Transitional forms illustrate these relationships and offer physical evidence of the process of speciation.

TRANSITIONAL FORMS

In examining the transition from *H. erectus* to *H. sapiens,* we need to cast a critical eye on the physical characteristics that distinguish the two species. *Homo*

erectus shares many physical features with modern humans; in fact, the postcranial skeletons are essentially the same, except for the generally heavier, more massive structure of *H. erectus* bones. The major differences between the two species appear in the skull. *Homo sapiens* skulls are high and vaulted, providing a large cranial capacity. In contrast, the skulls of *H. erectus* feature a **postorbital constriction,** meaning that the front portion of the skull is narrow and the high forehead of *H. sapiens* is absent. Lacking the high, vaulted cranium of *H. sapiens,* the skull of *H. erectus* is widest toward the base.

Homo erectus also exhibits a prognathic face, an attribute of early hominids that is absent in *H. sapiens*. The anterior teeth of *H. erectus* are relatively small compared to those of earlier *Homo* species but large in comparison to those of modern humans. Other distinctive characteristics of *H. erectus* make scientists believe that these creatures had strong jaw and neck muscles. These traits include a slight ridge at the back of the skull and heavy eyebrow ridges, structural features that have disappeared in modern humans.

Transitional forms bearing various combinations of *H. erectus* and *H. sapiens* features have been discovered in Europe, Asia, and Africa. The mosaic of physical characteristics found in some specimens has sparked debate over how to designate species most appropriately. This debate can be illustrated by the Petralona cranium, uncovered in eastern Greece in 1960 (Day, 1986). Scientists dispute the age of the find (claims of 1 million years old have been made), yet the consensus among paleoanthropologists leans toward an age of between 400,000 and 350,000 years. The species designation of this fossil has also been contested. The Petralona cranium exhibits certain *H. erectus* characteristics, including thick bones, pronounced browridges, and a low cranial vault. However, the cranial capacity is estimated at approximately 1,200 cc, placing it at the uppermost limits of *H. erectus* and the lower range of *H. sapiens*.

THE EVOLUTION OF *HOMO SAPIENS*

Although paleoanthropologists generally agree that *H. erectus* evolved into *H. sapiens,* they disagree about how, where, and when this transition oc-

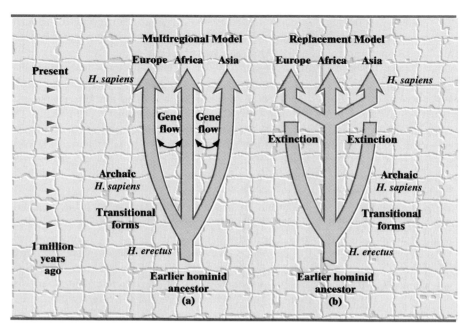

FIGURE 5.4 Two different interpretations of the emergence of *H. sapiens*. The multiregional evolutionary model (a) suggests regional continuity and the gradual evolution of all *H. erectus* and archaic *H. sapiens* populations into modern humans. In contrast, supporters of the replacement model (b) see modern humans as evolving in one world area and spreading out, replacing earlier hominid populations.

curred. Early interpretations were based on limited information and often emphasized the uniqueness of individual finds. Recent researchers have offered a number of different theories (Howells, 1976; Mellars, 1988, 1989; Sussman, 1993). These fall into two overarching models representing opposing perspectives: the multiregional evolutionary model and the replacement model.

MULTIREGIONAL EVOLUTIONARY MODEL

According to the **multiregional evolutionary model,** the gradual evolution of *H. erectus* into modern *H. sapiens* took place in many regions of Asia, Africa, and Europe at the same time, as illustrated in Figure 5.4(a). Through natural selective pressures and genetic drift, local *H. erectus* populations developed particular traits that varied from region to region. However, as characteristics of *H. sapiens* appeared in certain areas, gene flow—the widespread sharing of genes—between populations prevented the evolution of distinct species. The emergence of *H. sapiens* was, therefore, a widespread phenomenon, although different regional populations continued to exhibit distinctive features.

Working from the multiregional evolutionary model, we would expect to see a great deal of regional genetic continuity, meaning that the fossil finds from a particular geographic area should display similarities from the first *H. erectus* to those of modern populations. Supporters of this model argue that such continuities do indeed exist. For example, skeletal remains of early *H. sapiens* from different regions of China, North Africa, and Europe resemble modern populations in those areas in some respects (Smith, 1984; Thorne & Wolpoff, 1992). Certain distinctive features can be identified in the cranium, dentition, jaws, and particular features of the postcranial skeleton.

REPLACEMENT MODEL

The second major paradigm to explain the evolution of modern humans is the **replacement model,** or the single-source model (Stringer, 1985). This model holds that *H. sapiens* evolved in one area of the world first and migrated to other regions, as illustrated in Figure 5.4(b). It is called a replacement model because it assumes that *H. sapiens* were contemporaries of the earlier *H. erectus* but eventually replaced them. Thus, although the modern and archaic species overlapped in their spans on earth, they were highly distinctive, genetically different evolutionary lineages. According to the replacement hypothesis, *H. sapiens* populations all descended from a single common ancestral group. Consequently, there is minimal diversity among modern humans.

Some researchers believe that fossil evidence supporting the replacement hypothesis may be found in the homeland of all hominids, Africa. Fossils of anatomically modern *H. sapiens,* provisionally dated to between 130,000 and 70,000 years ago, have been found in eastern and southern Africa (Stringer & Andrews, 1988). In Omo, Ethiopia, hominid remains consisting of a mandible and postcranial bones have been classified as *H. sapiens.* In addition, some intermediate fossils with both archaic and modern traits have been found in North Africa. These African fossil finds may represent the earliest examples of modern humans found anywhere in the world. Some advocates of the replacement model contend that after evolving in Africa, early *H. sapiens* migrated to other regions, replacing earlier hominid populations that had arrived in those same regions hundreds of thousands of years before.

However, the fossil evidence from Africa has a number of limitations. Scientists cannot pinpoint the precise stratigraphic position of some of the finds, and the ages of the sites where the remains were located pose problems as well (Deacon, 1992; Klein, 1992; Mellars, 1989). The finds may be younger than present dating indicates. If so, they would not predate finds from other areas, thereby providing no support for the hypothesis that all humans diffused from Africa. Other researchers have postulated that *H. sapiens* may have originated in Southwest or East Asia. But these hypotheses, too, cannot be confirmed through available fossil evidence.

It is also possible that the emergence of anatomically modern humans was more complex, and encompassed more variables, than can be neatly wrapped up in either of the two overarching models (Lahr & Foley, 1994). Emergent human populations may have incorporated a great deal of physical diversity—as well as behavioral, social, and linguistic differences. It is unlikely that migrations ("Out of Africa" and elsewhere), were nice unidirectional affairs involving the movement of homogeneous populations. Many different migrations via different routes, recolonization of previously occupied territories, and gene flow with other populations was

more probable. Understanding of such variables may provide insight into not only the emergence of modern humans, but also the source of the diversity underlining present-day populations.

MITOCHONDRIAL DNA RESEARCH

Researchers have also brought biochemical techniques to bear on the question of the origins of modern humans. Working at the University of California, Berkeley, in the 1980s, a team of researchers studied the mitochondrial DNA of modern women (Cann, 1987; Stoneking et al., 1987). On the basis of the studies, which were widely publicized, they argued that modern humanity could be traced back to a single African female who lived between 200,000 and 130,000 years ago. Although seemingly providing an important means of validating fossil evidence, the technique proved to be fraught with problems.

The apparent strength of the technique lies in the distinctive characteristics of *mitochondrial DNA (mtDNA).* This type of DNA is located in the portion of the cell that helps convert cellular material into the energy needed for cellular activity. In contrast to nuclear DNA (see Chapter 3), mtDNA is not carried by the sperm when it fertilizes the egg. The genetic code embedded in mtDNA, therefore, is passed on only through the female. Thus, each of us inherits this type of DNA from our mother, our mother's mother, and so on, along a single genealogical line.

The study by the Berkeley team focused on the mtDNA of 147 women from Africa, Asia, Europe, Australia, and New Guinea (Cann et al., 1987). The accumulation of random mutations in the different populations displayed distinctive patterns. The mtDNA of the African women tended to be more diverse, or *heterogeneous,* suggesting that mutations present had a long time to accumulate. In other populations, the mtDNA was more uniform, or *homogeneous,* a sign that they had not had as much time to accumulate mutations. Using a set mutation rate, the researchers inferred a maternal line in Africa dating back to between 200,000 and 130,000 years ago. This interpretation of the mtDNA data favored a replacement model, with Africa as the place of origin of all later human populations.

Although mitochondrial dating initially received a great deal of media attention, aspects of the research were soon challenged by other researchers. Some researchers have noted that mtDNA may mutate rapidly and at irregular rates. If so, studies of mtDNA could not yield any precise dates for the separation of different human populations. Even more problematic is the computer program that was used to create the model. The computer created a human family "tree" by making the fewest number of branches that could be used to explain the genetic relatedness of the individuals studied. Unfortunately, the computer could produce millions of equally simple trees, with no guarantee that it had identified the correct model. Other computer runs produced trees that supported Asian or European origins. Although more recent studies have addressed some of these issues, the debate over the validity of the technique continues (Frayer et al., 1993; Moontain, 1998; Sussman, 1993; Templeton, 1993; Vigilant et al., 1991; Wilson & Cann, 1992).

ARCHAIC *HOMO SAPIENS*

Although debate still rages over the classification of certain hominids dating between 200,000 and 400,000 years ago, there is much more agreement over later finds. For the most part, all hominid fossils dating to the last 200,000 years are classified as *H. sapiens.*

This is not to say that *H. sapiens* populations of 200,000 years ago were identical to modern humans. Anatomically modern humans did not appear until between 130,000 and 40,000 years ago, earlier in some regions than others. Even these fossils do not have all the characteristics of modern populations. However, the distinctive features noted in these early *H. sapiens* populations are all considered within the range of variation found in a single species. Individual fossil finds are sometimes given subspecies names to signal a particular distinction, as, for example, *Homo sapiens steinheimensis,* found in a gravel pit in Steinheim, Germany on the River Murr in 1933. To simplify our discussion, hominids of the last 200,000 years can be divided into two categories: **archaic *Homo sapiens*** and **anatomically modern *Homo sapiens.***

Archaic *H. sapiens* follow the evolutionary path leading through transitional forms of *H. erectus* to modern *H. sapiens.* Indeed, some finds dating before 200,000 years ago, including the Petralona cranium, have been labeled archaic *H. sapiens.* At the

other extreme, hominid fossils with archaic *H. sapiens* features overlap with clearly modern humans.

The mosaic of features that characterizes archaic *H. sapiens* takes clear shape in remains unearthed at the Broken Hill Mine in Kabwe, Zambia. These finds consist of a cranium and the postcranial bones of three or four individuals, dated to at least 125,000 years ago (Day, 1986). On the one hand, the thickness of the bone, heavy browridges, and sloping forehead of the cranium are characteristic of *H. erectus.* Also like *H. erectus,* the Kabwe skull is widest at

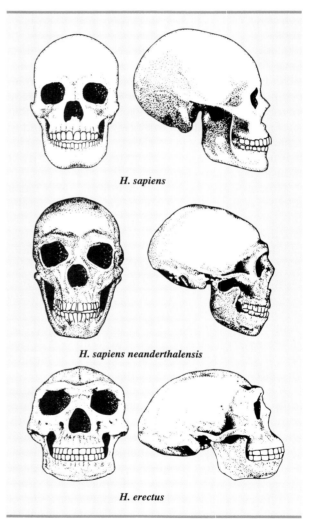

H. sapiens

H. sapiens neanderthalensis

H. erectus

FIGURE 5.5 A comparison of the skulls of *H. erectus, H. sapiens neanderthalensis,* and modern *H. sapiens.* The most distinctive feature of the latter is the high, vaulted forehead and the prominent chin.

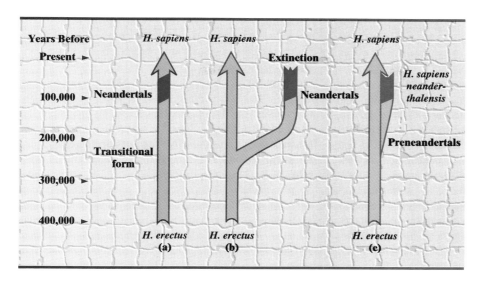

FIGURE 5.6 Three interpretations of the evolutionary relationships between Neandertals and modern humans: (a) unilinear evolution, (b) separate lineages, and (c) preneandertals.

the base. On the other hand, the cranial capacity is large (1,280 cc), and the postcranial skeleton bears a strong resemblance to *H. sapiens* (Kennedy, 1984). The Kabwe find once again raises the question of which features best differentiate species.

HOMO SAPIENS NEANDERTHALENSIS

One of the best-known examples of archaic *H. sapiens* is *H. sapiens neanderthalensis,* also known popularly as "Neandertal man." Neandertal fossils dating between 130,000 and 35,000 years ago have been discovered in Europe and the Middle East. In the past, climatic conditions in this area spanned a more extreme range than they do today. The southern regions had warmer, milder climates, and the northern regions were partially glaciated and extremely cold.

The Neandertal physique has become the model for the "cave men" portrayed in the media. They have often been portrayed as second rate hominids, swept to extinction by quicker-thinking modern humans (Brace, 1964; for readable, fictional portrayals, see Auel, 1981; Golding, 1981). This depiction stems, in part, from an early find of a skeleton of an elderly individual whom scientists later determined had suffered from arthritis. In fact, Neandertals were quite literally thick-skulled and had the heavy browridges seen in *H. erectus.* In the classic Neandertal, the midportion of the face protruded as if the nose and surrounding features were pulled forward (Figure 5.5).

The front teeth of Neandertals were larger than those of modern humans. Often, Neandertal teeth bear evidence of heavy wear (some were actually worn down to stubs), which leads researchers to believe that the teeth may have been used by Neandertals in much the same way as tools.

The image of the Neandertal as an entirely brutish creature is misleading. The large Neandertal cranial capacity ranged from 1,200 to 2,000 cc and could accommodate a brain as large as, or even larger than, that of a modern human. Moreover, relatively recent studies of Neandertal endocasts indicate that the structure and intellectual capacities of the Neandertal brain mirrored those of modern humans (Holloway, 1985). Artifacts used by these populations reflect a much more complex range of adaptive skills than do those of pre-*H. sapiens* hominids (see Chapter 7).

NEANDERTALS AND MODERN HUMANS Ever since the first Neandertal skulls were found in the nineteenth century, scientists have pondered the links between Neandertals and modern humans (Figure 5.6). Early interpretations that viewed Neandertals as an intermediate ancestor between *Homo erectus* and anatomically modern humans have been discarded. Their restricted geographic range and distinctive physical characteristics makes this scenario unlikely. Neandertals also appear to have coexisted with anatomically modern humans until the relatively recent past, perhaps as little as 30,000 years ago. A growing consensus among anthropologists holds that Neandertals were *H. sapiens* with distinctive physical features, but no one has come up with a cogent, widely accepted theory to explain which selective pressures produced these features. Paleoanthropologists tend to favor the hypothesis that a "pre-Neandertal" *H. sapiens* population, possibly originating in another region and migrating to the classic Neandertal area, underwent severe natural selection in response to the cold environment of Europe. In this view, natural selec-

tion and lack of gene flow with other *H. sapiens* populations produced the distinctive Neandertal characteristics. Such an interpretation might be consistent with recent molecular testing of genetic material extracted from Neandertal bone. These data suggest that Neandertals did not contribute to the mitochondrial DNA pool of modern populations (Krings et al., 1997). Despite some unique features, Neandertals remained *H. sapiens* and were eventually absorbed by later human populations with whom they interbred.

 ## SUMMARY

The hominids, a distinct family in the order Primates, include modern humans and their immediate ancestors. Although hominids share certain general features with all primates, they also evince a number of distinct characteristics. These include a fully bipedal posture; reduction of the face, jaw, and anterior teeth; and (in the genus *Homo*) increasing cranial capacity. Fossil evidence indicates that bipedalism stands as the earliest hominid trait, evolving sometime between 4 million and 10 million years ago. Changes in other features appear more recently, primarily during the past 2 million years.

Hominid fossils belong to a number of species and can be divided into two genera: *Australopithecus,* which includes species that may be ancestral to humans as well as extinct side branches, and *Homo,* the genus that includes modern humans. Paleoanthropologists believe that *A. afarensis* is at the base of the hominid family tree, with at least two branches that lead to later australopithecines on one side and one branch leading to the genus *Homo* and modern humans on the other.

Homo sapiens, the species that includes modern humans, evolved from *H. erectus* populations between 400,000 and 200,000 years ago. Skeletal remains from this time span often reveal a mix of characteristics from both species. These transitional hominid forms, which are difficult to classify, can be viewed as either advanced *H. erectus* or early *H. sapi-*

ens. Anthropologists point to these transitional forms as physical evidence of the process of speciation.

A general consensus holds that *H. erectus* populations gave rise to *H. sapiens* by 200,000 years ago; yet how, when, and where this transition occurred are topics of heated debate. The opposing perspectives fall into two overarching models. Advocates of the multiregional evolutionary model believe that *H. sapiens* emerged in Asia, Africa, and Europe more or less concurrently. According to this theory, a high degree of regional continuity marks the evolution of humans from the first arrival of *H. erectus* up to the present. In contrast to the multiregional perspective are several theories united in their view of a replacement, or single-source, model of *H. sapiens* origins. These theories argue that *H. sapiens* evolved first in one region of the world, later diffusing into other regions and replacing earlier hominid populations. Some researchers have concluded that Africa was the point of origin of *H. sapiens,* with expansion outward from there. Although this theory may hold a great deal of promise, fossil evidence and genetic studies must be evaluated cautiously.

Hominid skeletons from throughout the world dating to the last 200,000 years are generally classified as *H. sapiens.* Yet we find in these forms a great deal of variation, prompting archaeologists to identify them by a number of subspecies names. For convenience, *H. sapiens* can be divided into two categories:

archaic and anatomically modern forms. Archaic *H. sapiens* retain some of the traits seen in earlier *H. erectus* populations. Anatomically modern *H. sapiens,* showing most of the characteristics associated with modern humans, appeared approximately 130,000 to 40,000 years ago in different world areas.

The best-known archaic *H. sapiens* forms, *H. sapiens neanderthalensis,* have been identified through hundreds of finds in Europe and the Near East. Because Neandertals have many distinctive physical features, anthropologists have advanced many theories regarding their relationship to anatomically modern humans. Current consensus tends to regard Neandertals as an archaic subspecies of *H. sapiens* that disappeared as a result of intensive selective pressures and genetic drift.

 QUESTIONS TO THINK ABOUT

1. What are the critical evolutionary tends that differentiated the hominids from other primates?

2. Controversy surrounds the issue of how many genuses and species of australopithecines are present in the fossil record. How many genuses and species do you think there are? Is *Homo habilis* just another australopithecine? On what basis are these distinctinos made?

3. Many interpretations of hominid evolution (phylogenetic or family trees) have been advanced through the years. Which of the scenarios given in the text do you think makes the most sense? Why?

4. Fossils are not the only evidence used to provide information on hominid evolution. What are some of the other sources of information and data that are used to reconstruct the phylogeny of the hominids?

5. Comment on this statement: The transition from *Homo erectus* to *Homo sapiens* is easily documented by numerous fossils that have been found throughout the world that show a gradual change in morphology that occurred over a period of only 5,000 years.

6. Contrast the multiregional evolutionary model with the replacement model for the emergence of anatomically modern humans. Can you think of a "critical tests" that would allow you to evaluate the validity of these theories?

 KEY TERMS

anatomically modern *Homo sapiens*
archaic *Homo sapiens*
bipedalism

foramen magnum
hominids
multiregional evolutionary model
postorbital constriction

replacement model
sagittal crest
transitional forms

 SUGGESTED READINGS

BLINDERMAN, C. 1986. *The Piltdown Inquest*. Buffalo, NY: Prometheus Books. An account of the "discovery" and uncovering of the Piltdown man fraud. Although it does not conclusively solve the mystery, the book discusses all the possible suspects in this scientific whodunit. Now more than four decades old, J. S. Weiner's *The Piltdown Forgery* published in 1955 still conveys some of the excitement and speculation that surrounded the unmasking of the forgery.

JANUS, CHRISTOPHER J., AND WILLIAM BRASHER. 1975. *The Search for Peking Man*. A gripping tale of the discovery, disappearance, and search for the still missing remains of Peking man.

JOHANSON, DONALD C., AND MAITLAND A. EDEY. 1981. *Lucy: The Beginnings of Humankind*. New York: Simon & Schuster. An exciting, firsthand account of Johanson's discovery of "Lucy" (*Australopithecus afarensis*) in Hadar, Ethiopia, and the role of this discovery in restructuring interpretations of human origins. The book also includes an excellent overview of the history and development of the field of paleoanthropology.

LEWIN, ROGER. 1993. *The Origin of Modern Humans*. New York: Scientific American Library. This volume provides a highly readable introduction to the questions of when and where modern humans originated. It is well illustrated and also includes a glossary of key terms.

TRINKAUS, ERIK, AND PAT SHIPMAN. 1994. *The Neandertals: Of Skeletons, Scientists, and Scandal*. New York: Vintage Books. A gripping look at the discovery and interpretation of Neandertal, presented in historical perspective from the first finds of the nineteenth century to today.

Chapter 6

HUMAN
VARIATION

CHAPTER OUTLINE

A S WE NOTED IN CHAPTER 5, *HOMO SAPIENS* populations migrated throughout the world, settling in all sorts of climatic and environmental settings. Even though modern humans live in more diverse environmental conditions than does any other primate group, we all bear a striking genetic similarity. Recent scientific assessments of population genetics and human variation confirm that although humans are widely distributed and have experienced a degree of reproductive isolation, no human population has become so genetically isolated as to constitute a separate species. Rather, all modern humans are the product of a tremendous amount of gene exchange.

Although modern humans represent a single species, we are not all alike. Differences in many physical characteristics such as height, skin color, hair texture, and facial features, are readily discernible. Many other differences among human populations are the result of human culture—how we dress, the kinds of houses we live in, our marriage customs—and are not genetically determined. Physical anthropologists study humans as a biological species, whereas cultural anthropologists attempt to unravel the myriad elements of cultural diversity. As we shall see, however, cultural practices also affect

genetic and physical variation by influencing gene flow or altering the environment. Human physical variation and the challenge of identifying the sources of this diversity are the focus of this chapter.

SOURCES OF HUMAN VARIATION

To understand variation among human populations we must consider three primary causes: (1) *evolutionary processes,* which affect genetic diversity within and between populations, (2) *environment*—the variation among individuals that springs from their unique life experiences in specific environments, and (3) *culture*—the variation stemming from disparate cultural beliefs and practices inculcated during the formative years and reinforced throughout life.

GENETICS AND EVOLUTION

As discussed in Chapter 3, a population's total complement of genes is referred to as a *gene pool.* In *Homo sapiens,* as well as in many animal populations, genes may have two or more alternate forms (or alleles)—a phenomenon called **polymorphism** (literally, "many forms"). These differences are expressed in various physical characteristics ranging from hair and eye color to less visible differences in blood chemistry. Many of these traits vary in their expression in different world areas. For example, we associate certain hair texture and skin color with populations in specific geographic areas. Species made up of populations that can be distinguished regionally on the basis of discrete physical traits are called **polytypic.**

The genetic variation present in the human species is the product of the four fundamental processes of evolution that were examined in Chapter 3: mutations, natural selection, gene flow, and genetic drift. *Mutations,* which are random changes in the genetic code, bring about changes in alleles. Mutations may result in evolutionary change only if they occur in the sex cells of individuals, enabling this change to be passed on to succeeding generations. Mutations are important in explaining human variation because they are ultimately the source of all genetic variation. They may be beneficial, detrimental, or neutral in terms of an organism's reproductive success. The evolutionary process that determines which new mutations will enter a population is *natural selection.* Through this evolutionary process, traits that diminish reproductive success will be eliminated, whereas those enhancing the ability to reproduce will become more widespread.

Although natural selection has favored certain traits in human populations, it does not explain all genetic variation. Some physical characteristics, such as eye color, confer no discernible reproductive advantages. We might expect such *neutral* traits to be evenly distributed throughout human populations as a result of gene flow, yet this is not the case. The nonrandom distribution of neutral traits illustrates *genetic drift,* the processes of selection that alter allele frequencies. Genetic drift is particularly useful in explaining differences among genetically isolated populations.

Consider the physical differences that distinguish the people of central Africa from those living in northern Arctic regions. Gene flow, or the introduction of new genes into a population's gene pool through interbreeding with another population, is highly unlikely for these geographically distant peoples. In addition, most human cultures maintain rules of *endogamy*—that is, marriage to someone within one's own group—thereby further restricting gene

flow. Scientists speculate that Paleolithic populations consisted of small bands of between thirty and one hundred individuals in which genetic drift may have been an important factor.

THE PHYSICAL ENVIRONMENT

The physical environment influences human variation by promoting or restricting growth and development. Physical differences among humans may arise as a result of how well requirements for growth are met. **Acclimatization** is the physiological process of becoming accustomed to a new environment. We can examine the effects of the physical environment by studying how individuals with similar genetic makeup develop in different environmental settings. If, for example, identical twins were separated at birth and reared far apart in different regions of the world, any physical variation between them could be attributed to their disparate physical environments. In fact, studies have demonstrated that humans are highly sensitive to changes in their physical surroundings.

CULTURE

Many of the features that distinguish human populations stem from *culture.* People differ in the customs and beliefs that guide the way they eat, dress, and build their homes. Such differences are primarily superficial. If a child born in one region of the world is raised in another culture, he or she will learn and embrace the customs and beliefs of the adopted culture. Culture may influence human genetic variation through religious beliefs, social organization, marriage practices, or prejudices that restrict intermarriage among different groups and thus inhibit gene flow. Cultural beliefs also determine diet, living conditions, and the environment in which people work; these effects, in turn, either promote or hamper human growth and development.

EVALUATING REASONS FOR VARIATION

Although we know that genetic, environmental, and cultural factors all contribute to human variation, it is often difficult to assess the relative importance of each (see the box "The Human Aggression: Biological or Cultural?"). All three influences, in combination, yield the characteristics found in an individual.

CRITICAL PERSPECTIVES
HUMAN AGGRESSION: BIOLOGICAL OR CULTURAL?

One of the challenges faced by anthropologists in their studies is delineating the specific biological, environmental, and cultural variables that result in the particular phenomena they are interested in. As seen in Chapter 4, some researchers have suggested that natural selection has acted on primate and human behavior in much the same way it has selected certain physical characteristics. Following this line of argument, some behaviors are seen to be adaptive because they lead to enhanced reproductive success.

One area of human behavior that has been examined from many different perspectives is aggression. Throughout history, humans have been confronted with questions about violence, warfare, and conflict. Indeed, it sometimes appears as if aggression is an inescapable aspect of life. In today's world of mass media, we are con-

stantly confronted with accounts of violence and aggression, ranging from sports brawls to assaults and murders to revolutions and wars. The seeming universality of aggression has led some people to conclude that humans are "naturally" violent. What are the causes of human aggression? Is violence inbred, or is it learned?

Not surprisingly, anthropologists and other social scientists have examined these issues for decades, often arriving at conflicting conclusions. Disputes concerning the origins of violence reflect the biology-versus-culture debate. Some psychologists and ethologists attribute warfare and violence to humans' psychobiological-genetic heritage. For example, ethologist Konrad Lorenz developed an elaborate biological hypothesis based on comparisons with animal behaviors. In his widely read book *On Aggression*

(1966), Lorenz proposed that during humanity's long period of physical evolution certain genes were selected that provide humans with an aggressive instinct, which has survival value for the species. He argued that this instinct evolved through natural selection because of intergroup warfare and competition.

Lorenz noted that nonhuman animals usually do not kill within their species and that aggression among males within species is highly ritualized and rarely leads to death. Male deer, wolves, and other social animals fight one another, but this fighting establishes a hierarchy of dominant and submissive males and therefore helps to ensure order within the group. Thus, nonhuman animals have an instinct for inhibiting aggression that is activated by the ritualized fighting behavior. Humans, in contrast, have evolved as physical-

We can see the intertwined nature of these sources of variation by examining body height. How tall a person grows clearly stems, in part, from his or her genetic makeup. This can be illustrated by certain African, Philippine, and New Guinean populations that have mean heights of less than 5 feet. This average is much lower than that of most other human populations. Studies indicate that the relatively short stature in these populations is caused by a deficiency in a hormone that stimulates growth, a genetic trend (Shea & Gomez, 1988).

At the same time, however, height varies significantly even among populations that are genetically similar. One way to account for this is to examine

variation in environmental factors, such as the amount of sunlight a person is exposed to, the average daily temperature, changes in the available food resources, and rates of exposure to disease. Consider seasonal changes in growth rates: Children living in temperate climates grow more quickly during the spring and summer than during the fall and winter, and children in tropical climates experience growth spurts during the dry season rather than during the rainy season (Bogin, 1978). In both instances, scientists conjecture that more rapid growth correlates with greater exposure to sunlight, although precisely how this works remains unclear. One theory holds that the increased sunlight in certain seasons

ly weak creatures without sharp teeth, claws, beaks, or tremendous strength. Therefore, the instinct for inhibiting violence was not selected for the human species. According to Lorenz, this accounts for the prevalence of warfare in human societies and makes humans highly dangerous animals. Compounding this loss of instinctual inhibitions against violence is the human technological capacity to produce deadly weapons.

Many anthropologists challenged this concept of a universal instinct for aggression. Citing ethnographic evidence from sociocultural systems that experienced little violence, Ashley Montagu, Marshall Sahlins, and others proposed that cultural factors are more important than biological instincts in determining aggression (Montagu, 1968). They argued that human behavior can be shaped and influenced in many ways. Humans can be extremely violent or extremely pacif-

ic, depending on the prevailing cultural values and norms.

During the 1980s, anthropologists such as Napoleon Chagnon hypothesized that aggression is related to reproductive success and inclusive fitness. Although they denied the existence of an aggressive instinct, these anthropologists agreed with the biological school that humans possess an innate capacity or predisposition for violence. Violent behavior can be triggered by a number of factors that threaten humans' capacity to survive and reproduce, including scarcities of resources, excessive population densities, and significant ecological developments. They do, however, acknowledge that the norms and values of a particular sociocultural system can either inhibit or enhance aggressive tendencies. Thus, the same pressure that would lead to warfare in one society might be resolved without violence in a society with a different set of values.

Anthropologists today generally concur that human aggression results from both biological and cultural factors, although they continue to disagree about the relative importance of each. Meanwhile, research into the origins of aggression has raised some basic questions concerning human behavior.

Points to Ponder

1. If cultural norms and values can promote violence, can they also eliminate or sharply reduce it?

2. Can you foresee a society (or a world) in which conflicts and problems are resolved peacefully? Or will violence always be with us?

3. Given the complex relationships among various cultural and biological factors, what concrete steps can a society take to reduce the level of aggressive behavior?

---

stimulates the body's production of vitamin D, which promotes bone growth.

Finally, cultural factors can also affect people's health and, as a consequence, their growth. In certain cultures, for example, some social groups have greater access than others to food, shelter, and protection against childhood diseases, all of which affect growth rates. Underprivileged children whose basic nutritional needs are often unsatisfied will not grow as tall as those born into a society with material abundance.

Because of the complex interrelationships among genetic, environmental, and cultural influences, the relative importance of each of these elements can

be deciphered only through detailed analysis of specific human populations.

THE CONCEPT OF RACE

Physical characteristics such as skin pigmentation, nose shape, and hair texture have prompted people throughout history to classify humans into different "races." To modern biologists, **races** constitute divisions within humankind based on identifiable hereditary traits (Brues, 1977). These divisions are an attempt to categorize the great variation among humans. Bi-

An Ainu elder male. The Ainu, an ethnic group in northern Japan, are generally distinguished from other Japanese by such physical features as lighter skin and higher-bridged noses—attributes frequently associated with European populations. The distribution of characteristics like these confounds attempts at racial classification.

ologists recognize these divisions for what they are: crude and roughly drawn boundaries, at best.

Although the diversity of human populations is undeniable, delineating specific races has little practical or scientific value in studying human variation (Gould, 1977). As we shall see, physical characteristics do not divide humans into readily discernible groups or races. Furthermore, classification of physical characteristics serves only to label particular categories of phenomena arbitrarily selected by the researcher. It does not explain the reason for the observed variation. Despite the scientific limitations of the concept of race, we examine early racial classifications because incorrect and faulty ideas stemming from some classifications are still widespread. Table 6.1 summarizes racial classification systems.

ANCIENT CLASSIFICATION SYSTEMS

Early racial classifications were *folk taxonomies,* informal and unscientific classifications based on skin color. In the fourteenth century B.C., the ancient Egyptians divided all human populations into one of four categories: red for Egyptians, yellow for people to the east, white for people to the north, and black for Africans to the south (Gossett, 1963). Later, in the biblical book of Genesis, a similar classification scheme appears in a tale chronicling the distribution of the human population: "And the sons of Noah that went forth from the ark were Shem, Ham and Japheth: . . . these are the three sons of Noah: and of them was the whole earth overspread" (Genesis 9:18–19).

The descendants of Shem (the Semites) were the ancient Israelites. The descendants of Ham ventured to the south and the east, and the descendants of Japheth moved north. The word *Ham* originally meant "black" and referred to the black soil of the Nile delta, but its meaning was eventually changed to describe the skin color of Ham's descendants. At the end of Genesis, the descendants of Ham are condemned to be "servants of servants unto [their] brethren" (Genesis 9:25). Many Westerners subsequently cited this passage as the justification for an entrenched system of racial discrimination (Leach, 1988).

By correlating physical characteristics with cultural differences, classification systems such as these assumed erroneously that populations that shared certain physical traits, especially skin color, also exhibited similar behaviors. These beliefs gave rise to many popular misconceptions and generalizations concerning the values, traditions, and behaviors of non-Western peoples.

EARLY "SCIENTIFIC" STUDIES OF RACE

During the eighteenth and nineteenth centuries, the body of scientific knowledge expanded, and scientific methodologies became more sophisticated in the West; scientists and philosophers thus began to apply scientific principles to the question of race. In one of the earliest scientific efforts to organize human variation into racial categories, the Swiss scientist Carolus Linnaeus constructed a taxonomy in 1758 that divided *H. sapiens* into four races based on skin color: Europeans (white), North American Indians (red), Asiatics (yellow), and Africans (black). Linnaeus is discussed in Chapter 4 for his systematic classification of all living creatures.

In 1781, a German scientist, Johann Blumenbach, devised a racial classification system that is still some-

TABLE 6.1   How Many Races Are There?

Examples of different racial classifications, and their basis, are contained in the table. Other researchers have suggested completely different races and definitions. The great disagreement among scientists over the number and characteristics of different races is a good indication of the limited usefulness of the concept. Most modern researchers focus their efforts on explaining why there is variation in particular traits.

Origin of Theory	Number of Races	Description	Basis of Classification
Ancient Egyptians, 14th century B.C.	4	Egyptians (red), Easterners (yellow), people from the north (white), and people to the south (black)	Skin color
Carolus Linnaeus, 1735	4	Europeans (white), North American Indians (red), Asiatics (yellow), Africans (black)	Skin color
Johann Blumenbach, 1775	5	Caucasian, Ethiopian, Mongolian, Malay, Native American	Skin color, hair color, facial features, and other physical traits
J. Deniker, 1900	29	Adriatic, Ainu, Assyroid, Australian, Berber, Bushman, Dravidian, Ethiopian, Littoral-European, Western-European, Northern-European, Eastern-European, Ibero-Insular, Indo-Afghan, Indonesian, Melanesian, Negrito, Negro, Polynesian, Semite, South American, North American, Central American, Patagonian, Eskimo, Lapp, Ugrian, Turkish, Mongol	Hair color and texture, eye color
William Boyd, 1950	6	European, African, Asiatic, American Indian, Australoid, Early European	Blood groups
Carleton Coon, Stanley Garn, & Joseph Birdsell, 1950	30	Murrayian, Ainu, Alpine, N.W. European, N.E. European, Lapp, Forest Negro, Melanesian, Negrito, Bushman Bantu, Sudanese, Carpentarian, Dravidian, Hamite, Hindu, Mediterranean, Nordic, N. American Colored, S. African Colored, Classic Mongoloid, N. Chinese, S.E. Asiatic, Tibeto-Indonesian, Mongoloid, Turkic, Am. Indian Marginal, Am. Indian Central, Ladino, Polynesian, Neo-Hawaiian	Evolutionary trends, body build, and special surface features such as skin color and facial structure
Stanley Garn, 1961	9	Africans, Amerindian (Native Americans), Asiatics, Australians, Europeans, Indians, Melanesian-Papuans, Micronesians, Polynesians	Geographic boundaries restricting gene flow
Walter Bodmer, 1976	3	Africans, Caucasians, Easterners (including Australians and Pacific Islanders)	Major geographical groups

times used in popular, unscientific discussions of race. He divided humans into five distinct groups—Caucasian, Mongolian, Malay, Ethiopian, and Native American—corresponding to the colors white, yellow, brown, black, and red, respectively. Blumenbach based his racial typology primarily on skin color, but he considered other traits as well, including facial features, chin form, and hair color.

LIMITATIONS OF CLASSIFICATION SYSTEMS

Because these typologies were created before Darwin and Mendel had published their findings, they did not incorporate the modern principles of natural selection, heredity, and population genetics. For example, Mendel's principle of independent assortment

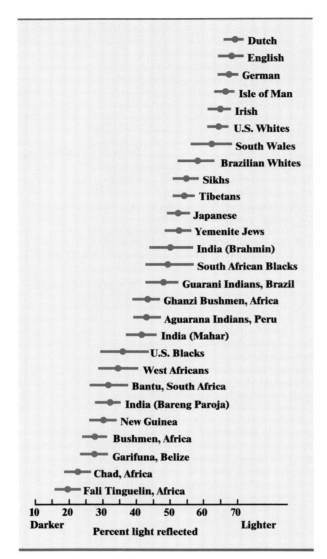

FIGURE 6.1 Variation in skin color, as measured by the amount of reflected light. The measurements cannot be divided into natural divisions, thus illustrating the arbitrary nature of racial classification.

Source: From *The Human Species: An Introduction to Biological Anthropology* by John Relethford. Copyright ©1990 by Mayfield Publishing Company. Reprinted by permission of the publisher.

holds that physical traits are not linked together in the process of reproduction and transmission of genetic material. In other words, there is no "package" of characteristics that is passed on to members of different "races." Thus, blond hair and blue eyes are not consistently found in tandem, just as a specific skin

color and hair texture are not linked to each other. Rather, these traits are independent of one another, leading to varying combinations in different individuals. Variation in the combination of traits makes it impossible to classify races according to well-defined criteria that hold for entire populations.

CONTINUOUS VARIATION Scientists encounter another fundamental problem in distinguishing races. Instead of falling into discrete divisions, many characteristics exhibit a spectrum from one extreme to another, a phenomenon called **continuous variation.** Figure 6.1 illustrates this concept by showing the overlap of different skin colors, as measured by reflected light. If skin color is to be used as the primary criterion for determining race, how, then, do we divide the races? Inevitably, any boundaries we draw are entirely arbitrary.

If races constituted fundamental divisions within the human species, such differences would be readily measurable; in fact, they are not. As scientific information has accumulated, the picture has become increasingly complicated and the boundaries more obscure. This is clearly illustrated by the disagreement among researchers over the number and characteristics of different races. Some people have attempted to explain continuous variation as a function of *mongrelization,* or interbreeding. This notion follows the logic that at some point in the past, the races were "pure," but the lines separating one race from another have become blurred by recent interbreeding. Such ideas reveal a naive understanding of the human past. Although gene flow may have been more restricted in some groups than in others, human populations have always interbred, so different races would have been impossible to distinguish during any time period.

RACISM At times, racial classifications have been used to justify **racism,** an ideology that advocates the superiority of certain races and the inferiority of others, and leads to prejudice and discrimination against particular populations. In many societies, including our own, racist beliefs were used to rationalize slavery and the political oppression of minority groups. Such racist views justified the African slave trade, which provided labor for plantations in the Americas until the Civil War. In the 1930s, the Nazi racist ideology, based on the pre-

The extermination of millions of Jews by the Nazis before and during World War II was justified by unscientific theories of race that had no basis in empirical fact.

sumed superiority of a pure "Aryan race," was used to justify the annihilation of millions of Jews and other "non-Aryan" peoples in Europe.

Racist beliefs have no basis in fact. Human groups never fit into such neat categories. Many Jewish people living in Europe during the Holocaust possessed the same physical features as those associated with Aryans. Even staunch advocates of Nazi ideology found it difficult to define precisely which physical characteristics supposedly distinguished one "race" from another.

GEOGRAPHICAL RACES

Stanley Garn (1971) took a different tack in classifying modern humans into races. Unlike earlier theorists, Garn did not rely on single, arbitrary characteristics such as skin pigmentation in developing his classification system. Instead, he focused on the impact evolutionary forces may have had on geographically isolated human populations. Garn divided modern humans into what he called *geographical races*, populations isolated from one another by natural barriers such as oceans, mountains, and deserts. He reasoned that because of these barriers most people in each population married within their own gene pool. Garn's taxonomy divided humans into nine geographical races: Amerindian, Polynesian, Micronesian, Melanesian-Papuan, Australian, Asiatic, Indian, European, and

African. These were further divided into smaller local races and microraces that reflected restricted gene flow among smaller populations.

In an innovative and substantial shift, Garn's approach sought to frame the classification of human races in evolutionary terms. However, critics have pointed out that even these divisions imply stronger differences among races than actually exist. In addition, some of Garn's designated races exhibit an enormous amount of variation. Consider, for example, the Mediterranean race, which extends, according to Garn, from Morocco in the far western Mediterranean to the Saudi Arabian peninsula thousands of miles to the east. It is difficult to imagine the culturally diverse groups included in this vast area as a discrete breeding population. Even more important, the degree of difference between this group and others is no greater than the variation within the group itself (Lewontin, 1972).

ALTERNATIVE APPROACHES TO HUMAN VARIATION

Taxonomies that classify humans into separate races, even those based on modern scientific evolutionary theories, fall short because they are too static to encompass the dynamic nature of human interaction and the consequences of varying environmental and evolutionary forces. Any criterion selected as the basis for classification is necessarily arbitrary. The physical characteristics that have historically been used to distinguish one race from another form an extremely small part of a human's total genetic make-up. There are so many variations among individuals within populations that the classification schemes tend either to break down or to become extremely blurred. For these reasons, the majority of physical anthropologists currently studying human variation steer clear of defining race. Instead, they focus on explaining variation in specific traits.

CLINAL DISTRIBUTION

Because many physical traits vary independently of one another, some researchers have found it useful to examine single traits, or unitary variables. In one contemporary approach to modern human variation, known as **clinal distribution,** scientists plot the distribution of individual traits on maps by zones known as **clines.** A clinal distribution map can be likened to a weather map: Rather than simply stating whether it is going to be hot or cold, weather maps detail the temperatures in different parts of the

country. Lines tracing temperatures identify approaching storm fronts, and special designations indicate areas experiencing heat waves. These graphic weather maps explain changing weather patterns beyond the local level. Similarly, plotting the distribution of individual traits may shed light on the genetic, environmental, and cultural factors that influenced their distribution. Using mathematical models to analyze evolutionary processes in a gene pool, scientists have tracked specific physical traits within a population.

Anthropologist Joseph Birdsell (1981) conducted a clinal distribution study of blond, or tawny, hair among Australian Aborigines. While conducting fieldwork, Birdsell noted that the majority of Aborigines had dark brown hair, but some had tawny hair. Significantly, the tawny hair trait was not evenly distributed throughout the Aborigine population, but was concentrated in certain areas. A map of the percentages of tawny haired individuals in each region revealed the spread of the trait (Figure 6.2). In some areas of the Western Desert, 100 percent of the people had tawny hair. Farther away from the western desert, fewer tawny-haired people were to be found. Birdsell speculated that a mutation or, more likely, repeated mutations produced light-colored

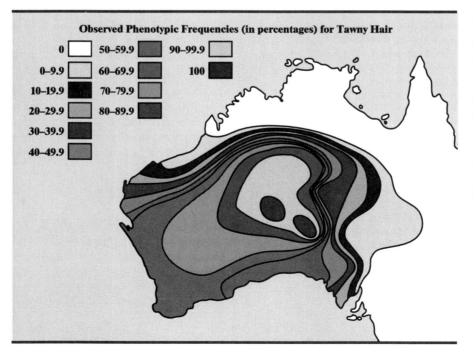

FIGURE 6.2 A clinal distribution map of tawny hair color in Australia. The trait probably orginated in the Western Desert, where it is most common. The percentage of tawny hair decreases in waves spreading out from this area. This is an example of a study focusing on one genetic trait.

Source: From *Human Evolution* by Joseph B. Birdsell. Copyright 1975, 1981 by Harper & Row, Inc. Reprinted by permission of Addison-Wesley Educational Publishers, Inc.

hair in some individuals. In certain areas the light-colored hair replaced the original dark brown hair color, for reasons that are unclear. Over time, through gene flow with surrounding groups, the new trait spread outward. The clinal distribution of tawny hair offers a graphic illustration of microevolutionary change over time within one human population.

MULTIVARIATE ANALYSIS

In contrast to univariate approaches that focus on a single trait, **multivariate analysis** examines the interrelationships among a number of different traits. Such studies are extremely complex, and scientists using this approach must decide which physical traits and which variables should be examined. Anthropologist R. C. Lewontin adopted a multivariate approach in his study of human variation. Lewontin (1972) probed the distribution of physical traits that vary in human populations, including those that have been considered distinctive to certain races, such as skin color and hair texture. In focusing on how the distribution of these traits compares to common divisions by race, he noted that traits used to identify races do not accurately reflect human variation. Observed differences among Africans, Caucasians, Mongoloids, South Asians, Oceanians, Australian Aborigines, and Native Americans (divisions that approximate Garn's geographical races) account for only about 6 percent of the total amount of variation in human populations. Almost 94 percent of human variation of physical traits occurs within each of these different "races." Lewontin's findings further underscore the highly limited usefulness of the concept of race.

ADAPTIVE ASPECTS OF HUMAN VARIATION

Natural selection has played a key role in the evolution of the human species, but what role has it played in variation among modern humans? How do we assess its effects? One way is to look at how different physical characteristics enable humans to adapt to disparate environmental conditions. If natural selection promoted these differences, there should be evidence to substantiate this assertion. As scientists probe human variation in its myriad forms, they are discovering that natural selection indeed underlies much of this variation.

SKIN COLOR

The genetic basis of skin color is complex. The current consensus among researchers is that it is a *polygenic* trait, that is, a consequence of variation in the alleles of more than one gene. However, the specific genetic loci involved and the precise manner of inheritance are unknown (Byard, 1981). Clearly, skin color varies among both individuals and groups. Multiple shades of skin color are found in different regions of the world. Whereas the classification of "white" refers to an ultra-pale skin color rarely seen in the world, the designations of "black," "yellow," and "red" each span many gradations of coloration.

Three substances combine to give skin its color: carotene, melanin, and hemoglobin (Poirer, Stini, & Wrenden, 1990). *Carotene,* an orange-yellow pigment that confers a yellowish tinge, is contained in certain foods, so people with a large amount of carotene in their diet may have an orange tone in their skin. However, the presence of carotene does not impart a yellowish cast to the skin of individuals of Asian descent. Rather, this skin tone is the result of a thickening of the exterior layers of skin.

Melanin, the dark pigment that is responsible for variations of tan, brown, and black skin color, primarily determines the lightness or darkness of skin. Melanin is produced by cells, known as *melanocytes,* in the bottom layers of skin. Interestingly, all modern human groups have about the same number of melanocytes. However, their arrangement and the amount of melanin they produce underlie variation in skin color; these factors are genetically controlled to some degree (Szabo, 1967).

Hemoglobin, a protein that contains iron, gives red blood cells their color. In people with less melanin in their skin, this red color shows through more strongly, tinting their skin pink.

Skin color is also directly influenced by the environment. Exposure to the ultraviolet radiation in sunlight stimulates the production of melanin, yielding what we call a tan. Thus, variation in skin color stems from the interaction of both genetic and environmental factors (Williams-Blangero & Blangero, 1992).

THE GEOGRAPHIC DISTRIBUTION OF SKIN COLOR
Analysis of the distribution of skin pigmentation

reveals a distinctive pattern. In most world areas, skin color is generally darker in populations closer to the equator (Birdsell, 1981). Further north and south of the equatorial zone, skin coloration is lighter. This observation was particularly true before the large population migrations of the last 500 years.

Scientists who have studied the distribution patterns of skin pigmentation hypothesize that natural selection played a decisive role in producing varying shades of skin color. Although interpretation remains open to question, several adaptive aspects of pigmentation suggest possible reasons why skin color may have been favored by natural selection. First, darker skin confers advantages in a tropical environment. Of the different pigments noted, only melanin provides protection from ultraviolet radiation, which can cause sunburn, sunstroke, and skin cancer. Originally, as *Homo sapiens* evolved in the tropical equatorial zones, a darker skin pigmentation most likely proved adaptive.

Darker skin is not advantageous in all environmental settings, however. As people moved into more temperate regions with less sunlight, other selective pressures, primarily the human need for vitamin D, conferred an adaptive advantage on lighter pigmentations. We know today that when human skin is exposed to the ultraviolet (UV) radiation in sunlight, the UV rays stimulate the synthesis of vitamin D. Insufficient levels of vitamin D can cause deficiency diseases like rickets, which makes bones soft and, ultimately, deformed. The fossil record indicates that some *Homo neanderthalensis* in northern Europe had the rickets disease (Boaz & Almquist, 1997). With this in mind, physical anthropologists conjecture that the production of optimal levels of vitamin D may have had a hand in the distribution of human skin color. That is, people who inhabited equatorial regions with ample exposure to sunlight evolved darker skin pigmentation to avoid overproduction of vitamin D. In contrast, individuals who lived in cold, cloudy climates and consequently wore heavy clothing improved their chances of survival if they had lighter skin, which absorbed higher levels of UV radiation and, thus, synthesized more vitamin D. Over time, therefore, in regions outside the tropics, natural selection favored lighter-skinned populations.

Other possible selective advantages of skin color have been suggested. For example, light skin may increase resistance to frostbite—another reason why lighter-skinned populations predominated in the colder northern regions. Studies of U.S. soldiers during the Korean War revealed that people with darker skin were more prone to frostbite (Post et al., 1975). Other studies have noted that melanin's resistance to UV radiation minimizes the body's destruction of folic acid, a member of the vitamin B complex critical to the production of red blood cells. Further studies have tentatively linked darker skin color with genetic resistance to certain tropical diseases as well (Polednak, 1974; Wasserman, 1965).

BODY BUILD

The influence of natural selection and environment on human body and limb forms is especially pronounced. These interrelationships were first noted by the nineteenth-century English zoologist Carl Bergmann. He observed that in mammal and bird populations, the larger members of the species predominate in the colder parts of the species range, whereas the smaller representatives are more common in warmer areas. This pattern also holds true for human populations.

Bergmann explained these findings in reference to the ways birds and mammals dissipate heat. *Bergmann's rule* states that smaller animals, which have larger surface areas relative to their body weights, lose excess heat efficiently and therefore function better at higher temperatures. Larger animals, which have a smaller surface area relative to their body weight, dissipate heat more slowly and are therefore better adapted to cold climates. The same applies to humans: People living in cold climates tend to have stocky torsos and heavier average body weights, whereas people in warmer regions have more slender frames, on the average.

Building on Bergmann's observations about heat loss, the American zoologist J. A. Allen did research on protruding parts of the body, particularly arms and legs. *Allen's rule* maintains that individuals living in colder areas generally have shorter, stockier limbs. Longer limbs, which lose heat more quickly, typify populations in warmer climates. Bergmann's and Allen's rules are illustrated by the contrasting body builds of a Masai warrior from Tanzania and an American Eskimo (see photographs on p. 131).

Bergmann's and Allen's observations can be partially explained by natural selection. However, ac-

A Masai warrior from Tanzania in East Africa (left) and a Native American Eskimo (right), illustrating how body weight and shape vary according to both Bergmann's rule and Allen's rule.

climatization also affects body and limb size, according to studies conducted among modern U.S. groups descended from recent migrants from Asia, Africa, and, primarily, Europe (Newman & Munroe, 1955). Researchers discovered that individuals born in warmer states generally developed longer limbs and more slender body types than did those from colder states. Because these developments occurred within such a short time (a few generations at most), they could not be attributed to evolutionary processes. Laboratory experiments with animals produced similar findings. Mice raised in cold conditions developed shorter, stouter limbs than did mice growing up in warmer settings (Riesenfeld, 1973).

CRANIAL AND FACIAL FEATURES

Because the human skull and facial features vary tremendously in shape, numerous theories explaining this variation have been advanced over the centuries. In the nineteenth century, many people embraced *phrenology,* the belief that a careful study

of the bumps of the cranium could be used to "read" an individual's personality or mental abilities or even the future. Other nineteenth-century theories posited a relationship among race, cranial shape, and facial features. None of these beliefs has withstood scientific scrutiny.

Why, then, do skull shapes vary? As with body build, the shape of the skull and face may represent adaptations to the physical environment. By examining hundreds of populations, researchers have found a close correlation between skull shapes and climate. People living in colder climates tend to have rounded heads, which conserve heat better, whereas people in warmer climates tend to have narrow skulls (Beals, 1972). Other studies have considered the environmental factors that may have favored specific nose types. Studies indicate that higher, narrower nasal openings have more mucous membranes, surfaces that moisten inhaled air. People living in drier climates tend to have more mucous membranes, regardless of whether the environment is hot or cold. Of course, these observations are generalizations; many individual exceptions can also be cited.

BIOCHEMICAL CHARACTERISTICS

Research on human variation has revealed less obvious differences among populations than skin color, body build, and facial appearance. Variation occurring in dozens of less visible features, such as blood type, the consistency of earwax, and other subtle biochemical traits, also illustrates evolutionary processes at work. It is easy to imagine how natural selection may have affected the distribution of some of these features. Consider, for example, resistance to disease. If a lethal illness were introduced into a population, individuals with a natural genetic resistance would have an enhanced chance of survival. With increased odds of reproducing, these individuals' genetic blueprints would quickly spread throughout the population (Motulsky, 1971).

History offers many tragic examples of one population inflicting disease on another that had no natural immunity. For example, when Europeans first came in contact with indigenous peoples of the Americas and the South Pacific, they carried with them the germs that cause measles and smallpox. Because these diseases had afflicted European populations for centuries, most Europeans had adapted natural immunities to them. When the diseases were introduced into populations that had never been exposed to them, however, plagues of catastrophic proportions ensued. Many Native American and Polynesian populations were decimated by the spread of diseases brought to their lands by Europeans.

BLOOD TYPES   Among the most studied biochemical characteristics are *blood types*. They are the phenotypic expression of three alleles—A, B, and O. A and B are both dominant, whereas O is recessive. These different alleles are a good illustration of polymorphism in a simple genetic trait. They are expressed in four phenotypes: type A (genotypes AA and AO); type B (genotypes BB and BO); type AB (genotype AB); and type O (genotype OO).

The three blood-group alleles are found throughout the world in varying frequencies from population to population. Type O is by far the most common, ranging from over 50 percent in areas of Asia, Australia, Africa, and Europe to 100 percent among some Native American groups. Type A occurs throughout the world but generally in smaller percentages than does type O. Type B has the lowest frequency. Believed to have been totally absent from native South American groups, type B is most common in Eurasia and can be tracked in a clinal distribution outward into Europe, in the west, and Asia, in the east.

Anthropologists, citing the nonrandom distribution of blood types, conclude that natural selection may have favored certain gene frequencies, keeping the percentage of individual alleles stable in particular populations. This natural selection might have something to do with resistance to disease. Each blood type constitutes a different antigen on the surface of red blood cells. An *antigen* is any substance that produces *antibodies,* proteins that combat foreign substances entering the body. The presence of these different antigens and antibodies is the reason doctors need to know a person's blood type before giving a blood transfusion. Type A blood has anti-B antibodies, and vice versa. Type O incorporates antibodies that fight against proteins in both type A and type B. People with blood type B (with anti-A antibodies) are better able to fight off diseases such as syphilis, which resemble type A antigens on a biochemical level. Similarly, scientists have posited links between blood types and resistance to many infectious diseases, including bubonic plague, smallpox, and typhoid fever. Before the advent of modern medical technology, natural resistance to these diseases would have conferred critical adaptive advantages.

SICKLE-CELL ANEMIA

By studying population genetics and evolutionary change within populations, scientists have gained important insights into certain genetic diseases, that is, those brought on by lethal genes that result in severe disabilities. One such disease, *sickle-cell anemia,* produces an abnormal form of hemoglobin, the blood molecule that carries oxygen in the bloodstream. In individuals with sickle-cell anemia, the abnormal hemoglobin molecules rupture and collapse into a sicklelike shape, inhibiting the distribution of oxygen. Individuals afflicted with sickle-cell anemia often die in childhood.

Why did natural selection fail to eliminate such a lethal gene? Geneticists and physical anthropologists investigating this question discovered that sickle-cell

A scanning electron micrograph of a deformed red blood cell (left) in sickle-cell anemia, a hereditary blood disease. To the right of the sickle cell is a normal, biconcave red blood cell.

anemia affects up to 40 percent of the population in regions where malaria, an infectious disease spread by mosquitoes, is prevalent. These regions include portions of Africa, the Mediterranean, the Arabian peninsula, and India. Investigators found that the blood of those who carry the sickle-cell gene is sufficiently inhospitable to the malaria parasite to confer on sickle-cell carriers an important genetic resistance over noncarriers in regions where malaria is rampant. Thus, although carriers may contract malaria, they are less likely to die from it.

It works like this: Recalling from Chapter 3 Mendel's principle of segregation, we note that there are three genotypes—homozygous dominant (AA), heterozygous (Aa), and homozygous recessive (aa). Because people who are homozygous for sickle-cell anemia usually die within the first year of life (Motulsky, 1971), only individuals who are heterozygous for the trait can transmit the disease to the next generation. Two heterozygous parents have a 25 percent chance of having a child who manifests the disease. Although these individuals may suffer from anemia, they are better suited to survive the threat of malaria than are individuals who do not carry the sickle-cell gene. Studies confirm that heterozygous carriers of sickle-cell anemia have higher fertility rates than noncarriers in regions where malaria is common (Livingston, 1971). Consequently, the survival of those with heterozygous genotypes balances the deaths of those that are homozygous recessive for the trait. The sickle-cell gene, therefore, is transmitted from generation to generation as an evolutionary adaptation in areas where malaria is prevalent.

BALANCED POLYMORPHISM In the case of sickle-cell anemia, a lethal, recessive gene confers partial protection against malaria. When homozygous and heterozygous genes exist in a state of relative stability, or *equilibrium,* within a population, this is known as *balanced polymorphism.* In equatorial Africa, 40 percent of the population carries the sickle-cell gene, constituting an evolutionary trade-off. Natural selection has created this balanced polymorphism to protect the African populations, but at a high cost: the deaths of some people.

By examining the sickle-cell gene, we also see an example of how natural selection acts against a harmful genetic trait. Approximately 2 to 6 percent of African Americans carry the sickle-cell gene—a greater percentage than that found in individuals of non-African origin in the United States but far lower than incidences of sickle-cell anemia among African populations (Workman et al., 1963). In part, this can be explained by gene flow between African Americans and other populations not affected by the sickle-cell gene. However, statistical studies point to another reason. Unlike Africa, the United States does not have high levels of malarial infection; therefore, the gene represents a severe liability. It is no longer favored by natural selection and is therefore gradually being eliminated from the gene pool.

LACTASE DEFICIENCY

Humans also vary in how well they digest particular foods. Most extensively studied is variation in the production of a digestive enzyme called *lactase,* which is responsible for the digestion of *lactose,* the sugar found in milk. All human infants can digest milk. However, the majority of humans lack the genetic coding that continues to produce lactase after about 4 years of age, a tendency also seen in other mammals. Without lactase, milk ferments in the intestine, causing diarrhea and cramps. This condition is referred to as **lactase deficiency.** As one researcher has noted, "Contrary to popular advertising, everybody does *not* need milk, at least not as adults. In fact, for millions of human beings, milk consumption leads to severe discomfort" (Nelson & Jurmain, 1988: 166). The majority of adults in Asian and African populations do not drink milk because they are not able to digest it properly.

Reasons for variation in lactase production among human populations are difficult to confirm, but researchers believe that certain conditions favored the ability to digest lactose. The capability is especially common among populations that have a history of *pastoralism,* the reliance on domesticated animals such as cows, sheep, and goats. Such animals provide plenty of milk to drink. In this cultural environment, natural selection favors individuals best able to make use of all available sources of nutrition. European populations, among the most *lactose-tolerant,* are partly descended from Middle Eastern pastoralists. African pastoralists such as the Fulani also produce significantly more lactase than do Africans who do not raise dairy animals (Relethford, 1997).

EFFECTS OF THE PHYSICAL ENVIRONMENT

We have highlighted the role of evolutionary processes in human variation, but we have also noted how differences in physical surroundings affect human variation. Think back to the differences between genotype and phenotype. The environment may produce vastly different appearances (phenotypes) in organisms of very similar genotypes. For example, if we take two plants with identical genet-

ic makeup and plant one in a fertile, well-irrigated field and the other in a stony, poorly watered area, the resulting plants will look completely different despite their genetic similarity. The physical environment plays a comparable role in causing differences in human populations.

HIGH-ALTITUDE ADAPTATIONS

Consider people living in high-altitude environments such as the Himalaya or Andes mountains. Because of the lower barometric pressure at high altitude, people take in less oxygen, making the air feel "thinner." So at high elevations, most humans experience dizziness and breathing difficulties, which are symptoms of *hypoxia,* or oxygen deficiency.

People raised in high-altitude environments, however, do not have these reactions. They have greater lung capacity, larger hearts, and more red cells in their blood, all of which promote greater oxygen exchange (Stini, 1975). We attribute this adaptation to high altitudes to acclimatization, because children born in lowland environments who are raised at higher elevations develop many of the same physical characteristics as those born in the latter environment (Frisancho, 1979).

CULTURAL DIVERSITY IN HUMAN POPULATIONS

As noted in Chapter 5, culture and society play key roles in our interaction with the environment. It is not surprising, therefore, that culture also influences human—and sometimes genetic—variation. Although humans can theoretically choose a mate from among any potential spouses within geographic limits, culture often circumscribes those choices. In the Middle East, for example, Christians, Jews, and Muslims live in close proximity to one another, yet most people marry within their own religious group. Sometimes these cultural sanctions take on the force of law. For example, at one time both South Africa and certain regions of the United States had laws prohibiting marriage between whites and blacks. Such cultural practices inhibit gene flow and contribute to genetic drift within a population. Other cultural practices actually alter the environment and can affect the allocation of resources. Nowhere is

Members of an Amish community. The Amish, a religious community in Pennsylvania and the Midwest, severely restrict interaction with other cultures. Religion, ethnicity, and perceived cultural differences can curb gene flow among human populations.

the impact of culture more pronounced than in modern urban societies.

THE IMPACT OF MODERN URBAN LIFE

Urbanization—the concentration of populations into large urban centers—has altered human lifestyles in dramatic and significant ways. Certain issues must be addressed whenever large numbers of people live together in a small area: How will they be supplied with food and water? How will sanitation needs be met? Will crowded living conditions enhance life or make it barely tolerable? Different cultures have worked through these issues with varying degrees of success. In some cities, overcrowding, combined with poor knowledge of sanitation, food storage, and personal hygiene, has contributed to nutritional deficiencies, reduced growth rates, and the spread of infectious disease. Daily life in modern American cities exposes people to air pollution, contaminated water, high noise levels, and other environmental hazards, all of which aggravate physiological stress.

Toxic waste, brought on by the improper disposal of hazardous chemicals, poses a problem of immense proportions in the United States. An example of the threat to human health and development is Love Canal, near Niagara Falls, New York, which was used as a dumping ground for chemical waste between 1940 and 1953 (Paigen et al., 1987; Vianna &

Polan, 1984). Studies have shown that women who lived close to the dump site gave birth to infants with lower average weights than women in other areas. Further research demonstrated that children raised near Love Canal were shorter than children of the same age raised elsewhere. This tendency was most pronounced in children who lived near Love Canal for the longest period of time.

Lower birth rates and reduced growth rates may be just the tip of the iceberg. As awareness of the threat of toxic waste increases, links to a host of other health hazards, neurological disorders, and cancer rates are being identified.

HEREDITY AND INTELLIGENCE

Intelligence is the capacity to process and evaluate information for problem solving. It can be contrasted with *knowledge,* which is the storage and recall of learned information. Heredity undoubtedly plays a role in intelligence. This is confirmed by the fact that the intelligence of genetically related individuals (for example, parents and their biological children) display the closest correlation. Yet other factors also come into play. In no area of study have the varying effects of genes, environment, and culture been more confused than in the interpretation of intelligence.

Following Darwin's publications on human evolution, many writers grounded allegedly "scientific," racist philosophies on misinterpretations of his theory. In nineteenth-century England, thinkers such as Herbert Spencer and Francis Galton believed that social evolution worked by allowing superior members of society to rise to the top while inferior ones sank to the bottom. These views reinforced the false belief that particular groups of people, or races, had quantifiably different intellectual capacities. In nineteenth-century France, Count de Gobineau developed a theory of history based on race. He argued that each race had its own, either high or low, intellectual capacity and that there were stronger and weaker races. Gobineau promoted the conquest of so-called weaker races by allegedly stronger ones. In his view, Caucasians, especially so-called Aryans, made up the strong race, the race responsible for the development of "advanced" civilization, which he equated with European civilization. Gobineau's views were revived by Nazi scientists during the 1930s and became the ideological basis for the genocidal policies of the Holocaust.

MEASURING INTELLIGENCE

Most scientists agree that intelligence varies among individuals. Yet it has been difficult to measure intelligence objectively because tests inevitably reflect the beliefs and values of a particular cultural group. Nevertheless, a number of devices have been developed to measure intelligence, most prominent among them the *intelligence quotient (IQ)* test devised by French psychologist Alfred Binet in 1905. Binet's test was brought to the United States and modified to become the Stanford-Binet test. The inventors warned that the test was valid only when the children tested came from similar environments; yet the IQ test is widely used today for tracking students in the U.S. educational system, sparking controversy among educators and social scientists alike. In a controversial book called *The Bell Curve: Intelligence and Class Structure in American Life* (1994), Richard Herrnstein and Charles Murray argue that research evidence supports the conclusion that race is related to intelligence. Utilizing a bell curve statistical distribution, they place the intelligence quotient of people with European ancestry at 100. People of East Asian ancestry exceed that standard slightly, averaging 103; people of African descent fall below that standard, with an average IQ of 90. Their findings imply that IQ scores are related to genetic differences among races.

A number of scientists have noted the faulty reasoning used by Herrnstein and Murray, as well as others who have attributed IQ differences between African Americans and European Americans to so-called racial groupings. If there truly were differences between African Americans and European Americans for IQ scores, then African Americans with more European ancestry ought to have higher IQ scores than those with less European ancestry. In a major IQ study of hundreds of African Americans based on blood testing to determine European ancestry, Scarr and Weinberg (1978) found no significant differences between IQ scores and the degree of European admixture.

In more recent studies of ethnic groups around the world, Thomas Sowell (1994, 1995), an African American social scientist, has demonstrated that most of the documented racial differences in intelligence are due to environmental factors. Sowell tracked IQ scores in various racial and ethnic categories from early in the twentieth century. He found that, on average, immigrants to the United States from European nations such as Poland, Italy, and Greece, as well as from Asian countries including China and Japan, scored ten to fifteen points below longtime U.S. citizens. However, people in these same ethnic categories today, Asian Americans, Polish Americans, Italian Americans, or Greek Americans, have IQ scores that are average or above average. Among Italian Americans, for example, average IQ rose by almost ten points in fifty years; among Polish and Chinese Americans, the rise was almost twenty points. It is obvious that as immigrants settled in the United States, their new cultural and environmental surroundings affected them in ways that improved their measured intelligence.

Among African Americans, Sowell noted, northerners have historically scored higher than southerners on IQ tests, by about ten points. Moreover, the IQ scores of African Americans who migrated from the South to the North after 1940 rose in the same way as they did among immigrants from abroad. Other studies indicate that middle-class African Americans score higher on IQ tests than do lower-class white Americans.

These test-score disparities indicate that cultural patterns matter. African Americans are no less intelligent than other groups, but, carrying a legacy of disadvantage, many contend with a cultural environment that discourages self-confidence and achievement. Most anthropological research on this topic indicates that when differences in socioeconomic status and other factors were controlled for, the difference between African Americans and European Americans was insignificant (Molnar, 1983, 1992). Additional studies show that educational enrichment programs boost IQ scores (Jensen, 1980; Loehlin, Lindzey, & Spuhler, 1975; Molnar, 1992).

WHAT DO INTELLIGENCE TESTS MEASURE?

Most psychologists agree that intelligence is not a readily definable characteristic like height or hair color. Psychologists view intelligence as a general capacity for "goal-directed adaptive behavior," that is, behavior based on learning from experience, problem solving, and reasoning (Myers, 1989). Though this definition of intelligence would be acceptable to most social scientists, we now recognize that some people are talented in mathematics, others in writing, and still others in aesthetic pursuits such as music, art, and dance. Because abilities vary from individual to individual, psychologists such as Howard Gardner question the view of intelligence as a single factor in human makeup. Based on cross-cultural research, Gardner (1983) has concluded that intelligence does not constitute a single characteristic but rather amounts to a mix of many differing faculties. According to Gardner, each of us has distinct aptitudes for making music, for spatially analyzing the visual world, for mastering athletic skills, and for understanding ourselves and others—a type of social intelligence. Not surprisingly, Gardner concludes that no single test can possibly measure what he refers to as these "multiple intelligences."

The majority of psychologists and other scientists concur with Gardner's findings that intelligence spans a wide and diverse range. The IQ test ranks people according to their performance of various cognitive tasks, especially those that relate to scholastic or academic problem solving. Yet it cannot predict how successfully a person will adapt to specific environmental situations or even handle a particular job. Throughout the world people draw on various forms of intelligence to perform inventive and creative tasks, ranging from composing music to developing efficient hunting strategies. Before we call someone "intelligent," we have to know what qualities and abilities are important in that person's environment. Different sorts of cognitive abilities lead to success in a hunting-and-gathering society, an agricultural society, or an industrial society.

 SUMMARY

Human beings throughout the world constitute a single species, *Homo sapiens.* Nevertheless, humans vary in specific physical traits, such as skin color, body build, cranial shape, facial features, and a variety of biochemical characteristics. Early scientists divided humanity into races based on these differences. However, further research has shown such divisions to be of little practical use and of no scientific utility; divisions into races are entirely arbitrary.

Individual physical traits, such as skin color, body build, and biochemical characteristics, are the products of the dynamic interaction of evolutionary forces, environment, and cultural variables. Recent research has focused on the distribution and study of specific traits and the explanation of the processes that may have produced them. Studies suggest that many genetically controlled traits may be the result of natural selection. For example, the protection against ultraviolet radiation afforded by the melanin in dark skin would be adaptive in tropical environments. Other advantageous characteristics seem to be associated with body build, cranial and facial features, and biochemical characteristics such as blood groups.

Human beings also display variation in intelligence, which can be defined as the capacity to process and evaluate information for problem solving. Although individuals vary in their intelligence, researchers generally agree that environmental and cultural factors influence intelligence much more than hereditary or genetic factors. A consensus among educators and social scientists now holds that rather than being a singular trait, intelligence is actually a mix of all sorts of different faculties, which cannot be measured by one test.

QUESTIONS TO THINK ABOUT _____

1. What are the basic sources of variation of human physical characteristics?
2. Why have racial classifications proven of little explanatory value to physical anthropologists?
3. Comment on this statement: "There are no races, just clines."
4. Consider variation in different physical characteristics in humans. What evolutionary factors explain these differences? Do the same evolutionary processes act on all characteristics?
5. What factors need to be considered when studying the relationship between race, language, intelligence, and culture?

 KEY TERMS _____

acclimatization
clinal distribution
clines
continuous variation

lactase deficiency
multivariate analysis
polymorphism
polytypic

races
racism

 SUGGESTED READINGS _____

BOAZ, NOEL T., AND ALAN J. ALMQUIST, 1997. *Biological Anthropology: A Synthetic Approach to Human Evolution*. Englewood Cliffs, NJ: Prentice Hall. An excellent introduction to most of the important topics on human variation explored by biological and physical anthropologists.

BRUES, ALICE M. 1977. *People and Races*. New York: Macmillan. Brues examines physical variation in human populations living in different geographic regions. Though somewhat dated, this introductory work contains a great deal of useful information about human variation.

GOULD, STEPHEN JAY. 1981. *The Mismeasure of Man*. New York: W. W. Norton. Highly readable, this book spot-lights the limitations of several attempts to correlate race with intelligence. The author looks at each theory and its flaws.

HARRISON, G. A., J. M. TANNER, D. R. PILBEAM, AND P. T. BAKER. 1988. Human *Biology: An Introduction to Human Evolution, Variation, Growth, and Adaptability*. Oxford: Oxford University Press. A thorough overview of human variation.

RELETHFORD, JOHN. 1996. *The Human Species: An Introduction to Biological Anthropology*. Mountain View, CA: Mayfield. This well-written and well-organized textbook on biological anthropology offers a good introduction to human evolution, microevolution, and variation.

Chapter 7

PALEOLITHIC CULTURES

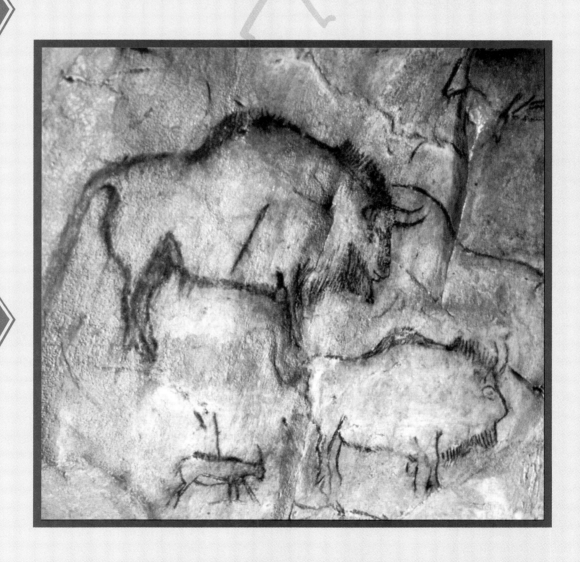

CHAPTER OUTLINE

IN THE PRECEDING CHAPTERS WE SAW HOW physical anthropologists have used fossils and biochemical characteristics to trace the evolution of humans as a biological species over the past 4 million years. During this period there were also changes in how humankind lived as well as in human social organization and culture. It is the archaeologist's job to interpret the record of these aspects of humans and of human ancestors.

In tracing the origins of human culture, researchers face a daunting task. Unlike the evolution of bipedalism and the increase in cranial capacity, which can be studied through fossil remains, the roots of human culture can be inferred only from the living areas, food remains, and tools the early hominids left behind. Thus archaeologists and other anthropologists must bring to the field a keen sense of imagination, a detective's acuity, and boundless patience to piece together the lifeways of our long-dead hominid ancestors.

LIFESTYLES OF THE EARLIEST HOMINIDS

Much of what we know about the lifestyles of the earliest hominids is based on the tools they left behind. As noted in Chapter 5, bipedalism freed hominids' hands for such tasks as food gathering, infant

care, and tool use. There is no question that tools had important consequences for early hominids, allowing them to exploit a wider range of food, defend themselves more effectively, and generally perform many tasks that they would not have been able to do otherwise. For example, tools allowed early hominids to cut through the tough hides of other animals, which could then be used as food. The first tools were very likely unmodified pieces of wood, stone, bone, or horn that were picked up to perform a specific task and then discarded. Observations indicate that chimpanzees use tools such as sticks or folded blades of grass to extract termites from their nests (Goodall, 1986). Early hominids probably used similar tools. Unfortunately, as noted in Chapter 5, artifacts of this kind are unlikely to be preserved or identified in archaeological contexts.

Archaeologists have studied stone tools extensively because they survive very well, even after being buried in the ground for millions of years. Researchers refer to the earliest stone tools produced by pre-*Homo sapiens* hominids as **Lower Paleolithic** implements, referring to the earliest part of the Old Stone Age (roughly between 2.4 million and 200,000 years ago).

THE OLDOWAN

The oldest known stone tools, dating back 2.4 million years ago, were first identified by Louis and Mary Leakey at Olduvai Gorge in Tanzania, East

Africa. They called this technology **Oldowan,** another version of the name *Olduvai.* Oldowan tools are basically river cobbles with sharpened edges. The cobbles, perhaps measuring as much as 4 inches across, were sharpened by using a hammer stone to break off chips, or *flakes*, in a process called **percussion flaking** [Figure 7.1(a)]. As flakes were removed, they left behind a sharp edge that could be used for cutting. Tools with flaking on one side are termed unifacial; those with flaking on two sides, bifacial. The tools most characteristic of the Oldowan are pebble tools, or choppers [Figure 7.1(b)]

Although pebble tools are the most distinctive Oldowan implements, other artifacts have also been found. Flakes, the chips of stone struck off during the manufacture of choppers, are very common. These were sometimes flaked themselves to sharpen their edges so that they, too, could be used as tools. Smaller tools like these were well suited for cutting and scraping. Other implements required no modification. For example, the hammer stones used to manufacture stone tools and possibly to crack bones to extract the nutritious marrow are no more than natural stones of convenient size and shape. Called *manuports,* some of these stones bear no obvious evidence of use, but archaeologists know that they were handled by early hominids because they are found together with other artifacts in places where such stones do not naturally occur.

HOW TOOLS WERE USED Because the Oldowan tools are highly primitive, even trained researchers often find it difficult to tell which stones were man-ufactured by early hominids and which were broken by natural forces. In general, archaeologists note that the flaking in manufactured tools follows a more regular, consistent pattern than in stones in which a sharp edge is produced by natural processes. Through **experimental studies,** paleoanthropologists may actually duplicate the process used by these protohumans to create their tools, striking flakes off a cobble to produce a chopper, for example. Experimenting with these replica tools, researchers have discovered that they can be used for a variety of tasks, including cutting through hides, dismembering animals, and whittling wood (Isaac, 1984).

Researchers also conduct **use-wear and residue studies** to help decipher how early tools were used. When a person cuts or shaves something with a stone tool, the tool's edge becomes nicked and dull. In addition, the material the tool has cut may leave a distinctive polish or residue on certain types of stone. In examining fifty-six stone artifacts from Koobi Fora, dating to 1.5 million years ago, Lawrence Keeley and Nicholas Toth (1981) determined that four of these tools had been used to cut meat and five others bore residues of plant matter.

Indirect evidence of tool use comes from the remains of butchered animals (Potts, 1988). By analyzing cut marks on animal remains, researchers have determined that tools were used to cut and scrape soft flesh off the bone. Chopping marks, resulting from the splitting of bones or the separating of joints, are rare. However, paleoanthropologists suspect that hominids, presumably with the help of stone tools, did break open bones to expose the marrow.

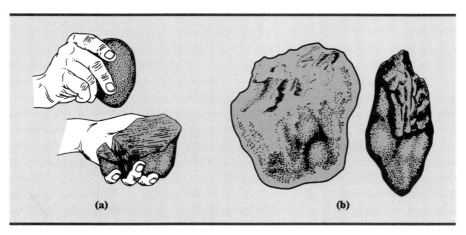

(a)　　　　　　　　　　**(b)**

FIGURE 7.1 Oldowan technology: (a) The percussion flaking method, in which a hammer stone is used to remove flakes from a stone to produce a chopper tool. (b) An Oldowan chopper.

LIVING SITES The Oldowan tools convey valuable information about the lifestyles of early hominids, particularly when they are found in an archaeological context and associated with other artifacts (see Chapter 2). At Olduvai Gorge, tools have been found in clusters along with discarded animal bones and flakes cast off from tool manufacture. In one case more than 40,000 bones and 2,600 stones were found together. In another instance, researchers came upon a ring of stone 14 feet in diameter, a feature that some experts have interpreted as evidence of a foundation for a simple structure or shelter (M. D. Leakey, 1971). Although most researchers are reluctant to make the theoretical leap from a circle of stone to an actual shelter, these archaeological sites have frequently been interpreted as living floors or occupation areas, where early hominids lived, prepared food, ate, and slept.

DECIPHERING THE OLDUVAI FINDS Piecing together the artifacts from the Olduvai sites, researchers have advanced various theories to explain the origins of human culture. During the 1960s and 1970s Glynn Isaac (1978) and Louis Leakey developed a model that focused on "man the hunter" (Isaac, 1978; Lee & Devore, 1968). According to this perspective, hunting—facilitated by bipedalism, increased cranial capacity, and tool use—was the key to early hominid life. In a model that assumed a great deal of social complexity, Isaac and Leakey conjectured that the Olduvai sites served as **home bases,** locations where hominids gathered. They argued that males brought back meat, particularly from large animals, to the home base to be shared with adults (especially females) and children who had remained in camp. In this view, food sharing, prolonged infant care, and the social interaction that occurred at the home base prefigured the kind of social arrangements that characterize modern human societies. Drawing an analogy to human hunting-and-gathering populations, such as the !Kung San of southern Africa, this model provides an idealized interpretation of human lifestyles stretching back 2 million years.

Many researchers viewed the man-the-hunter model as problematic, however. Critics pointed out that its idealized version of the hominid past puts far too much emphasis on the role of the male and too little on that of the female (Dahlberg, 1981; Fedigan, 1986; Tanner, 1981). Drawing on observations of modern hunter-gatherers, they argued that plants—fruits, seeds, and vegetables—and not meat were the major food source for early hominids. Because women generally gathered plants, they played the key role in providing food. In addition, use of the hands was as important to females (who cared for infants) as to males (who made tools and hunted). Because hominid infants had lost the ability to cling to the mother with their hands and feet, the mothers' hands had to be free to carry their offspring.

Emerging perspectives have produced very different interpretations of early hominid life. Recent scholarship has criticized the preceding theories for assuming that early hominid behavior paralleled that of modern humans (Binford, 1985; Gould, 1980; Potts, 1988). More importantly, these theories are not supported by archaeological evidence. Using innovative analytical techniques, researchers have proposed a number of alternative theories. For example, Lewis Binford (1985) suggests that much of the bone found at hominid sites is not the product of hunting forays but rather the remains of animals killed by other predators, which were then scavenged by hominids. Binford's interpretation gains support from microscopic studies of bones showing that, at least in some cases, cut marks from tools overlie carnivore tooth marks (Shipman, 1986b). In addition, studies of wear patterns on hominids' teeth indicate that fruit was integral to their diet. At this point, however, we cannot state with any degree of certainty the relative importance of gathered plant foods, scavenging, and hunting in early hominid societies.

An emerging consensus among paleoanthropologists and archaeologists holds that the Olduvai sites do not represent home bases along the lines suggested by Glynn Isaac and Louis Leakey. Rather, archaeological evidence indicates that these places were visited by large predatory animals and probably were not safe for early hominids. Current theories do not assume the high degree of social complexity posited by the home-base model. However, paleoanthropologist Richard Potts (1988) notes three key factors suggesting that early hominids lived in social groups of some kind: "(1) Social groups are a general part of higher primate life; (2) tool use and manufacture were undoubtedly learned in a social setting; and (3) the feasibility of site production would seem to depend on communal use of stone materials at sites" (p. 304).

A young female chimpanzee uses a stick to dig insects out of a fallen tree. Study of nonhuman primate behavior may provide clues to the behavior of early hominids.

PRIMATE MODELS OF HUMAN BEHAVIOR

To obtain insights into what the social life of early hominids may have been like, some researchers have turned to the study of living nonhuman primates. Although the behavior of human ancestors did not mirror that of modern chimpanzees or baboons—or any living primate—studies of modern primates shed light on how environmental factors shape social behavior, leading researchers to develop models of early hominid behavior (Kinsey, 1987). The research has been made possible by long-term studies of nonhuman primates such as those conducted by Jane Goodall and Dian Fossey, noted in Chapter 4. The mother-infant bonds, friendships, and complex patterns of social interaction found in higher primates offer useful insights into early hominid behavior. Of the living nonhuman primates, two groups—the baboons and the chimpanzees—have received particular attention.

BABOON BEHAVIOR   Although baboons constitute a distinct evolutionary branch within the primates, they resemble hominids because they evolved in an arboreal environment, but today they are primarily terrestrial. Many preliminary generalizations about primate social interaction were based on observations of baboon behavior simply because they were among the earliest primate groups to be studied (Strum & Mitchell, 1987). Baboons exhibit clear patterns of competition, dominance, and aggression, so these features figured prominently in hypotheses

about early hominid behavior (Ardrey, 1961; Morris, 1967), with scientific speculations stressing the adaptive role of aggression as a defense against predators and as a mechanism for mate selection. These initial conjectures also portrayed hominid society as male dominated. In recent years, however, more complete information on baboons, as well as descriptions of the behavior of other primate species, have prompted researchers to conclude that primate behavior varies considerably, even within species. For example, baboons forge cooperative ties and friendships as well as acting out aggressive tendencies (Smuts, 1987; Strum & Mitchell, 1987).

CHIMPANZEE BEHAVIOR   Fieldwork by Jane Goodall inspired several models of hominid behavior based on the activities of wild chimpanzees. According to one model, proposed by Nancy Tanner (1987), chimpanzee behavior supports the argument that females played a key role in obtaining plant foods among early hominids. Chimpanzees are the only apes that regularly use tools to crack nuts and to extract insects from nests. Field observations have shown females to be more consistent and more proficient tool users than males. The interactions between mothers and their offspring have piqued scientists' curiosity as well: Female chimps commonly share food with their young, and mothers have been observed rewarding their daughters for learning how to use tools. Tanner suggests that early hominids may have engaged in similar dynamics in tool use and strategies for the acquisition of food and the feeding of young.

HOMO ERECTUS CULTURE

In contrast to the scant traces of earlier hominid species, the archaeological record of *Homo erectus* is more substantial and the tools more readily recognizable. Physically, *H. erectus* resembled modern humans, with a cranial capacity less than that of today's humans but greater than that of earlier hominids. This larger brain no doubt contributed to innovations in behavior, technology, and culture. In addition, the shape of *H. erectus* skulls leads researchers to believe that these protohumans may have been the first hominids with both the physical and mental capacities for speech.

THE ACHEULIAN

Lower Paleolithic tool traditions changed very slowly. The simple Oldowan choppers remained essentially unmodified 1 million years after their first appearance. Approximately 1.5 million years ago, some changes began to appear, including tools with more advanced flaking that may have served as drills and protobifaces. These increasingly sophisticated implements are referred to as signs of the Developed Oldowan. Also during this period a new tool tradition emerged, the **Acheulian** (Figure 7.2). Acheulian tools have been found with the remains of *H. erectus,* the hominid credited with producing them.

The Acheulian is named after the town of St. Acheul, France, where some of the first finds were made. Like the Oldowan choppers, Acheulian tools were produced by percussion flaking, but exhibit more complexity. Most characteristic of the Acheulian is the hand ax, a sharp, bifacially flaked stone tool shaped like a large almond, which would have been effective for a variety of chopping and cutting tasks [Figure 7.3(a)]. Unlike Oldowan choppers, which consisted of natural cobbles with a few flakes removed, the hand ax was fashioned by removing many flakes to produce a specific form. In other words, the toolmaker had to be able to picture a specific shape in a stone. Late Acheulian tools were produced through a more refined brand of percussion flaking, the **baton method,** pictured in Figure 7.3(b). In this technique a hammer, or *baton,* of bone or antler was used to strike off flakes. The baton al-

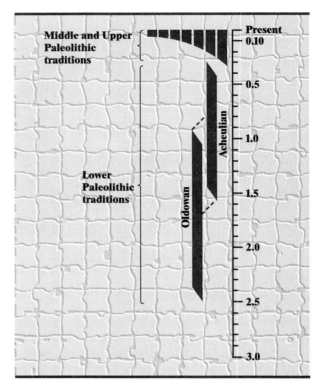

FIGURE 7.2 A chart giving historical dates for the Oldowan and Acheulian technologies. The Lower Paleolithic traditions were relatively simple and long-lasting. They were followed by Middle and Upper Paleolithic industries, which were more complex and varied.

lowed for more accurate flaking and produced shallower, more delicate flakes than a hammer stone.

Archaeological sites bearing signs of *H. erectus* life reveal a material culture that was more elaborate than that of earlier hominids. Cave sites such as Zhoukoudian China, which yielded the remains of Peking man, also preserved an array of animal bones, stone tools, and possibly traces of cooking fires (Binford & Ho, 1985; Clark & Harris, 1985). Studies showing bite marks on some hominid fossils indicate that predators also frequented these sites, raising questions about how the cave deposits were formed and about hominid activities at the site.

Acheulian tools have also been found in sites dating between 400,000 and 200,000 years ago—a period of time associated with *transitional forms* of hominid intermediate between *H. erectus* and *H. sapiens.* In southern Europe, open-air sites such as Tor-

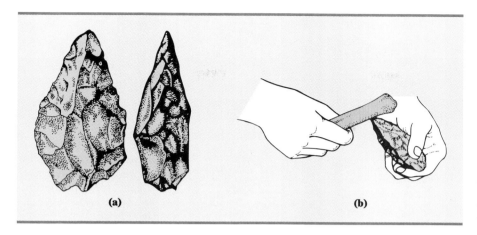

FIGURE 7.3 In later
Acheulian technology, the
characteristic hand axes (a)
were made through the
baton method (b).

ralba and Ambrona in Spain and Terra Amata on the
French Riviera have produced large assemblages of
tools (Laville et al., 1980; de Lumley, 1969; Villa, 1983).
Interpreting the Terra Amata finds, Henri de Lumley
suggested they may include evidence of simple shelters (see Figure 7.4 on page 146). The occupants of
Terra Amata enjoyed a varied diet, consuming such
large animals as extinct species of deer, elephant,
boar, rhinoceros, and wild ox. Other food sources,
drawn from the Mediterranean, included mollusks,
such as oysters, mussels, and limpets, as well as fish.

THE APPEARANCE OF *HOMO SAPIENS*

The archaeological record associated with *H. sapiens* becomes increasingly sophisticated and diverse.
Tools and dwellings become more complex, and
display a wide array of local variation. The time
span associated with European and Southwest
Asian archaic *H. sapiens* (beginning approximately 200,000 years ago) is called the **Middle Pale-**

*The archaeological site of Olorgesailie, Kenya, one of the most informative Acheulian sites in
Africa, dates to approximately 700,000–900,000 years ago. A walkway at the site allows
visitors to view Paleolithic tools exposed on the surface. A cleaver or cutting tool is shown on
the right.*

Source: Courtesy of Pamela Willoughby, University of Alberta.

FIGURE 7.4 A hypothetical reconstruction of a shelter of stones and saplings as it may have appeared approximately 300,000 years ago at Terra Amata, France.

Source: Adapted from Henry de Lumley, "A Paleolithic Camp at Nice." Copyright © 1969 by Scientific American, Inc. All rights reserved.

olithic. The term *Middle Stone Age* is used to refer to the Middle Paleolithic in Africa and Eurasia. The time roughly associated with modern *H. sapiens* (after 40,000 years ago) is called the **Upper Paleolithic,** or *Late Stone Age.* These periods are separated on the basis of variations in tool types and manufacturing techniques. Exemplifying the Lower Paleolithic are the Oldowan and Acheulian traditions; the Middle and Upper Paleolithic periods, saw a burst of creative energy that brought forth myriad innovations.

TECHNOLOGICAL ADVANCES

Percussion flaking gained some refinement in the Middle Paleolithic with the *Levalloisian technique.* Using this method, the toolmaker first shaped a stone into a regular form by striking flakes off the exterior. Additional pieces were then struck off from one side to form a flat striking platform. From this prepared core the toolmaker struck the edge to produce flakes that were fashioned into scrapers, points, and knives. The Levalloisian technique created longer cutting edges relative to the size of the stone, allowing for a more economical use of the stone. Tools were more standardized and could be produced in less time than with earlier methods.

The archaeological record points to an increasing amount of regional variation in toolmaking tech-

niques, most likely reflecting the needs of groups who were adapting to different environments. We have gathered more information about the Middle Paleolithic in Europe than in other areas because more archaeological research has been conducted there. Early stone tool traditions have been discovered in other regions but they have not yet been described in detail. However, regional innovations in stone tool technology can be seen in Middle Paleolithic finds throughout the world. In Africa, for example, rapid innovation in tool technology occurred during this period. J. Desmond Clark (1970), who devoted years of study to the prehistory of Africa, notes that this change "is in marked contrast to the very slow development of the Acheulian tool tradition during the preceding half a million years, or more" (p. 108).

Some of the most sophisticated Middle Stone Age percussion-flaking techniques from Africa can be seen in the Howieson's Poort industry tools from southern Africa, tentatively dated between 90,000 and 50,000 years ago (Phillipson, 1993). These tools reveal a trend away from larger, cruder forms to smaller, more carefully flaked implements—such as small scrapers and shaped flakes—made of finer-grained lithic materials. Some archaeological evidence indicates that more refined toolkits may have facilitated the first settlement of portions of Africa that had not been extensively occupied earlier, including the arid regions of northeastern Africa (Bailey et al., 1989; Clark, 1970: 107).

NEANDERTAL TECHNOLOGY

Neandertals, the archaic *H. sapiens* who inhabited Europe and the Middle East between 130,000 and 35,000 years ago, also fashioned implements whose versatility far surpassed earlier technologies. The Middle Paleolithic stone tool industry associated with Neandertal populations is known as the **Mousterian,** which is named after a rock shelter at Le Moustier, France, where it was first described. Produced by the Levalloisian technique, Mousterian implements

Although it has often been assumed that tool traditions can be associated with particular hominid species, such associations are difficult to demonstrate. For example, tool kits recovered from some Neandertal sites are strikingly similar to tools associated with occupations of anatomically modern Homo sapiens.

THE FAR SIDE   By GARY LARSON

© 1983 FarWorks Inc. All Rights Reserved

"Neanderthals, Neanderthals! Can't make fire!
Can't make spear! Nyah, nyah, nyah . . .!"

could have been used for cutting, leather working, piercing, food processing, woodworking, hunting, and producing weapons (Binford & Binford, 1966; Bordes, 1968).

The Neandertals were probably the first humans to adapt fully to the cold climates of northern Europe. Their technology must have included the manufacture of cloth; otherwise they would not have survived the cold winters. Clues to Neandertal life have come from the archaeological record: Scientists have discovered substantial evidence that Neandertals occupied caves and rock shelters, as well as open-air sites that may have served as temporary camps during the summer months. Archaeologists cite remains of charcoal deposits and charred bones as indications that Neandertals utilized fire not only for warmth but also for cooking and, perhaps, for protection against dangerous animals. Efficient hunters, Neandertals stalked both small and large game, including such extinct creatures as European elephants, elk, bison, and huge bears that stood between 9 and 12 feet tall and weighed up to 1,500 pounds. Their lifestyle must have been carefully attuned to the seasonal resources available. In the cold climate of Ice Age Europe, the storage of food for the winter months was probably of great importance.

NEANDERTAL RITUAL BELIEFS

Study of Neandertal sites has also given archaeologists the first hints of activities beyond hunting-gathering and the struggle for subsistence—possible evidence that Neandertals practiced rituals. Regrettably, much of this evidence, portrayed in countless movies, novels, and caricatures, is far more circumstantial than archaeologists would like. Finds that have been examined include both bear bones and Neandertal artifacts. From Drachenlock Cave in Switzerland, it was even reported that twenty bear skulls had been found in an arrangement of stone slabs—a discovery interpreted as a crude shrine. Some writers have used these discoveries to paint a complex picture of Neandertal ritual.

Despite the romantic appeal of a Neandertal "cave bear cult," however, these interpretations lack the most important thing archaeologists need to glean insights into such complex issues as prehistoric ritual beliefs: clearly documented archaeological context (Chase & Dibble, 1987; Trinkaus & Shipman, 1994). In the absence of clear associations between the bear bones and the tools, this evidence suggests only that Neandertals visited a cave in which bears may have hibernated and occasionally died. The Drachenlock Cave finds were not excavated by trained archaeologists, and no plans or photographs of the discovery were made at the time of excavation (Rowley-Conwy, 1993). Without this information, interpretation of a Neandertal shrine remains entirely speculative.

More convincing than the evidence for a bear cult are discoveries suggesting that Neandertals were the first hominids to intentionally bury their dead. Finds at a number of sites—including Shanidar, Iraq; Teshik-Tash, Uzbekistan; La Chapelle-aux-Saints, France; and Kebara, Israel (Rowley-Conwy, 1993; Solecki, 1971)—have been interpreted as burials. Of these finds, the evidence for intentional burial is most compelling at the French and Israeli sites. In both instances, the skeleton of a Neandertal man was found in a pit that seems to be too regular in shape to have been formed naturally.

Other skeletal evidence indicates that Neandertals cared for individuals with disabilities. At the Shanidar site, for example, archaeologists identified the remains of one individual who had the use of only one arm—the result of an accident or a birth defect. Despite that disability, this individual lived a relatively long life. Although no set of ritual beliefs can be inferred on the basis of these finds, they clearly do indicate the growing group communication, social complexity, and awareness that distinguish humans.

MODERN *HOMO SAPIENS* AND THEIR CULTURES

The first anatomically modern *H. sapiens* began to appear about 40,000 years ago. Between 40,000 and 10,000 years ago, these populations migrated throughout the globe, adapting both physically and culturally to conditions in disparate regions.

Physically, these *Homo* populations resembled modern humans in most respects. Their fossilized skeletons do not have the heavy, thick bones, large teeth, and prominent browridges seen in the Neandertals and other archaic forms. The high, vaulted shape of the cranium is modern, too, with dimensions similar to those of present-day humans. From the cold climates of northern Asia to the deserts of Africa, groups of *H. sapiens* shared similar characteristics as part of one species. Like modern populations, however, these early groups likely had various physical traits, such as body build and facial features, that represent adaptations to local environmental conditions and selective pressures (see Chapter 6).

THE TECHNOLOGY OF *HOMO SAPIENS*

Sites associated with anatomically modern *H. sapiens* display a flowering of cultural expression—in everything from toolmaking to home building, so-

Shanidar Cave in northern Iraq is the site of some of the most fascinating discoveries of Neandertal remains.

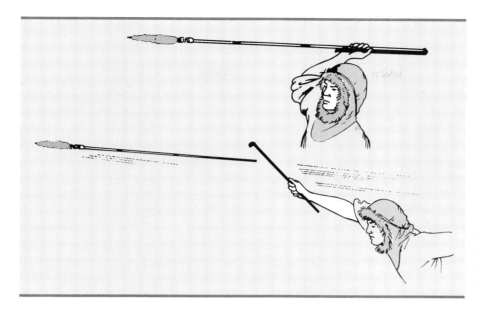

FIGURE 7.5 An innovation of the Upper Paleolithic was a spear thrower, a device that extended the hunter's arm, enabling him to make a more powerful throw.

cial arrangements, and subsistence strategies. As modern *H. sapiens* populations crafted increasingly complex tools and strategies to deal with their varied environments, a number of different stone tool traditions sprang up.

European archaeologists divide the Upper Paleolithic period into the Chatelperronian, Aurignacian, Gravettian, Solutrean, and Magdalenian stone industries, which encompass tremendous variation in stone tool types. Stone tool production made a major technological advance with increasingly fine techniques of producing *blades*—long, narrow flakes that had all sorts of uses as knives, harpoons, and spear points. Among the most striking examples of Upper Paleolithic percussion flaking are Solutrean projectile points, dated to 20,000 years ago. These implements, often measuring several inches long, probably functioned as spear points. Yet the flaking is so delicate and the points so sharp that it is difficult to imagine them fastened to the end of a spear. Some researchers have ventured a guess that they may have been made as works of art, not tools for everyday use.

TOOLS Upper Paleolithic peoples produced a number of specialized stone tools as well, including *borers,* or drills, and *burins,* chisellike tools for working bone or ivory. Tools like these facilitated the manufacture of the bone, antler, and ivory artifacts that be-

came increasingly common during the Upper Paleolithic.

Spear throwers—(Figure 7.5), long, thin pieces of wood or ivory that extended the reach of the hunter's arm,—were invented during this period, too. A particularly important innovation, spear throwers enabled Upper Paleolithic hunters to hurl projectiles much faster than they could by hand.

This period also marked the debut of *composite tools,* implements fashioned from several pieces. For example, consider the harpoon, which might consist of a wooden shaft that is slotted for the insertion of sharp stone flakes. Additional signs of technological progress at Upper Paleolithic sites ranged from needles for sewing clothing and fibers for making rope to evidence of nets and trapping equipment.

Like Lower Paleolithic groups, Upper Paleolithic technology indicates that early *H. sapiens* were efficient hunters. Many archaeological sites contain bones from mammoths, giant deer, musk ox, reindeer, steppe bison, and other animals. In addition, piles of animal bones have been discovered at the bottom of high cliffs. In reconstructing the meaning of these finds, archaeologists conjecture that *H. sapien* hunters had stampeded the animals off cliffs to be killed and butchered by waiting hunters below. Archaeologists have also found the remains of traps that Upper Paleolithic hunters used to snare animals.

Reconstruction of an Upper Paleolithic dwelling made with mammoth bones, such as those found at the 15,000-year-old site of Mezhirich in Ukraine.

Upper Paleolithic people gathered plants to supplement their diet and probably for medicinal purposes. However, because of the small size of Upper Paleolithic living areas and the limited amount of plant remains recovered from archaeological sites, we can sketch only an incomplete picture of the diet during that period.

SHELTERS In the Upper Paleolithic period, technology advanced to the point where people were proficient at building shelters, some quite elaborate. Among the more spectacular are five shelters, from a 15,000-year-old site at Mezhirich in the Ukraine, constructed from bones of mammoths, an extinct species of elephant (Gladkih et al., 1984). The mammoths' jaws formed the shelter's base, and ribs, tusks, and other bones were used for the sides. Inside, archaeologists discovered work areas, hearths, and accumulations of artifacts. Storage pits were located between the structures—indications that the shelters were inhabited for long periods. Scientists speculate that the settlement may have been occupied by more than fifty people.

To accomplish the technological and social innovations of the Upper Paleolithic, *H. sapiens* had by this time fully acquired the skills needed to accumulate and transmit knowledge, that is, the rudiments of language. In addition, the inhabitants of these settlements had developed highly efficient subsistence strategies that gave them free time for experimentation and innovation (Figure 7.6).

ETHNICITY AND SOCIAL ORGANIZATION

Advances in toolmaking took different forms around the world. Some of this regional variation stemmed from specialized technologies suited to particular environments; in other cases, toolmaking technology was driven by the specific types of stone available in a region. Regional differences may also reflect patterns of culture, ethnicity, and individual expression. Archaeologist James Sackett (1982), who has studied the classic Middle and Upper Paleolithic finds in France, notes that tools serving the same function seem to exhibit a great deal of variation. Many Upper Paleolithic artisans fashioned their stone tools in distinctive styles that vary from region to region, possibly signaling the first traces of ethnic and cultural divisions in human populations. Just as we often associate particular styles in dress, decoration, and housing with specific ethnic groups, archaeologists rely on expressions of ethnic identity preserved in material remains to piece together the lifestyles of earlier peoples. A sense of imagination comes through in Upper Paleolithic artifacts as well. Compared to the Middle Paleolithic, there are more nonutilitarian objects, including items for personal adornment (White, 1982). In addition, because some of these artifacts were obtained from distant sources, archaeologists believe that trade networks had arisen by this time. The term archaeological cultures refers to the lifeways of past peoples reconstructed from the distinctive artifacts these cultures left behind.

To glean insights into the culture and social organization of Paleolithic peoples, researchers have looked at modern hunter-gatherers. Contemporary hunter-gatherer societies, with their relatively small groups, low population density, highly nomadic subsistence strategies, and loosely defined territorial boundaries, have social organizations that tie kin (related individuals) together and foster unity within and among groups. Constant circulation of material goods in such societies not only enhances and maintains kin ties through mutual obligations but also inhibits the accumulation of wealth by any individuals in the society. This enables these societies to remain *egalitarian, —* societies that have very small differences in wealth among individuals. There are no rich or poor in these types of societies.

The most common form of political organization among ethnographically documented hunter-gatherer societies is the **band,** a fairly small group of people

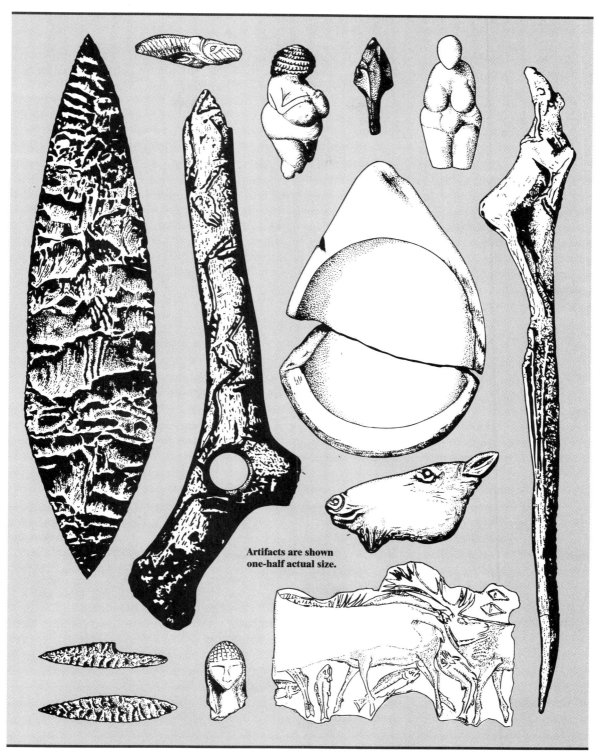

Artifacts are shown
one-half actual size.

FIGURE 7.6 The archaeological record of the Upper Paleolithic becomes progressively more elaborate, including more sophisticated stone tools and nonutilitarian items such as those pictured here.

Source: From W. J. Sollas, *Ancient Hunters and Their Modern Representatives,* Macmillan, 1911.

tied together by close kinship relations. A band is the least complex form of political system—and most likely the oldest. Each band is politically independent of the others and has its own internal leadership. Most of the leaders in the bands are males, but females also take on some important leadership roles. Leaders are chosen because of their skills in hunting, food collecting, communication, decision making, or other personal abilities. Political leaders, however, generally do not control the group's economic resources or exercise political power as they do in other societies, and there is little, if any, social stratification between political leaders and others in the band. In other words, band societies are highly egalitarian, with no fundamental differences between those with and those without wealth or political power. Thus, leaders of bands must lead by persuasion and personal influence rather than by coercion or withholding resources. Leaders do not maintain a military or police force and thus have no definitive authority.

Although it is tempting to draw a similar picture of Paleolithic hunters, the analogy is not without limitations. The archaeological record of the Paleolithic is consistent with small kin-based groups. Yet archaeological information on Paleolithic hunter-gatherers suggests that their subsistence strategies varied substantially and probably included the beginnings of some of the complexity, socioeconomic inequality, and more sedentary lifestyle that characterize more recent periods (Price & Brown, 1985).

THE UPPER PALEOLITHIC IN EUROPE

In Europe, the Upper Paleolithic marked changes that mirrored developments in much of the rest of the world. Best known among the European *H. sapiens* of the Upper Paleolithic are the *Cro-Magnon,* dated at about 40,000 years ago, whose name comes from the Cro-Magnon site in Dordogne, France. The fragmentary remains of five or six individuals, recovered from a rock shelter in 1868, provided the first evidence for the entry of anatomically modern *H. sapiens* into Europe. At the various Cro-Magnon sites, artifacts reveal an elaborate technology. Cro-Magnon stone workers were particularly adept at producing blades, which varied according to the different activities they were used for, the region, and possibly the social group.

By 17,000 years ago, some Cro-Magnon groups had specialized in procuring game from migrating herds of reindeer. In the summer, these bands es-

Upper Paleolithic painting of bison and other animals from a cave in France. Although interpretations of the meanings vary, paintings such as these convey the changing nature of the Paleolithic.

tablished encampments on rises that offered a clear view of the herds. For shelter in the warm weather, scientists speculate that they may have used lightweight tents. As the weather turned cold, they would relocate in smaller groups back to the warmth and shelter of the caves. In a major technological advance, Cro-Magnons completely harnessed fire, using such materials as iron pyrite to make sparks to ignite dry tinder. Archaeologists have found evidence for this mastery of fire in sites as widespread as Belgium and the former Soviet Union. Hearths have been uncovered in caves and in open-air locations bearing remnants of shelters. Charred wood and large quantities of bone ash at these sites indicate that Cro-Magnons used fire for cooking as well as for heat.

CARVING AND CAVE PAINTING

In addition to their other technological accomplishments, the Upper Paleolithic peoples of Europe created an impressive array of artwork. Sculptures in bone, ivory, and stone, found throughout Europe, depict human figures, including some speculatively dubbed "Venus" fertility goddesses. Magnificent abstract and naturalistic renderings of animals, and humans dressed in the hides of animals, decorate the walls of caves in Spain and France, including the Lascaux caves. These murals, or cave paintings, may have been drawn to celebrate a successful hunt or to ensure a better future. Because some of them are located deep in underground caves, researchers have speculated that this art held profound spiritual and religious significance for the Cro-Magnons. Imagine how awe inspiring a religious celebration or initiation ceremony would have been in a dark underground chamber adorned with beautiful paintings. Still, these evocative cave murals could have been painted solely as an expression of the artist's view of life (Halverson, 1987).

THE MIGRATION OF UPPER PALEOLITHIC HUMANS

Upper Paleolithic hunter-gatherers developed a number of specialized technologies that allowed them to adapt to diverse environments in ways their precursors could not have imagined. By harnessing fire to cook and heat shelters, they could eat a greater variety of foods and live in all sorts of climates, including the frigid north. In addition, their technology allowed them to produce simple watercraft. At home in such a wide variety of environments, *H. sapiens* increased their populations and settled in all parts of the globe, including North and South America and Australia, continents that had previously been unoccupied by hominids.

Changes in world climatic conditions during the past 100,000 years allowed the movement of modern *H. sapien* populations into new areas. During the latter part of the Pleistocene, or Ice Age, when climatic conditions were much cooler and wetter than they are now, Asia, North America, and the northern regions of Europe were covered by huge sheets of ice, or *glaciers,* which extended from the polar ice caps. The vast amount of water frozen in these glaciers lowered sea levels around the world by hundreds of feet, exposing vast tracts of land, known as *continental shelves,* that were previously (and are presently) beneath the sea. Many regions of the world surrounded by water today were connected to other land areas when these low-lying landmasses were exposed by lower water levels. These "land bridges" allowed *H. sapiens* to migrate from one region to another over land that is now covered by seawater.

UPPER PALEOLITHIC HUNTERS IN THE AMERICAS

Archaeologists believe that the first humans came to the Americas from Siberia into modern Alaska over a low-lying land bridge that is now the Bering Strait. Following herds of large game animals, these people migrated southward into North and South America. Today the Bering Strait is a frigid body of water connecting the Pacific and Arctic oceans. Between 75,000 and 10,000 years ago, however, glaciers that lowered sea levels transformed this region into a landmass, known as Beringia, that was more than 1,000 miles wide (Hopkins, 1982). A natural land bridge, Beringia linked Europe and Asia to the Americas.

Physical similarities between Asians and Native Americans, bolstered by historical linguistic analyses, clearly establish the Asian origin of Native American populations. Historical linguistic studies by Joseph

Clovis spear points found at various sites in North America. First described at a site near Clovis, New Mexico, these spear points are characteristic of New World Paleo-Indian cultures.

Greenberg of Stanford University divide Native American languages into three broad groups: Aleut-Eskimo, Na-Dene, and Amerind. Greenberg (1986) argues that these different linguistic groups represent three migrations from Asia. Because Amerind is the most widespread and diversified, the linguist contends that it has had more time to evolve. Aleut-Eskimo, in contrast, is confined to a comparatively small number of Native American groups, an indication that these are the most recent arrivals.

Although most anthropologists agree that Native Americans migrated from Asia, they disagree about when this migration took place. Contemporary scholars embrace one of two major perspectives concerning the peopling of the Americas—the *Clovis-first hypothesis* and the *pre-Clovis hypothesis*. The **Clovis-first hypothesis** holds that the first humans arrived in the Americas about 12,000 years ago. The **pre-Clovis hypothesis** pegs this migration at a much earlier date, perhaps 30,000 to 40,000 years ago.

THE CLOVIS-FIRST HYPOTHESIS The term *Clovis* refers to a distinctive type of stone spear point first found at Blackwater Draw, near Clovis, New Mexico. Clovis spear points, measuring several inches long, have flaked channels, or *flutes,* down the middle, which may have made them easier to attach to the spear shaft. Clovis points have been recovered from a number of North American sites and are closely dated to between about 11,200 and 10,900

years ago (Haynes, 1991). In the Clovis-first hypothesis, these tools are interpreted as the work of Paleo-Indians—the first migrants to enter North America from Asia. They are presumed to have arrived in the Canadian Yukon approximately 12,000 years ago, a figure based on estimates of how long it would have taken migratory hunters to reach the Clovis sites identified in the United States. The Clovis people, in turn, gave rise to later Native American cultures.

Clovis spear points and associated artifacts clearly establish that human settlements dotted America after 12,000 years ago. Yet archaeologists are puzzled by the rapid spread of early Paleo-Indian cultures throughout the Americas. Could humans have migrated from Alaska and put down roots in such disparate areas as Maine and the southernmost tip of South America in a mere 1,000 or 2,000 years? Recent studies indicate that the rate of migration proposed by Clovis-first proponents is unrealistically fast (Whitley & Dorn, 1993). Questioning this accelerated time frame, some scientists believe that humans must have arrived in the Americas before 12,000 years ago in order to have settled throughout the Americas by 10,000 to 11,000 years ago.

THE PRE-CLOVIS HYPOTHESIS Advocates of the hypothesis that humans inhabited the Americas earlier than 12,000 years ago are building a strong but still controversial case. Some archaeologists have dated sites in North and South America to between 12,000 and 40,000 years ago, but because dating these sites is extremely difficult, many archaeologists do not accept them as evidence of pre-Clovis settlement. Most of the artifacts found at these sites fail to meet one of two key criteria: (1) They do not demonstrate conclusively that humans lived at these sites, or (2) they cannot be accurately dated to a period before 12,000 years ago (Carlisle, 1988; Dillehay & Meltzer, 1991; Meltzer, 1993; Taylor, 1995). Two pre-Clovis sites serve to illustrate the types of finds represented. These are the Meadowcroft Rock Shelter, near Pittsburgh, Pennsylvania, and Monte Verde in Chile.

MEADOWCROFT ROCK SHELTER Excavations at Meadowcroft were undertaken by archaeologist James Adavasio over a period of six years (Carlisle & Adavasio, 1982). Artifacts uncovered here provide evidence of early human occupation, including

Paleo-Indian bison kill in the western United States. Early humans in North America probably depended on a wide variety of plants and animals. Kill sites such as these, however, suggest a heavy reliance on large animals.

Paleo-Indian tools dated to 15,000 years ago. Deeper levels of the excavation produced dates of 19,000 years ago, but artifacts pegged to these dates are limited; the earliest is a fragment of bark that the excavators believe was trimmed to make a basket. Critics also say that the flora and fauna remains represented are not consistent with the supposedly late Pleistocene age of the early levels.

MONTE VERDE A pre-Clovis site has been identified in the Monte Verde forests of Chile, where, beginning in 1976, archaeologist Tom Dillehay (1989) has conducted research and excavations. Because of the moist soil in the area, preservation is much better than is generally seen at other Paleo-Indian sites. Dillehay found the remains of forty-two plants and the bones of rodents, fish, and larger animals that had been hunted by human settlers, as well as flaked stone tools. There were also traces of a row of huts, built of wooden frames with hide coverings, which may have housed between thirty and fifty people. Some of the stone tools had been brought from 60 miles away. Most significant of all, radiocarbon dating has established the age of the Monte Verde artifacts at between 12,000 and 13,000 years ago. Subsequent excavations of simpler stone tools from earlier strata may yet push this date back beyond 30,000 years ago.

South America has yielded other archaeological signs of pre-Clovis human settlements, but the finds await further analysis and verification. Through recent experimental dating techniques on rock art and artifacts from the western United States, archaeologists have also pegged as pre-Clovis a number of these items (Dorn et al., 1986; Dorn & Whitley, 1988). Finds such as these seriously challenge Clovis-first interpretations.

HOMO SAPIENS IN ASIA, AUSTRALIA, AND OCEANIA

The occupation of Asia by hominids extends back several hundred thousand years, as we saw with the discoveries of *H. erectus* in Java and China. But this settlement did not spread to Japan or the island continent of Australia; instead, these two areas began to be populated only in the Upper Paleolithic with the advent of *H. sapiens*.

Like the settlement of the Americas, the first arrival of human populations in Japan is still an open question. Lower sea levels during the Pleistocene produced expanses of dry land linking Japan to the Asian mainland and allowed migration. Yet few early sites have been located. Archaeologists excavating at Fukui Cave in northern Kyushu have discovered several stone tools and flakes dated to more than 31,000 years ago (Aikens & Higuchi, 1981).

THE INITIAL SETTLEMENT OF NEW GUINEA AND AUSTRALIA

Archaeological research pins the earliest human settlement in Australia to 50,000 years ago, perhaps somewhat earlier. At that time, Australia was connected by a continental shelf to New Guinea, but there were no land bridges between the Australia–New Guinea landmass and the Southeast Asian mainland. Scientists therefore speculate that the first humans arrived in Australia by simple rafts or boats, but to date no evidence of these vessels has been uncovered. There are signs of human occupation by 35,000 to 40,000 years ago over most of the continent, with the possible exception of Australia's central desert (White, 1993; White & O'Connell, 1982). The earliest finds from the island of Tasmania, located south of Australia, are somewhat more recent, dating to between 30,000 and 35,000 years ago.

On the basis of stone tool industries and skeletal remains, some Australian archaeologists have postulated that the continent was settled by two waves of immigrants. The earliest of the arrivals are represented by the finds at Bobongara on New Guinea's Huon Peninsula, dated to about 45,000 years ago. The diagnostic stone tools at this site consist of crude axes, weighing 4 to 5 pounds, that were made by striking large, irregular flakes off river cobbles (White, 1993).

The Bobongara finds can be contrasted with later tools (dated after about 26,000 years ago) from sites such as Nawamoyn and Malangangerr in northern Australia and Nombe in the New Guinea highlands. Some of the stone tools from these sites were partly shaped by grinding the edges against other stones. As will be seen in Chapter 8, this *ground stone technique* does not become common in other world areas for another 10,000 years. Citing these discoveries, some researchers have argued that these different stone tool technologies might represent different immigrant populations.

Searching for support of this theory, researchers have turned to physical anthropology and skeletons of early Australians. The earliest human remains in Australia, dating back some 20,000 years, come from Lake Mungo in the southeastern interior of the continent. The skulls of the two individuals represented are described as delicate. They have frequently been contrasted with finds at another site, Kow Swamp, which dates somewhat later. Scientists have identified more than forty burials at Kow Swamp, and compared to the Lake Mungo remains, the skulls are generally heavier and have sloping foreheads.

On the basis of these differences, some researchers postulate that these remains reflect two different migratory populations. This view can only be classified as highly speculative, however, since the Lake Mungo remains represent only two individuals. The two collections do share many similarities, and perhaps if more individuals had been found at Lake Mungo the two populations might appear to be closer in heritage (Habgood, 1985).

PACIFIC FRONTIERS

The Pacific Islands, or Oceania, were the last great frontier of human habitation. The Pacific Ocean covers one-third of the earth's surface, broken only occasionally by islands and coral reefs. The islands, spread over thousands of miles, are usually divided into three groups: Melanesia, which includes the smaller islands immediately north and east of Australia; Micronesia, which is farther north and includes the islands of Guam, Naura, Marshall, and Gilbert; and Polynesia, which includes widely scattered islands such as Tahiti and Hawaii east of the preceding groups. The achievements of the early settlers of these regions are all the more remarkable when it is realized that many of the islands are not visible from other bodies of land. Nevertheless, humans had settled hundreds of the islands long before the arrival of the first European mariners in the sixteenth century.

Not surprisingly, the earliest evidence for human settlement comes from the larger islands of Melanesia, closer to New Guinea. Archaeological evidence from the Bismarck and Solomon islands places the first evidence for human occupation at about 30,000 years ago. Like the early settlements of New Guinea and Australia, this evidence consists of stone tools associated with hunting-and-gathering adaptations. The settlement of the smaller islands of Melanesia, as well as those of Micronesia and Polynesia, was substantially later, perhaps beginning no more than 3,000 or 4,000 years ago (Irwin, 1993). These settlers were farmers, who brought domesticated animals and root crops with them (see Chapter 8). They also brought a pottery and shell tool technology.

 SUMMARY

The lifestyles of the earliest hominids remain an enigma. Much of the insight we have gleaned comes from the study of the stone tools and butchered animal bones they left behind. Interpretations of these finds vary considerably. Some models liken hominid behavior to that of human hunter-gatherers. Other hypotheses stress the importance of females in child rearing, social learning, and food gathering. Still other studies draw parallels between early hominids and the behaviors of modern primates.

Although early hominids undoubtedly lived in social groups and learned how to make tools in a social setting, the growing consensus among researchers is that they exhibited distinctive behavior patterns unlike any modern primates, including humans.

In contrast to earlier hominids, the activities and tools of *H. erectus*—the species that precedes *H. sapiens*—are more readily identifiable. The Acheulian tool technology is characterized by the hand ax, a bifacially flaked tool that could be used for a variety of cutting and chopping tasks. Technological innovations, including the development of more elaborate stone tools, and probably the use of fire and the ability to make simple shelters, may have prompted the movement of *H. erectus* out of Africa into cooler climates.

Homo sapiens created stone tools in the Middle and Upper Paleolithic. Stone tools and other elements of technology became increasingly more complex, particularly during the Upper Paleolithic, when all sorts of new stone tools were added to the material inventory. These implements helped in the manufacture of ivory, bone, and horn artifacts. The Upper Paleolithic also saw the debut of artistic expression in cave painting, sculpture, and engraving.

Regional stone tool industries reflect increasingly specialized adaptations to local environmental conditions. Learning how to make use of more varied resources allowed Upper Paleolithic populations to expand into world regions that had previously been unoccupied by hominids. Areas first settled during the last 50,000 years include Japan, Australia, and the Americas. Migrations into these areas were facilitated by lower sea levels, which exposed land that had been, or is currently, underwater.

Archaeological information on these migrations remains scanty, leading to disagreements about when they occurred. However, consensus has emerged among researchers about how the Americas were settled. Scientists contend that the first hominid inhabitants of the Americas arrived via a land bridge linking North America and Asia that was created by low sea levels during the late Pleistocene. Substantial evidence confirms human settlement from 12,000 years ago on, represented by the Clovis tradition, but disagreement persists over the possibility of an earlier human presence. Recent research suggests that the date for the first human occupation of the Americas may be extended farther back in time. Archaeologists have also dated the sequence of migrations of *Homo sapiens* into Asia, Australia, and Oceania.

 QUESTIONS TO THINK ABOUT

1. Given the information we have available, describe what might have been a typical day in the life of an early hominid family. Where did they live? What kind of society did they have? What did they eat? What tools did they use?

2. Describe the specific changes in chipped stone tools that characterized the evolution of technology from the Oldowan to the Acheulian, Mousterian, and Upper Paleolithic periods. What do you think were the consequences of these technological changes?

3. What do the changes in stone tools over time reveal about changes in the humans who were making them? How were the styles of tools and their methods of manufacture affected by changes in human behavior?

4. What are the principal arguments for and against the idea of a pre-Clovis migration to the New World? What is the nature of the archaeological evidence that supports or refutes the pre-Clovis hypothesis?

5. When was the continent of Australia first populated and by whom? What kind of archaeological evidence do we have for the first native Australians? (Be sure to mention specific archaeological sites.)

 KEY TERMS

Acheulian
band
baton method
Clovis-first hypothesis
experimental studies

home bases
Lower Paleolithic
Middle Paleolithic
Mousterian
Oldowan

percussion flaking
pre-Clovis hypothesis
Upper Paleolithic
use-wear and residue studies

 SUGGESTED READINGS

GAMBLE, CLIVE. 1986. *The Paleolithic Settlement of Europe.* Cambridge: Cambridge University Press. This book surveys current issues in the interpretation of the European archaeological record, including the first evidence of *H. erectus* through the appearance of *H. sapiens.*

KINSEY, WARREN G., ed. 1987. *The Evolution of Human Behavior: Primate Models.* Albany: State University of New York Press. A collection of scholarly essays illustrating the relevance of modern, nonhuman primate behavior to interpretations of early hominid activities. The text includes useful overviews, as well as case studies of particular primate groups.

LEAKEY, RICHARD, AND RICHARD LEWIN. 1992. *Origins Reconsidered: In Search of What Makes Us Human.* New York: Doubleday. In this volume, internationally known paleoanthropologist Richard Leakey attempts to reach beyond the fossil remains to the origins of mental and cognitive abilities that distinguish human beings. Touching on philosophy, anthropology, biology, and linguistics, the book above all provides a personal narrative of some of the major hominid discoveries of all time.

PHILLIPSON, DAVID W. 1993. *African Archaeology,* 2nd ed. New York: Cambridge University Press. A survey of African archaeology ranging from an overview of human origins up to the late prehistoric period. The text includes an excellent overview of developments during the Middle and Late Stone Ages.

WHITE J. PETER, AND JAMES O'CONNELL. 1982. *A Prehistory of Australia, New Guinea and Sahul.* Sydney: Academic Press. A definitive and well-written presentation of the information available on the settlement of Australia and New Guinea.

Chapter 8

The Origins
of Domestication
and Settled Life

CHAPTER OUTLINE

The Late Pleistocene: Changes in Climate and Culture
Mesolithic and Archaic Technology

Origins of Food Production: The Neolithic Period
Plant and Animal Domestication

Why Did Domestication Occur?
The Oasis Theory / The Readiness Hypothesis / Population Models / Human Selection and the Environment / Coevolution / Agricultural Origins in Perspective

Domestication in Different Regions of the World
Southwest Asia / Europe / East Asia / Africa

The Origins of Domestication in the Americas
Mesoamerica / South America / North America

Consequences of Domestication
Population Growth / Health and Nutrition / Increasing Material Complexity / Increasing Social Stratification and Political Complexity

I N THE PAST 15,000 YEARS, HUMANS HAVE undergone minimal changes in physical characteristics; in contrast, human cultural adaptations have grown substantially more sophisticated. The most significant of these cultural shifts relates to subsistence, the manner in which humans obtain food and nourishment.

As seen in Chapter 7, Upper Paleolithic populations were probably relatively mobile, *nomadic* people who followed the migrations of the herd animals they hunted. These hunter-gatherers drew extensively from their environment but likely made no conscious effort to alter or modify it intentionally.

Beginning late in the Pleistocene epoch, approximately 15,000 years ago, this pattern gradually began to change in some parts of the world. Rather than moving around in pursuit of large animals, humans started to make more intensive use of smaller game animals and wild plants in one area. Fishing and gathering marine resources also yielded valuable food sources as people became less mobile and increasingly focused their energies on the exploitation of plants and animals within particular local environments. In time, they also started to experiment with planting crops and raising wild animals in captivity, practices that set the stage for food production. This chapter explores the shift from food gathering to food production.

THE LATE PLEISTOCENE: CHANGES IN CLIMATE AND CULTURE

Between the late Pleistocene and the early Holocene (the current geologic epoch), a gradual warming of the earth's temperature caused the great glaciers of the Pleistocene to melt. Sea levels rose in coastal areas, and lands that had been compressed under the glaciers rose. As the earth's climate changed, many species of plants and animals became extinct. For example, Pleistocene megafauna like the mammoth disappeared. Yet many others adapted to the new conditions and even expanded in the new environments. In both North America and Europe, as the ice sheets melted, thick forests replaced the *tundra,* the vast treeless plains of the Arctic regions. These climatic changes enabled human populations to migrate to northern areas that previously had been uninhabitable.

The reshaping of the earth's environments prompted new patterns of technological development. As large game became extinct in Europe and North America, for example, humans captured smaller game, learned how to fish, and gathered plants to satisfy nutritional needs in a strategy that represented a subtle change, one to **broad-spectrum collecting.** Because

160

of variation in local environments, many specialized regional patterns and technologies developed, making it increasingly difficult to generalize about developments worldwide. These new subsistence strategies have been referred to as the **Mesolithic** in Europe, Asia, and Africa and the **Archaic** in the Americas.

MESOLITHIC AND ARCHAIC TECHNOLOGY

The transition to broad-spectrum collecting began in different regions at different times and had varying consequences. In some areas relatively permanent settlements emerged, whereas in other regions people maintained mobile, nomadic lifestyles. In general, however, percussion-flaked Mesolithic and Archaic tools differ markedly from those of the Paleolithic: Typically they are much smaller and more specialized than Paleolithic implements. Some of the most common Mesolithic tools are known as **microliths,** small flakes of stone that were used for a variety of purposes, including harpoon barbs and specialized cutting tools. The bow and arrow appeared in the Upper Paleolithic, and both Mesolithic and Archaic peoples made extensive use of this technological innovation, which allowed hunters to kill game from a greater distance and with more accuracy than did spears.

A new type of stone tool, *ground stone,* also became common in many societies. Some of these implements were probably unintentional products of food processing. To make seeds and nuts more palat-able, people pulverized them between a hand-held grinding stone and a larger stone slab or even a large rock. This activity shaped the hand stones and wore depressions, or grooves, into the stone slabs. Using a similar grinding process, Mesolithic peoples intentionally made some stones into axes, gouges, and *adzes* (specialized tools to shape wood). Tools with similar functions had been produced by percussion flaking during the Paleolithic, but ground-stone tools tend to be much stronger.

The increasingly sophisticated stone-working technology that characterized the Mesolithic and Archaic periods allowed for a great many innovations in such areas as the harvesting of resources and the shaping of wood for building. Although watercraft were developed during the Upper Paleolithic, ground-stone tools made it easier to cut down logs and hollow out the inside to make dugout canoes. Vessels of this type improved mobility and enabled people to exploit more diverse ocean, lake, and river resources. Ground-stone sinkers and fishhooks made from shell, bone, or stone also attest to the importance of aquatic resources in this era.

THE EUROPEAN MESOLITHIC   European Mesolithic sites display a variety of subsistence strategies, reflecting a range of adaptations to various local conditions. Northern European sites such as Meiendorf and Stellmoor in northern Germany include extensive evidence of reindeer kills. On the other hand, sites in France, Britain, and Ireland display reliance on a greater diversity of animals (Price & Brown,

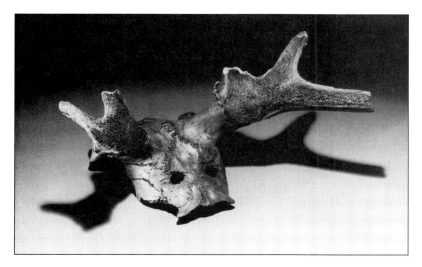

Antler mask from the British Mesolithic settlement of Star Carr, dating to approximately 10,000 years ago. Careful excavation also produced a host of other remains that aided the interpretation of the activities and diet of the inhabitants of the site.

1985). Archaeological evidence for Mesolithic subsistence strategies and how archaeologists interpret this information can be illustrated by the well-studied British site of Star Carr, dating to about 10,000 years ago. Meticulous excavation by archaeologist J. G. D. Clark between 1949 and 1951 produced traces of many different activities (Clark, 1979). Large piles of refuse, or *middens,* contained bones of wild game such as elk, pig, bear, and especially red deer. Discovered tools included projectile points fashioned from elk and red deer antlers and stone microliths, which may have been used for scraping hides or shaping antlers. The careful excavations also produced a host of organic remains.

Though researchers agree that Star Carr was a small, intermittently occupied camp, opinions about its precise function have varied. Clark saw it as a winter settlement of several families of men, women, and children, who used a period of semisettled existence to replace equipment expended during other, more mobile seasons. More recent evidence supports the view that the site had a more specialized function, perhaps serving as a hunting camp or butchering station used for very short periods at different times of the year (Andresen et al., 1981). The camp may have been optimally located to exploit game trails. Such seasonal occupation may be characteristic of other Mesolithic sites.

THE ARCHAIC IN THE AMERICAS   The changes and diversity in adaptations noted at European Mesolithic sites also characterize Archaic sites in the Americas. The specific technologies and resources involved are different, but the experimentation and intensive exploitation of local resources are similar. American Archaic sites are usually categorized on the basis of geographic divisions, such as the Western Archaic, the Great Lakes Archaic, and the Northern Archaic (Jennings, 1989). In turn, these categories are subdivided into regional traditions (distinguished by different tool complexes) and temporal periods (that is, Early, Middle, and Late).

The Koster site in the Illinois River valley illustrates Archaic traditions. The Early Archaic remains found at this site, dating from about 9,500 years ago, indicate that the occupants of the site ate fish, freshwater mussels, and nuts, and hunted white-tailed deer and a variety of smaller animals. These remains likely indicate there were seasonal camps at the site, repeatedly occupied by small bands of hunters and

gatherers over several centuries. The Early Archaic finds contrast dramatically with the Middle Archaic finds, which date from around 7,000 years ago, also discovered at Koster. This period is characterized by larger and much more permanent settlements, including sturdy houses made of log supports covered with branches and clay. The occupants subsisted on many of the same foods as in the preceding period, but they concentrated more on easily obtainable resources found within a few miles of their settlement. The site may have been occupied for most, if not all, of the year.

Some North American Archaic sites seem to prefigure the technological and social complexity typical of later food producing societies. For example, the Poverty Point sites in the lower Mississippi Valley, which are dated to between 3,300 and 2,200 years ago, include the remains of naturally available foods and stone tools typical of other Archaic sites, but also include fired clay balls. These clay balls were likely used as *cooking stones* that could be heated in a fire and dropped into wooden bowls or baskets to cook food. These balls provide the earliest evidence of fired clay in North America. Even more striking are the vast earthworks of five concentric octagonal ridges and mounds found at the Poverty Point site in Louisiana. The earthworks provide evidence of organizational capabilities and ritual expression more typical of later food-producing societies.

ORIGINS OF FOOD PRODUCTION: THE NEOLITHIC PERIOD

The **Neolithic,** or New Stone Age, beginning before 10,000 years ago in some world areas, marks one of the most pivotal changes in human history: the shift from food gathering to food production. Like the change from the Paleolithic to the Mesolithic, the transition to the New Stone Age occurred gradually. During the Mesolithic period, human populations experimented with new types of subsistence activities, including the practice of growing plants, known as **cultivation.** Some groups deliberately collected seeds for planting, not just for consumption. In addition, certain populations began to tame animals like wild dogs or wolves to have as companions and to help in hunting. Other groups sought to capture wild varieties of sheep, goats, cattle, and horses and to travel with these animals to suitable pastures as the seasons changed.

Eventually, some populations came to rely on certain cultivated plants more than on others. They also concentrated their energies on raising particular animals. In other words, some of these groups engaged in **artificial selection,** a process similar to natural selection in which people encourage the reproduction of certain plants or animals and prevent others from breeding. In effect, these human populations were modifying the reproductive patterns of certain plants and animals to propagate certain characteristics better suited to their own needs. Gradually, this process yielded plants and animals that were distinct from wild species and dependent on humans. This process is referred to as **domestication.**

Through artificial selection, humans modified the reproductive patterns of certain plants and animals to characteristics better suited to their own needs.

THE FAR SIDE   By GARY LARSON

"It's Bob, all right ... but look at those vacuous eyes, that stupid grin on his face — he's been domesticated, I tell you."

To some extent, the domestication of plants and animals may have occurred in an unplanned way. When people gathered wild seeds, the larger seeds on the stem were easier to pick. Similarly, people kept more docile, easily tamed animals rather than the more aggressive members of the species. In some world areas about 10,000 years ago, these processes of artificial selection promoted societies that placed great emphasis on domesticated plants and animals. Because people had to remain in certain areas to tend their crops, they began to put down roots for a more permanent home.

PLANT AND ANIMAL DOMESTICATION

Much of what we know about domestication comes from the archaeological record. Because wild and domesticated species differ physically, researchers can trace the transition to domestication by examining plant and animal remains (Cowan & Watson, 1992; Struever, 1970; Ucko & Dimbleby, 1969). For example, wild species of grains such as wheat and barley have fragile *rachis,* the portion of the stem where the seeds are attached. This feature is advantageous in the wild because the seeds easily fall to the ground, where they can germinate and reproduce. In contrast, on domesticated plants, seeds tend to cling to the stems, attached with firm rachis. This feature facilitates harvesting by humans. Domesticated plants also have larger edible parts, as a rule, something early farmers would have favored.

Increasing knowledge about both plant domestication and the exploitation of wild species is a result of intensifying awareness among researchers of the need to recover plant remains from excavations through more refined recovery techniques. A great deal of information has been obtained by the use of a technique known as **flotation.** When placed in water, soil from an excavation sinks, whereas organic materials, including plant remains, float to the surface. These can then be skimmed off and examined for identifiable fragments. Other information may be obtained by studying the stomach contents of well-preserved bodies and *coprolites,* or fossilized feces.

Although archaeologists can easily distinguish plant species in the wild from those that were domesticated, the domestication of animals is more difficult to discern from archaeological evidence, even though many features distinguish wild from domesticated animals. Unlike their wild counterparts, domesticated cattle and goats produce more milk than

Domesticated species of plants and animals differ physically from wild varieties. In this photograph, the oldest maize cobs can be readily distinguished from the larger, more recent examples.

their offspring need; this excess is used by humans. Wild sheep do not produce wool, and undomesticated chickens do not lay extra eggs. Unfortunately, however, the animal remains—primarily skeletons—found at archaeological sites often exhibit only subtle differences between wild and domesticated species. Researchers have traditionally considered reduction in jaw or tooth size as an indication of domestication in some species, for example, the pig. Other studies have attempted to identify changes in bone shape and internal structure. Although providing possible insights, such approaches are problematic when the diversity within animal species is considered: The particular characteristics used to identify "domesticated" stock may fall within the range found in wild herds.

A different approach to the study of animal domestication is to look for possible human influence on the makeup and distribution of wild animal populations, for example, changing ratios in the ages and sexes of the animals killed by humans. Archaeological evidence from Southwest Asia shows that Paleolithic hunters, who killed wild goats and sheep as a staple of their lifestyle, initially killed animals of both sexes and of any age. However, as time went on, older males were targeted, whereas females and their young were spared. This has been interpreted as an intentional move to maximize

yield. Some sheep bones dating back 9,000 years have been found in Southwest Asian sites far away from the animals' natural habitat, suggesting that animals were captured to be killed when needed (Perkins, 1964).

Observations such as these may suggest human intervention and incipient domestication, but conclusions need to be carefully assessed (Wilson et al., 1982). Recent research has pointed out that sex ratios and percentages of juvenile individuals vary substantially in wild populations. Moreover, all predators, not just humans, selectively hunt. Finally, information on the ancient distribution of animal species is unknown.

OTHER ARCHAEOLOGICAL EVIDENCE In the absence of direct evidence from plant and animal remains, archaeologists attempting to identify Neolithic communities at times indirectly infer a shift to domestication. For example, because the food-processing requirements associated with food production, as opposed to hunting and gathering, necessitated specific technological innovations, food-processing artifacts such as grinding stones are found more frequently at Neolithic sites. In addition, Neolithic peoples had to figure out ways to store food crops, because agricultural production is seasonal. Thus, during the Neolithic, structures used as granaries became increasingly common, allowing for the stockpiling of large food supplies against periods of famine. Agricultural peoples constructed large and small granaries or storage bins and pits out of such diverse materials as wood, stone, brick, and clay. Remnants of these storage structures are found archaeologically.

Sherds of pottery, too, often give clues to Neolithic communities. Whereas nomadic hunters and gatherers could not easily carry heavy clay pots in their search for new herds and food sources, the settled agrarian lifestyle encouraged the development of pottery, which makes it easier to cook and store food.

Generalizations about farming cannot be made solely on the basis of indirect evidence such as pottery, however, as the same artifact inventory is not associated with the transition to domestication in all cultural settings. In many instances evidence for domestication precedes the use of pottery. For example, in some sites in Southwest Asia, domesticated barley appears before the use of pottery (Miller, 1992). Conversely, some of the earliest pottery yet discovered, some 10,500 years old, was produced by the Jomon culture of Japan, a sedentary hunting-and-gathering society (Aikens & Higuchi, 1981).

DISTRIBUTION OF WILD SPECIES Archaeology does not provide all of the answers. Plant and animal remains are often poorly preserved or nonexistent. Furthermore, finding early plant or animal remains at a particular site does not necessarily mean that the plant was domesticated there. The species may have originally been domesticated elsewhere and introduced to the area where the remains were discov-

ered. One way researchers trace the provenance of a domesticated species is to pinpoint the areas where related *wild* species are currently found. Because wild species served as the breeding stock for domesticated varieties, domestication probably occurred in these areas. For example, wild species of tomato have been identified in South America, making that a prime candidate as the site of initial cultivation. Domesticated lettuce, which is presently grown in many regions of the world, probably derived from prickly lettuce, a wild species of the same genus that is native to Asia Minor. The origin of domesticated cattle has been traced to *Bos primigenius,* the wild ox, an animal native to Europe, Asia, and Africa that became extinct in Europe in the early seventeenth century (Fagan, 1998).

ETHNOGRAPHIC STUDIES We can also discover clues to domestication in ethnographic data. Modern cultures make varying use of wild, as well as domesticated, plants and animals.

The cross-cultural study of plant use in modern populations is termed *ethnobotany;* the study of the interrelationship between ancient plants and human populations is called **paleoethnobotany.** In one example of this kind of approach, archaeologist Merrick Posnansky (1984) examined what he refers to as "the past in the present." He observes that in addition to domesticates, modern farmers in Ghana, West Africa, harvest more than 200 species of wild plants on a seasonal basis. The farmers also trap wild animals such as the grass cutter, or cane rat, to supplement their diet. Posnansky draws a parallel between these practices and similar activities by early farmers,

Jomon pottery from Japan. Pottery, produced from fired clay, is generally associated with the Neolithic period. However, in the case of the Jomon culture, it was initially developed by sedentary hunter-gatherers.

based on finds at nearby archaeological sites some 3,500 years old.

As the early inhabitants of the region faced similar subsistence challenges in a comparable environmental setting, Posnansky speculates that studies of modern farmers may provide insights into the options available to ancient farmers and the mechanisms that eventually led to domestication.

WHY DID DOMESTICATION OCCUR?

Today we take food production for granted. The vast majority of the world's population depends for sustenance on crops and animals domesticated and cared for by humans. In contrast, hunting and gathering accounts for a comparatively small part of our diet. However, scientists have not determined conclusively why domestication initially took place.

At first glance the transition to reliance on domesticated foods does not seem to be a wise choice. In contrast to hunting and gathering, agriculture takes much more time and energy. The soil has to be tilled, seeds must be planted, and the crops need protection from weeds and pests. Moreover, agriculture demands patience. Several months may elapse from the time a crop is planted until it can be harvested. Tree crops like bananas and plantains may not bear fruit until almost a year after planting. In addition, agricultural production is a risky enterprise. Despite the considerable effort that people invest in planting, weeding, and protecting crops, the harvest may still fail, producing enormous hardships for the entire society.

Hunting-gathering, in comparison, represents a highly successful subsistence strategy. Compared to the farmers' investment in labor, hunter-gatherers spend a comparatively limited amount of time procuring food. Ethnographic studies of groups like the Bushmen and San of southern Africa indicate that they may invest only between twelve and nineteen hours a week in the search for food (Diamond, 1987; Lee, 1969). Although figures may vary, this clearly affords them a great deal more leisure time than their agrarian neighbors. Undoubtedly, as seen in Chapter 7, humans were hunter-gatherers for the vast majority of their history; as recently as 500 years ago, 15 percent of the world's populations still subsisted by that means (Murdock, 1968).

The disadvantages of agricultural production, then, would appear to outweigh the benefits, yet most of our hunting-and-gathering ancestors made the transition to agriculture. Diverse explanations have been posited, but no consensus has emerged. An early theory of the nineteenth century credited a solitary (unknown) genius with suddenly coming up with the idea of planting seeds; this innovation, then, led to agricultural civilizations. Such a simplistic scenario is clearly unlikely. Gathering information from a number of world regions, archaeologists have formulated several provocative theories to explain the transition to agriculture.

THE OASIS THEORY

In the 1930s, V. Gordon Childe (1936, 1952) advanced one of the first scientific theories concerning the move to domestication. Childe suggested that at the end of the Pleistocene a major climatic change transformed the environment in regions like Southwest Asia and made new subsistence strategies necessary. Severe droughts forced humans to move to isolated fertile areas called *oases* and to take up agriculture. According to Childe, who called this presumed period of dramatic change the "Neolithic revolution," agriculture enabled humans to maintain a reliable food supply in extreme conditions. Once invented, the concept of food production spread rapidly to other regions.

Popularly known as the *oasis theory*, Childe's theory was readily accepted by the archaeological community for a number of years. However, subsequent archaeological and geological research has not confirmed his interpretations. Thus far, little evidence has surfaced to prove that such a dramatic change took place in the environment of Southwest Asia following the Pleistocene epoch or that populations clustered around isolated, fertile oases as a result.

THE READINESS HYPOTHESIS

A different theory, developed by archaeologist Robert Braidwood (1960) of the University of Chicago, was based on finds excavated in Southwest Asia during the 1940s and 1950s. Braidwood noted that climatic conditions comparable to those at the end of the Pleistocene had existed in this region at several time periods dating back at least 75,000 years. If agriculture grew out of environmental pressures, as

Childe suggested, why hadn't domestication occurred earlier? Braidwood (1960) hypothesized that after a long period human populations became increasingly familiar with the plants and animals around them:

Around 8000 B.C. the inhabitants of the hills around the fertile crescent had come to know their habitat so well that they were beginning to domesticate the plants and animals they had been collecting and hunting. At slightly later times human cultures reached the corresponding level in Central America and perhaps in the Andes, in southeastern Asia, and in China. From these "nuclear" zones cultural diffusion spread the new way of life to the rest of the world. (p. 6)

Braidwood's statement may present a plausible description of agricultural origins, but like Childe's theory, Braidwood's hypothesis does not really explain what prompted hunters and gatherers to adopt agriculture as a way of life. Underlying his model is a sweeping assumption about human nature or psychology—that earlier peoples were not ready to innovate or develop agriculture for some unexplained reason. This theory, sometimes referred to as the *readiness hypothesis,* does not answer two key questions: how and why domestication originated when it did.

POPULATION MODELS

More recent models of the origins of agriculture have been influenced by economist Ester Boserup's (1965) theories about the relationships among population, labor, and resources. Although Boserup initially set out to explain changes in complex agricultural practices, her ideas can be applied equally well to the origins of domestication. She speculated that societies will intensify their cultivation practices only when they are forced to by rising population pressure on the available resources. Making the transition from simple to intensive agriculture requires such a substantial increase in labor that the results may not warrant the effort. History attests to this statement. Many hunting-and-gathering societies were familiar with intensive agriculture but did not adopt these practices because of the vast increase in labor needed to succeed. In this view, then, agricultural production would not make sense for populations who enjoy reliable food resources and experience limited pop-

ulation growth. At some point, however, population pressures may force people to adopt food-production techniques. Researchers differ in their interpretations of what factors may have caused these pressures.

DEMOGRAPHIC STRESS Archaeologist Lewis Binford (1968) linked increasing population pressure to environmental change. Binford noted that at the end of the Pleistocene period, sea levels began to rise with the melting of the world's glaciers in the temperate regions. He reasoned that rising sea levels would have forced coastal peoples to migrate into interior regions, where other populations had already settled. In Binford's view, this movement would lead to population increases and demographic stress. In response to these demographic and environmental shifts, populations would begin systematically to cultivate the land so there would be adequate food supplies for the expanding population. Thus, Binford contended, population pressure prompted the development of agriculture.

POPULATION GROWTH Archaeologist Mark Cohen (1977) formulated another hypothesis that attributes domestication to population growth. Cohen pointed out that by the end of the Paleolithic era, hunting-and-gathering societies had spread to all parts of the world. During their migrations they gradually expanded the amount and variety of wild food resources they could draw on for sustenance. Eventually, these populations were using nearly all of the naturally available food. As populations continued to grow and territorial expansion left very few unpopulated areas for nomadic hunters and gatherers to explore, the need to feed greater numbers of people gave rise to agrarianism.

Recent archaeological research delving into the origins of agriculture has unearthed more controversy than consensus. Some researchers, including Fekri Hassan (1981), have criticized the population-pressure models proposed by Boserup, Binford, and Cohen, arguing that population pressures alone would not make people abandon hunting and gathering in favor of intensive agriculture. Furthermore, although most archaeologists would agree that population densities did increase at the end of the Pleistocene, they have no clear sense of whether this population surge occurred before or after the transition to food production.

HUMAN SELECTION AND THE ENVIRONMENT

Many researchers have focused on particular local conditions and cultural settings that may have precipitated or affected different patterns of domestication. Carl O. Sauer (1952) was among the first archaeologists to stress how human adaptations may have paved the way for domestication. His research on the early cultivation of root crops in East Asia yielded the conclusion that plants were first domesticated by successful, sedentary food collectors who had the opportunity to observe the plants' growth cycle. Sauer considered Southeast Asia to be a major center of domestication, and he suggested that domestication spread outward from there.

Studying data from Mesoamerica and Southwest Asia, archaeologist Kent Flannery (1965, 1973) approached the question from a different perspective. He argued that an important push for domestication came when humans introduced plants to environmental zones outside the areas where the plants normally flourished. Why did they do this? According to Flannery's hypothesis, transplantation might have stemmed from population growth or from a human desire to exploit certain resources on a more permanent basis. Under these circumstances, humans would have had to invest extra time to nurture plants removed from their natural environment. This, then, eventually resulted in domestication.

COEVOLUTION

One archaeologist, David Rindos (1984), has examined the question of domestication in a biological evolutionary framework. Rindos criticizes other interpretations of the origins of domestication for placing too much emphasis on conscious or intentional human selection. He argues that humans unintentionally promoted the survival and dispersal of certain types of plants through such activities as weeding, storing, irrigating, and burning fields. As particular species of plants became more common because of these practices, human reliance on them increased. This, in turn, led to further changes in plant distribution and morphology. According to Rindos, human agricultural practices and biological changes in cultivated plants evolved simultaneously. Human agents unintentionally selected and germinated specific varieties of plants and adopted cer-

tain behaviors that created a link between agricultural yields and population growth.

Rindos's research underscores the role of unconscious human choice in the process of domestication. However, other archaeologists emphasize that learning, cognition, culture, and conscious processes are as crucial in explaining the origins of agriculture as are unconscious choices. For example, Michael Rosenberg (1990) points out that cultural norms regarding property and territorial arrangements in hunter-gatherer populations were a conscious societal development. Similarly, he believes that such deliberate cultural choices affected transition to agriculture.

AGRICULTURAL ORIGINS IN PERSPECTIVE

Contemporary anthropologists tend to credit a complex interplay of factors for the transition from food gathering to food production as human populations evolved. Some schools of thought hold that our predecessors consciously decided to gamble on food production after considering the advantages and drawbacks of agricultural enterprise. According to this view, early humans weighed population pressures and the need for reliable food sources against the intense labor and the uncertainties of growing their own food plus the likelihood that naturally occurring resources would meet all their needs.

Other theoretical approaches suggest that a less intentional process was at work. These models hold that particular plants took well to the soil in areas where humans had settled, and people just followed their instincts, making the shift from hunting-gathering to agriculture with little thought of the consequences.

Although archaeologists have reached no consensus concerning the exact reasons for domestication, all agree that the process was much more intricate and gradual than the "revolution" envisioned by early theorists.

DOMESTICATION IN DIFFERENT REGIONS OF THE WORLD

Researchers looking into the Neolithic have traditionally spotlighted Southwest Asia and, to a lesser extent, China and Mesoamerica, where some of the

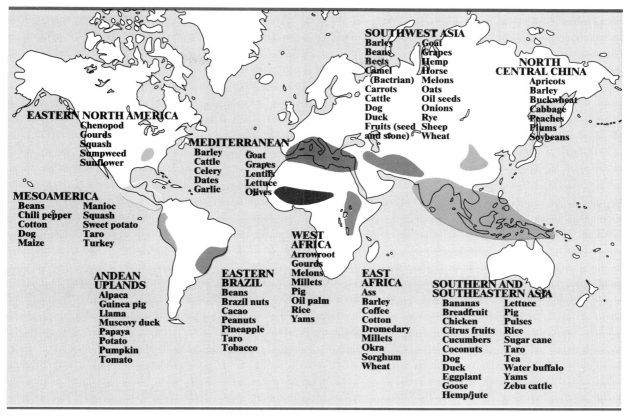

FIGURE 8.1   A world map showing centers of domestication.

Source: Adapted from Arthur Getis, et al., *Introduction to Geography,* 5th Ed., copyright © 1995 The McGraw-Hill Companies. Used by permission of the McGraw-Hill Companies.

first archaeological evidence pertaining to domestication was recovered (Harlan, 1971). However, the complex issues underlying the origins of domestication are worldwide in scope. Although domestication occurred earlier in some areas than in others, it took place independently in many regions and it involved a vast number of species (Figure 8.1). We now turn to how the transition was made from hunting and gathering to food production in different areas of the world.

SOUTHWEST ASIA

In Israel, Jordan, Syria, Turkey, Iraq, and Iran—the area known as the *Fertile Crescent*—scientists have chronicled a gradual trend toward the exploitation of a variety of resources. This area—which extends along a curve from the Red Sea, north along the east-

ern edge of the Mediterranean Sea, then southeast through the Zagros mountains of Iran and Iraq, down to the Persian Gulf (Figure 8.2)—includes a number of distinct environmental zones, ranging from high mountains to the fertile Tigris and Euphrates River valleys. Here, early hunters and gatherers found a wide range of natural resources.

THE NATUFIANS   The best-known Mesolithic people of Southwest Asia are the Natufians, who lived in the eastern portion of the Fertile Crescent in what is today part of Israel and Jordan, subsisting on wild animals and wild plants (Henry, 1984; Mellaart, 1975). Approximately 13,000 to 11,400 years ago, the Natufians settled in villages, where they cultivated wild grains and cereal grasses. Archaeologists have discovered a number of items that confirm these new dietary practices, from mortars, pestles, and ground-

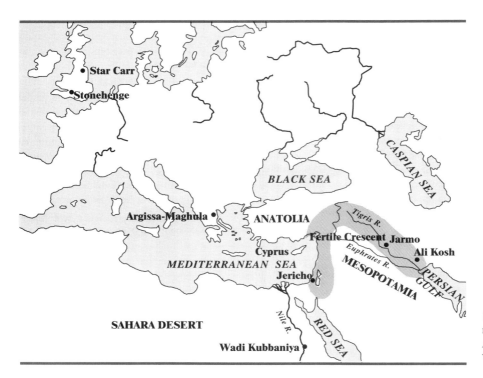

FIGURE 8.2 Sites of domestication in Southwest Asia, Europe, and the Nile Valley.

stone bowls to sharp flint blades that were inserted into bone handles and used to cut grains (Figure 8.3). Archaeologists know how these blades were used because a distinctive residue, called **silica gloss,** was left by the plant stalks. Among the plants exploited by the Natufians were wild barley and various species of legume, which provided a rich source of protein (Miller, 1992).

Natufian society also demonstrates the increasing elaboration associated with more permanent settlement. Whereas early Natufian sites were established in natural rock shelters, some later settlements were quite substantial, incorporating houses with stone foundations, paved floors, and storage chambers. Variation in the material wealth found in graves suggests that there was some differentiation in social status. Imported items such as seashells, salt, and obsidian (a volcanic glass) attest to the Natufians' expanding trade contacts as well. These data suggest greater social complexity than that of earlier societies.

The manipulation of wild plants by the Natufians and other Mesolithic populations paved the way for later farming communities, which emerged more than 11,000 years ago, and display a reliance on various combinations of domesticated plants and ani-

mals. After several thousand years, a distinct pattern of village life—based on wheat, barley, peas, beans and lentils, sheep, goats, pigs, and cattle—appeared. These Neolithic societies continued the elaboration of material culture initiated in earlier periods.

One of the most famous of these Neolithic settlements was Jericho, which was flourishing long before the time of Joshua and the Israelites as described in the Bible (Kenyon, 1972). The earliest settlement at Jericho was most likely a temporary seasonal camp that gradually became a more permanent settlement. By 10,000 years ago Jericho was a sizable town with permanent mud-brick structures containing finely plastered floors and walls. A massive stone wall with towers and a defensive ditch were built around the settlement. The long occupation at Jericho, consisting of substantial villages built on the same spot over thousands of years, provides an excellent illustration of *tell,* a type of archaeological site that became increasingly common during the Neolithic.

THE EASTERN FERTILE CRESCENT Remnants of the Neolithic transition have been found in other parts of the Fertile Crescent as well. In the valley steppe-land areas of the Zagros mountains and surrounding

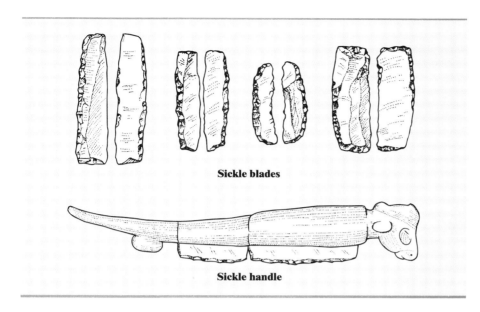

Sickle blades

Sickle handle

FIGURE 8.3 A hypothetical reconstruction of a Natufian sickle, showing stone blades set in a bone handle. Traces of silica gloss, a distinctive polish resulting from cutting plant stems, have been found on stone blades, suggesting how they were used.

Source: Adapted from D. A. E. Garrod and D. M. A. Bate, *The Stone Age of Mount Carmel* (New York: Oxford University Press, 1937); and James Mellaart, *The Earliest Civilizations of the Near East* (London: Thames and Hudson, 1975).

hills, archaeologists have uncovered extensive evidence of a way of life based on cultivation and pastoralism, a subsistence strategy that has as its core the rearing of domesticated animals. By 10,000 years ago, Neolithic farming villages dotted the entire region.

Agriculture spread rapidly to the lowland area of the Mesopotamian plain, which was flooded each spring by the Tigris and Euphrates rivers. The annual flooding enabled crops to mature before the soil was completely dried out, and it gave rise to one of humankind's most important innovations—irrigation systems to maintain crop growth from season to season. At Ali Kosh, one of the earliest farming sites in this region, archaeologists have discovered sequential phases of human occupation dating back 9,500 years. The earliest settlements may have been seasonal, the people relying for sustenance on wild animals, fish, and waterfowl. Over time, however, the region's inhabitants gradually began cultivating or domesticating plants and animals, including sheep, emmer wheat, and barley. Signs of irrigation and domesticated cattle appear about 7,500 years ago.

EUROPE

Although considered a discrete area here, Europe actually comprises a variety of different climatic, topographical, and cultural regions, ranging from the mild Mediterranean climates in the south to the frigid areas of Sweden, northwestern Russia, and northern Poland. Mesolithic hunting-and-gathering populations exploiting resources in these varied climes, probably initially used crop agriculture only as a minor adjunct.

Traditional views held that early farmers expanded in a wave from southwestern Asia and Anatolia, displacing earlier Mesolithic populations (Evans & Rasson, 1984; Renfrew, 1987). Although current evidence suggests that most crops, including barley, peas, lentils, vetch, and rye, were introduced, there was a great deal of variation in the manner of adoption. In addition, the local domestication of some crops, such as oats and peas, cannot be ruled out.

Surveying the available data, we can discern three different patterns of agricultural adoption (Dennell, 1992). First, in some areas such as southeastern and central Europe, evidence for domestication appears suddenly and is frequently associated with village settlement and pottery. In the second pattern, the evidence for farming appears gradually; there is no clear distinction between Mesolithic and Neolithic settlements. This pattern holds true for western Europe, the British Isles, and much of the Mediterranean coast. Domesticated grains, legumes, and livestock turn up in different combinations and at different times, eventually giving rise to true farming communities only after hundreds of years. Third, in

Excavated remains of the great tower of early Jericho, a thriving Neolithic settlement that developed from a temporary camp used by Mesolithic hunter-gatherers.

some areas such as Sweden and Finland, a shift to agriculture gradually occurs, but the evidence for Mesolithic subsistence patterns continues, suggesting that the agricultural practices were not initially successful in the northern climates.

MEGALITHS By 6,000 years ago, agriculture was thriving throughout most of Europe. One of the most interesting and enigmatic developments of Neolithic Europe was the appearance of large stone structures, or **megaliths,** Greek for "large stones." Archaeological sites dating to the period show traces of large houses, fragments of well-made pottery, and evidence of trade contacts with other regions. Distinctive burial complexes also characterize this period: Elaborate tombs with chambers for several

bodies were constructed from large stone slabs and boulders. These structures were the earliest megaliths. Later megaliths were much more intricately designed, consisting of circular patterns of huge, upright stones.

The most famous megalithic structure is *Stonehenge,* located on Salisbury Plain in southern England. Popular wisdom credits the massive stone monuments to the Druids, a religious sect in Britain during the first century B.C. In fact, Stonehenge far predates the Druids. Archaeological research reveals that work on Stonehenge first began around 3000 B.C. and spanned the next 2,000 years. As illustrated in the chapter-opening photo, a circle of thirty vertical stones, each weighing more than 25 tons, marks the center of the site. Running across the tops of these stones was a series of thirty stones, each weighing about 6 tons.

No one has yet pinpointed why Stonehenge and other megaliths were constructed, but studies over the past several decades have indicated that many of the stone structures may mark astronomical observations (Hawkins, 1965; Hoyle, 1977). For example, viewed from the center of Stonehenge, outlying stones indicate the summer and winter solstices, whereas other alignments point to positions of the moon and possibly other celestial bodies. The knowledge of the heavens expressed in Stonehenge may indicate the importance of the stars in interpreting the seasons of the year. These imposing monuments also clearly indicate a growing social complexity in Neolithic Europe. They may have served to reinforce the social, economic, and religious authority of the people who directed their construction.

EAST ASIA

Agricultural practices unrelated to those in Southwest Asia emerged in the far eastern part of the continent. As in other world areas, the advent of domestication was preceded by a period of ambitious experimentation with naturally occurring plants and animals (Gorman, 1969, 1977). Researchers disagree about when people in the Asian region initially domesticated plants. Some, citing evidence from Thailand, argue that root crops, particularly various species of yam, were domesticated very early, perhaps as far back as 11,000 years ago (Solheim, 1971). Crops like these are propagated by dividing and replanting living plants, an activity referred to as

vegiculture. Unfortunately, however, it is very difficult to distinguish domesticated from wild root crops, as Carl Sauer (1952), an early researcher on the origins of Asian domestication, has noted. In addition, archaeological remains needed to clear up this mystery are fragmentary. So whereas experts agree that root crops were exploited very early in Southeast Asia, no one can specify with certainty the precise time period of domestication.

The earliest evidence for the domestication of rice, too, is uncertain, but archaeologists believe it occurred somewhat later than the domestication of root crops. Yet, many areas have not been studied in detail, and much of the available information is inconclusive. In one major site in northeastern Thailand, known as Spirit Cave, archaeologist Chester Gorman identified artifacts that hint at rice cultivation by 9,000 years ago, but no actual grains were found. Rice husks dating back 7,000 years have been found at Khok Phanon Di, Thailand. Other sites in the region have yielded evidence of domestication that may date even earlier (Fagan, 1998).

CHINA The origins of domestication in China remain sketchy. Some researchers have asserted that Chinese agricultural beginnings date to some time before 9,000 years ago (Chang, 1970, 1975; Crawford, 1992). Sites with pottery dating before 8,500 years ago are generally assumed by archaeologists to be early Neolithic, although actual plant remains are absent. The earliest direct evidence for domestication comes from northern China between 8,500 and 7,000 years ago. Villages of this period supported a mixed economy dependent on hunting-gathering, fishing, and plant and animal husbandry. Domesticates include foxtail millet, broomcorn millet, and Chinese cabbage. Domesticated animals are represented by the pig, the dog, and possibly the chicken.

These early societies were the forerunners of later cultures such as Yangshao, which flourished throughout much of the Yellow River valley between 5000 and 3000 B.C. Yangshao sites are typically more than five times the size of those of the preceding period (Crawford, 1992). In addition to earlier domesticates, domesticated animals such as cattle, sheep, and goats were raised. Significantly, these societies were not entirely dependent on domestication; hunting, gathering, and fishing continued to supplement their diet. Rice, an important staple today, did not become important in northern China until much more recently. Yangshao farming villages comprised fairly substantial dwellings that were built partly below ground. Special areas were designated for the manufacture of pottery, and cemeteries were located outside the village. Many characteristics of modern Chinese culture can be identified in these communities.

From these early centers of domestication in China and Southeast Asia, archaeologists have documented the spread of agriculture from the Asian mainland into Korea, through Japan, and into the islands of Southeast Asia and the Pacific.

In contrast to developments in China, broad-spectrum collecting continued in Korea until 4,000 years ago and even later, after 2,400 years ago, in Japan. Subsequently, rice cultivation predominated in these regions, culminating in the evolution of complex agricultural societies.

The peoples of Southeast Asia also concentrated on intensive rice growing, which propelled the development of societies throughout Malaysia and the Indonesian Islands. The Pacific Islands (Melanesia, Micronesia, and Polynesia) staked their agricultural claims on taro, yams, and tree crops like breadfruit, coconut, and banana.

AFRICA

Archaeological information on the growth of agriculture in Africa is not as well documented as it is for other regions of the world. Only limited research has been undertaken there, and the tropical climates that predominate on the continent do not preserve plant and animal remains well over time. This is unfortunate because modern African societies maintain a wide variety of agricultural practices, making the region a key area of study for researchers probing the transition to food production (Clark & Brandt, 1984). More than forty genera of domesticated plants have been identified, including many crops that are still of considerable importance in Africa. Based on these findings, researchers believe that several centers of domestication developed independently.

Changes in subsistence strategies similar to those that heralded the advent of the Mesolithic and the Neolithic in the Near East and Asia took place very early in Africa. Upper Paleolithic populations in Egypt's Upper Nile Valley were harvesting, and perhaps cultivating, grains 18,000 years ago (Wendorf

Archaeologists working at Wadi Kubbaniya used fine screens to sift excavated soil for plant and animal remains. Improved recovery techniques such as this one have dramatically increased available information on ancient diet and the origins of domestication.

& Schild, 1981). Evidence for these activities comes from excavations at such sites as Wadi Kubbaniya on the Nile River, where archaeologist Fred Wendorf recovered grinding stones and sickle blades with silica gloss. By 8,000 years ago, wetter environmental conditions had allowed the expansion of people into the Sahara, and semipermanent and permanent settlements dating to this period have been identified in Egypt's western desert (Wendorf & Schild, 1984). Excavations at these sites have yielded remains of domesticated cattle, barley, well-made pottery, and storage pits. Other sites in Egypt, of more recent age, have produced emmer wheat, flax, lentils, chickpeas, sheep, and goats. Whereas the domesticates represented suggest a southwestern Asian origin, the associated toolkits are distinctly African (Harlan, 1992).

Archaeological finds in sub-Saharan Africa reveal many regional traditions that effectively made use of a host of wild plant and animal resources. In contrast to northern Africa, however, the transition to food production seems to have occurred more slowly. The thick tropical rainforests of western and central Africa impeded the diffusion of agricultural technology to this region. Furthermore, the plethora of naturally occurring animal and plant resources were more than sufficient to meet the subsistence needs of the indigenous hunting-and-gathering populations.

Some 4,500 years ago, climatic conditions in the Sahara desert were less arid and more suitable for

human settlement than they are today. The primacy of pastoralism in this region is confirmed by archaeological evidence that domesticated cattle, goats, and sheep spread southward from northern Africa beginning about 7,000 years ago and expanded into eastern and western Africa about 4,000 years ago (A. Smith, 1984). At the earliest sites, artifacts such as grinding stones suggest that plant foods were being processed, even though few actual plant remains or residue have been recovered. In the western Sahara, the earliest trace of the extensive use of domesticated plants dates after 3,200 years ago. The most telling evidence comes from Dhar Tichett, an archaeological site where impressions of bulrush millet and sorghum were preserved in pottery sherds.

Plant remains of probable African domesticates such as sorghum, pearl millet, African rice, and yams do not appear until much later. However, it is likely that domestication had occurred earlier: Vegiculture based on the cultivation of indigenous yam species may have begun in sub-Saharan West Africa between 7,000 and 6,000 years ago (Phillipson, 1993). Yet speculation stems entirely from circumstantial evidence: Pottery and ground stone appear in the archaeological record of this time period. More intensive agricultural practices in forest regions, beginning approximately 2,000 years ago, may have been facilitated by the advent of ironworking, when iron tools made clearing land more labor efficient (Fagan, 1998).

THE ORIGINS OF DOMESTICATION IN THE AMERICAS

In the Americas, the transition from the Paleolithic to the Archaic began around 7000 B.C. This shift in subsistence strategies encompasses a wide range of cultures and tool technologies. Many American cultures—reflecting numerous regional variations and spanning a great many environmental zones—developed subsistence strategies based on domesticated crops that flourished locally. In describing these adaptations, archaeologists generally use specific names rather than the generic *Neolithic*.

MESOAMERICA

Mesoamerica extends from the northern boundary of present-day Mexico to the southern borders of Nicaragua and Costa Rica. As in the past, the region today encompasses a great deal of environmental and cultural diversity. Early cultivation techniques and domesticates also varied, a factor that needs to be considered when reaching general conclusions about the advent of agriculture in the region. At pre-

sent, all of the early evidence for plant cultivation and domestication comes from Mexico, much of it from dry cave sites; these data, then, can represent only a small portion of the variation probably present (Flannery, 1985; McClung de Tapia, 1992).

One of the major archaeological studies of this region was carried out by archaeologist Richard Mac-Neish, who examined the sequence of agricultural developments in the Tehuacan Valley in Mexico. Through the excavation of twelve sites, MacNeish found traces of human occupation extending back 12,000 years, with early settlers subsisting on the meat of wild animals. Beginning around 10,000 years ago, however, the population increasingly turned to wild plants as a source of food, eventually cultivating wild species. By 7,000 years ago, these native peoples had domesticated maize, chilis, avocados, and gourds (MacNeish, 1970). Gradually, the population of the Tehuacan Valley came to rely almost exclusively on domesticated crops, particularly maize (Figure 8.4). Similar sequential developments—from hunting to cultivating plants and, finally, a complete reliance on domesticates—have been documented in other areas.

Archaeologists have tracked changes in Mesoamerican subsistence strategies by examining artifacts such

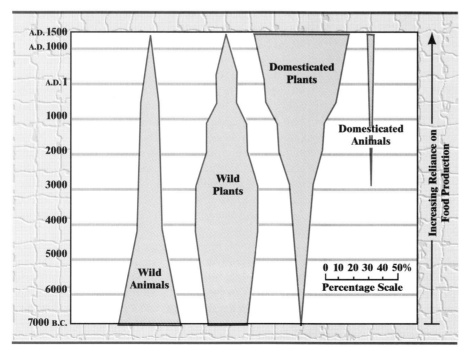

FIGURE 8.4 A graph illustrating changes in the relative importance of wild and domesticated food resources exploited by people of the Tehuacan Valley, Mexico, as determined from archaeological remains. The figure traces the gradual transition from an almost exclusive reliance on natural resources to dependence on domesticated crops. Graphs of other world areas would follow a similar pattern.

Source: Adapted from R. S. MacNeish, Q. Nelken-Turner, and I. W. Johnson, *The Prehistory of the Tehuacan Valley,* Vol. 2, University of Texas Press, 1967. Copyright ©1967 Robert S. Peabody Museum of Archaeology. Phillips Academy, Andover, Mass. All rights reserved.

as stone-grinding slabs, locally called *metates,* and hand-grinding stones, known as *manos.* They have also noted the construction of food storage facilities and the permanent settlement of large populations in village farming communities. This pattern of intensive agriculture eventually took hold in various regions of Mesoamerica and spread to North America.

SOUTH AMERICA

As in Mesoamerica, South America encompassed a number of diverse cultural and ecological settings for early agriculture (Pearsall, 1992). These included at least three distinct systems: first, a suite of practices adapted to the low-altitude cultivation of manioc (a root crop), maize, and beans; second, a mid-altitude system focused on maize, peanuts, and beans; and third, a high-altitude Andean system involving a number of minor tubers and cereals but dominated by the potato. There is no complete picture for the domestication of any one crop.

Indications of early agricultural developments in South America have been found in a number of areas, but many of the better-known sites are located in Peru. Near the foothills of the Andes on the Peruvian coast are fertile river valleys that are ideal for the cultivation of food. Following the Pleistocene, people congregated in coastal regions, finding sustenance in fish and other marine resources. By about 10,000 years ago, these sedentary communities were experimenting with a variety of wild plant species. The earliest cultivated species—including the bottle gourd, manioc, chiles, beans, squash, and potatoes—appeared before 4500 B.C. Evidence for peanuts, jack beans, and achira (a starchy root) dates slightly later. Notably, the evidence for the use of many crops precedes the presence of ceramics. In fact, the abundant remains of cotton dating between 2000 B.C. and 1500 B.C. has given this period the name the "Cotton Preceramic."

Because the highland terrain includes steep mountain slopes that cut through a variety of microenvironments, the subsistence strategies of the early inhabitants of the Peruvian highlands differed markedly from those of coastal people (Fagan, 1989). Here the transition to agriculture is documented by archaeological finds at such sites as the Guitarrero and Pikimachay caves, which preserve a record extending back about 12,000 years. Early highland hunting-and-gathering populations first subsisted on animals like the giant sloth and smaller species of wild game, plus a variety of wild plants. Finds of more recent age from a number of sites indicate an Archaic subsistence based on the hunting of camelids and deer and the gathering of wild plants (Pearsall, 1992). The earliest evidence for plant cultivation dates to 10,000 years ago. Many of the earliest finds of cultivated plants are from the highlands, supporting interpretations that many of the crops involved may have originated there.

Species of maize, squash, potatoes, beans, gourds, and other crops became common between 8,000 and 5,000 years ago. Archaeologists believe that the llama, a South American pack animal, and the guinea pig had been domesticated by 5,500 years ago.

NORTH AMERICA

Researchers have tended to view North America as peripheral to the domestication center located in Mesoamerica. Although knowledge about important plant crops such as corn, squash, and beans appears to have filtered northward from Mesoamerica, inhabitants of North America also domesticated other plants and animals on their own (Smith, 1989). Archaeologists have uncovered evidence that various starchy and oily seeds were domesticated in what has been termed the *Eastern Agricultural Complex* (Ford, 1985; Keegan, 1987). As in other American areas, the manipulation of wild plants began early in the Archaic, with some domesticated species recovered from sites dating between 4,000 and 3,000 years ago.

The cultivation of Mesoamerican domesticates such as maize and beans spread into North America during the period between A.D. 800 and 1100 (Smith, 1989). Gradually, maize became the most important crop in farming communities throughout North America. In what is now the southwestern United States, maize cultivation led to specialized dry-land farming practices in Pueblo cultures such as the Hohokam, the Anasazi, and the Mogollon. These practices included irrigation methods and precise planting and harvesting schedules organized around rainy seasons.

An array of agricultural technologies, which promoted the cultivation of maize, beans, and squash, diffused throughout the Midwest, the Southeast, and eventually the Northeast, culminating in the Adena and Hopewell cultures (in the eastern United States) and the immense Mississippian cultural complex with its center at Cahokia, now in Illinois. Artwork and

Native American cultures developed a wide variety of agricultural practices suitable for many different environments. Farming communities, such as those at the Cliff Palace ruins in Colorado, flourished for hundreds of years by making effective use of the limited water available.

other artifacts generated by these societies indicate contacts with Mesoamerica across the Gulf of Mexico and through the river systems connecting the Mississippi with the Ohio, Missouri, and Illinois rivers.

CONSEQUENCES OF DOMESTICATION

Although no one can pinpoint the precise reasons for the domestication of plants and animals, these activities clearly had far-reaching consequences, especially on the environment. All farming involves modification of the natural environment by clearing the land of naturally occurring vegetation. Some agriculturalists do periodically allow the land to lie fallow (uncultivated), but it is eventually cleared again, preventing the regrowth of wild plants. Domesticated animals also alter the environment by grazing and preventing the regrowth of plants. Larger settlements and more intensive land use by humans frequently contribute to soil erosion and a decline in soil fertility.

POPULATION GROWTH

As agriculture transforms the landscape, it also gives rise to increases in human populations by making food supplies more stable and reliable. Even more significant, agriculture yields more food per acre of land, which allows a given region to support a larger population. The surge in world population during the Neolithic period constituted a major demographic shift in human history. The annual population growth rate rose dramatically, leading to a tenfold population increase from the end of the Paleolithic, when scientists estimate the total world population at 30 million. By the year A.D. 1, some researchers conclude that the world population stood at approximately 300 million (Hassan, 1981).

HEALTH AND NUTRITION

Although agricultural developments promoted population growth, they did not necessarily improve the quality of life in agricultural societies. In fact, in a number of areas, the advent of domesticated crops actually contributed to a decline in human health (Cohen & Armelagos, 1984; Larsen, 1995). The larger settlements of the Neolithic increasingly brought people into contact with one another, facilitating the spread of infectious disease. In some cases people also became dependent on particular domesticated plants like corn, to the exclusion of most other foodstuffs. This restricted diet did not fulfill nutritional requirements as well as that of hunter-gatherers, which encompassed a wide variety of both plants and animals. Reliance on one crop rather than on a variety of wild resources also boosted the risk of famine.

Archaeologists can study the impact of Neolithic life by studying human skeletons (Goodman et al., 1984). Poor nutrition, which causes arrested growth and disease, leaves identifiable traces on the bone. Signs of physiological stress brought on by food shortages show up in *Harris lines,* lines in long bones indicating periods of arrested growth, and in *enamel hypoplasias,* deficiency in tooth enamel. Calculations of prehistoric people's average height and age of death also shed light on changes in general health.

In a survey of worldwide data, Anna Curtenius Roosevelt (1984) concluded that Paleolithic and Mesolithic populations did experience food stress. However, she found more signs of stress and other health and nutritional problems in sedentary Neolithic communities:

> It seems that a large proportion of most sedentary prehistoric populations under intensive agriculture underwent chronic and life-threatening malnutrition and disease, especially during infancy and childhood. The causes of the nutritional stress are likely to have been the poverty of the staple crops in most nutrients except calories, periodic famines caused by the instability of the agricultural system, and chronic lack of food due to both population growth and economic expropriation by elites. The increases in infectious disease probably reflect both a poorer diet and increased interpersonal contact in crowded settlements, and it [infectious disease] is, in turn, likely to have aggravated nutritional problems. (pp. 573–74)

INCREASING MATERIAL COMPLEXITY

Technological advances during the Neolithic brought about dramatic changes in food production and other economic and cultural activities. Archaeologically, the Neolithic takes shape through an explosion of artifacts. Most Neolithic settlement sites have huge trash mounds, or middens, containing food remains, broken tools, and other garbage. Sorting through these artifacts and the detritus of these societies reveals an increasingly sophisticated material culture. Clay was shaped into vessels of many forms and was also used to make smoking pipes, lamps, and sculptures. Plants cultivated by humans included cotton and flax, which were then woven into

cloth. Interestingly, many Neolithic artifacts resemble material goods familiar to modern-day humans. For example, some sites contained the remains of chairs, tables, and beds similar to those used today (Clark & Piggott, 1965). Ritual structures and ornamentation also became more elaborate.

Innovations in transportation technology also occurred in the Neolithic. In Southwest Asia, people used the wheel to construct transportation vehicles. American civilizations knew how to make wheels (they were found on toys in Mesoamerica), but they did not use the wheel in any vehicles, most likely because, unlike peoples in Europe, Asia, and Africa, they had not domesticated oxen or cattle to pull vehicles. Moreover, in the mountainous regions of the Andes where the llama was domesticated, wheeled transportation was inefficient.

As populations settled permanently in villages and urban areas, they built durable dwellings of mud, brick, stone, and mortar, depending on locally available materials. As a sign of growing divisions on the basis of wealth, prestige, and status in these societies, some houses had many rooms with private courtyards and rich furnishings, whereas others were very modest.

INCREASING SOCIAL STRATIFICATION AND POLITICAL COMPLEXITY

As the preceding discussion of the Neolithic illustrates, during the past 10,000 years human societies became more and more individualized. Some populations successfully pursued hunting-and-gathering subsistence strategies up through modern times. For others, relatively simple agricultural methods sufficed. Looking at how human societies evolved through history, however, we see a gradual but unmistakable progression from egalitarian social arrangements in which all members of a society have roughly equal access to power and prestige to more stratified societies. Paleolithic cultures, with their emphasis on hunting-gathering by small bands, accorded people of the same sex and capabilities more or less equal shares in the benefits of social living, even as they acknowledged that some were better hunters or more gifted leaders than others. In contrast, during the Mesolithic and the Neolithic, there was a clear trend toward greater social stratification all over the world. Certain members of these societies acquired more influence than others in societal decision making, such as how to allocate agricultural surpluses, and were thus able to accumulate more wealth. Another marked change was the emergence of full-time craft specialists, individuals who concentrated on the manufacture of tools and other goods. These developments set the stage for all sorts of momentous changes in human social and political life, as we shall see in the next chapter.

 SUMMARY

The transition from a reliance on naturally occurring plants and animals to food production occurred in many world areas at different times. This shift took place in several stages. First, in the period known as the Mesolithic or Archaic, humans started to exploit plants and animals intensively in particular environments. In some cases, permanent or semipermanent settlements were built around these resources, and specialized tool technologies appeared for processing plant foods, for hunting, and for fishing. Many Mesolithic and Archaic populations also started to plant wild seeds and capture wild animals. The manipulation of wild species was the first stage in the transition to food production. After a period of human selection, some plants and animals became domesticated—physically distinct from wild varieties and dependent on humans for reproduction.

The change in subsistence to domesticated species marks the beginning of the Neolithic period. Yet, why domestication took place at all remains speculative. On the one hand, domestication provides more regular food supplies, allowing for growth in human populations. On the other hand, hunter-gatherers actually invest much less time in subsistence activities than do food producers. Domesticated crops must be planted, weeded, and watered for several months before they can be harvested, and the chance of crop failure makes agriculture a risky investment.

Archaeologists have offered many hypotheses to explain the origins of food production. Some cite climatic changes at the end of the Pleistocene and population growth as key elements. Others tend to view the shift to food production as a complex process involving fundamental changes in the way humans interacted with the environment. Archaeological evidence indicates a great deal of regional and local variation in the plant and animal species involved and in the specific subsistence practices.

Food production clearly had important consequences for human history. Because people had to stay near their crops, Neolithic people became more sedentary than earlier hunting-and-gathering populations. This transition, however, was not without disadvantages: Diseases associated with larger population concentrations and the tendency to concentrate on growing one crop gave rise to poorer health and deficiency diseases.

In addition to a dramatic increase in population, the Neolithic period is also marked by increasingly sophisticated material culture, social stratification, and political complexity. These trends set the stage for the emergence of states in many world areas.

 QUESTIONS TO THINK ABOUT

1. What were the principal effects of climatic change on human populations at the end of the Pleistocene period? What aspects of human existence would likely have been affected by climate change?

2. Discuss what is meant by the terms *Archaic* and *Mesolithic*. In what ways are they both different from and similar to the periods that preceded and followed them?

3. Childe, Braidwood, Binford, Cohen, and Rindos provide very different models for the origins of domestication. What are the strengths and weaknesses of each? Based on what you have learned, which theory do you favor?

4. Why is information about the size and density of human populations relevant to studies of the domestication of plants and animals?

5. Compare and contrast the origins of agriculture in Asia, Europe, and the Fertile Crescent. When and where does the first evidence for agriculture appear? What were the most important agricultural crops, and where were they domesticated?

6. In what ways have the ability to produce food through agriculture and the adoption of a sedentary lifestyle affected the human experience? Discuss specific ways that these changes affected human health and reproduction?

 KEY TERMS _____

Archaic

artificial selection

broad-spectrum collecting

cultivation

domestication

flotation

megaliths

Mesolithic

microliths

Neolithic

paleoethnobotany

silica gloss

vegiculture

 SUGGESTED READINGS _____

CLARK, J. DESMOND, AND STEVEN A. BRANDT, eds. 1986. *From Hunters to Farmers: The Causes and Consequences of Food Production in Africa.* Berkeley: University of California Press. An edited collection of essays probing plant and animal domestication. Using environmental, archaeological, ethnographic, and linguistic data, these essays explore developments in a wide range of areas.

CLARK, J. G. D. 1979. *Mesolithic Prelude.* Edinburgh: Edinburgh University Press. A survey of European Mesolithic sites and an overview of interpretations of the findings.

COHEN, MARK, AND GEORGE J. ARMELAGOS, eds. 1984. *Paleopathology at the Origins of Agriculture.* New York: Academic Press. This volume focuses on the consequences of domestication for human nutrition and health in different parts of the world.

COWAN, C. WESLEY, AND PATTY JO WATSON. 1992. *The Origins of Agriculture: An International Perspective.* Washington, DC: Smithsonian Institution Press. This edited volume surveys evidence for domestication in many areas, including Far East Asia, Southwest Asia, Africa, and the Americas. It provides the most comprehensive and readable synthesis currently available.

FAGAN, BRIAN M. 1998. *Peoples of the Earth: An Introduction to World Prehistory,* 9th ed. New York: HarperCollins. A broad introductory survey that spans human evolution through the origins of agriculture, placing developments in different regions of the world in perspective.

RINDOS, DAVID. 1984. *The Origins of Agriculture: An Evolutionary Perspective.* New York: Academic Press. An in-depth look at the origins of agriculture from an evolutionary viewpoint. Also included are brief critiques of other theoretical approaches.

Chapter 9

THE RISE
OF THE STATE
AND COMPLEX SOCIETY

CHAPTER OUTLINE

A S WE SAW IN CHAPTER 8, THE ADVENT OF domestication during the Neolithic brought many changes to human societies, including more settled communities, population growth, increased social stratification, and growing political complexity. Gradually, as permanent settlements expanded, people in many regions of the world developed such techniques as irrigation, plow cultivation, and the use of fertilizers, which allowed fields to be cultivated year after year. These intensive agricultural practices produced food surpluses, which contributed to further changes in human subsistence, and political organization. Beginning approximately 5,500 years ago in some areas, these developments coalesced in the appearance of complex societies commonly called *civilizations*.

One aspect of the growing social complexity was institutionalized government, or the *state,* run by full-time officials. The intensification of agriculture was accompanied by the rise of agrarian states in Mesopotamia, Egypt, India, China, and later in Greece, Rome, Byzantium, Southeast Asia, sub-Saharan Africa, and feudal Europe. The agrarian states in the Americas included the Teotichuacán, Mayan, and Aztec empires of Mesoamerica and the Incan Empire of Peru. The location of some of these states is highlighted in Figure 9.1. In this chapter we examine how archaeologists study the development of social and cultural complexity in early societies and some of the theories that explain why states initially formed.

THE STATE AND CIVILIZATION

V. Gordon Childe, whose theories on domestication were discussed in Chapter 8, also wrote on the origin of complex societies. He believed that the rise of civilization could be easily defined by the appearance of a specific combination of features, including urban centers of between 7,000 and 20,000 inhabitants; a highly specialized division of labor, with economic roles other than those pertaining to agricultural production; a ruling class of religious, civil, and military leaders; extensive food surpluses; monumental architecture; the use of numbers and writing systems for record keeping; developments in arithmetic, geometry, and astronomy; sophisticated art expressed in a variety of materials; long-distance trade; and an institutionalized form of political organization based on force—the state (Childe, 1950).

Although definitions such as Childe's incorporate many of the features that are popularly used to characterize civilization, such definitions have little use for anthropologists. A survey of the archaeological and historical information that is now available reveals that such neat definitions as Childe's are too rigid to define the diversity in the societies under study. In fact, all of the features noted by Childe are rarely present in the earliest societies termed civilizations. Different interpretations of what charac-

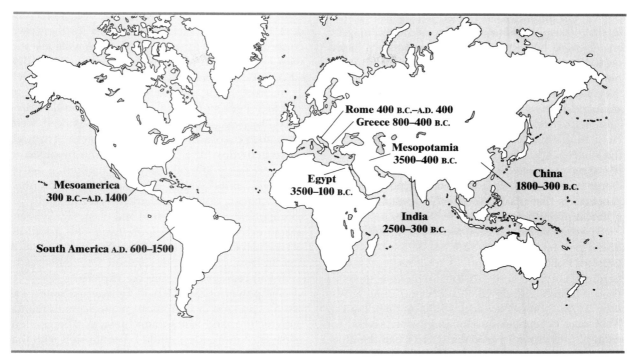

FIGURE 9.1   Early states present an effervescence of technology, craft specialization, and artistic achievement. The states shown here represent some of the earliest states in the world.

teristics are the key ingredients of civilization create more problems than they solve. In this text, the term *civilization* is used in a general way to indicate a complex society with some of the features noted by Childe.

TYPES OF POLITICAL SYSTEMS

In studying the development of early societies, today's anthropologists generally focus on specific features rather than all-encompassing definitions. One of the most important distinctions researchers make concerns different kinds of political organization. Many anthropologists use variations of a four-fold classification system first developed by anthropologist Elman Service (1971), which divides societies into bands, tribes, chiefdoms, and states. As we saw in Chapter 7, a *band* is the least complex—and, most likely, the oldest—form of political system. It is the most common form among hunter-gatherer societies, and is based on close kinship relations within a fairly small group of people. **Tribes**

are more complex societies with political institutions that unite larger groupings of people into a political system. **Tribes** do not have centralized, formal political institutions, but they do have *sodalities*, groups based on kinship, age, or gender that provide for a more complex political organization. **Chiefdom** political systems are more complex than tribal societies in that they are formalized and centralized. Chiefdoms establish centralized authority over many communities through a variety of complex economic, social, and religious institutions. Despite their size and complexity, however, chiefdoms are still fundamentally organized by kinship principles. Although chiefdoms have different levels of status and political authority, the people within these levels are related to one another through kinship ties. Eric Wolf (1982) has referred to bands, tribes, and chiefdoms as *kin-ordered societies*.

The **state** is structurally distinguished from other societies on the basis of an institutionalized bureaucracy or government. States are political systems with centralized bureaucratic institutions to establish power and authority over large populations in different territories. While the ruler of a state may be an inherited position, states systems are not based on kinship. Because early states were more complex and highly organized than prestate societies, they could not rely solely on kinship for different status positions. Landownership and occupation became more important than kinship in organizing society. In the highly centralized agricultural societies, the state itself replaced kin groups as the major integrating principle. State bureaucracies govern society on behalf of ruling authorities through procedures that plan, direct, and coordinate highly political processes.

This classification system, too, is not without some of the problems faced in Childe's definition of civilization: Political organizations form more of a continuum than four neat divisions and some societies do not fit into an individual category (Cohen & Service, 1978; Johnson & Earle, 1987). Archaeologically, it may be especially challenging to assess, for example, whether a particular society had the characteristics of a chiefdom or a state. The classification serves, like any classification scheme, to facilitate comparison and organize information, but it should be used flexibly, with the dynamic nature of human societies in mind.

EARLY STATES

Early states were typically characterized by a high degree of social inequality. The creation of substantial food surpluses, along with better food-storage technologies, led to new forms of economic relations. Mastery was primarily based on the control of agricultural surpluses, often administered by a complex administrative system. For this reason the term **agricultural states** is often used to describe these early societies. Many people were freed from working in the fields to pursue other specialized functions. Hundreds of new occupations developed in the urban centers. Craftsworkers produced tools, clothing, jewelry, pottery, and other accessories. Other workers engaged in commerce, government, religion, education, the military, and other sectors of the economy.

This new division of labor influenced both rural and urban areas. Farm laborers were not only involved in essential food production, they also turned to crafts production to supplement their income.

Over time, some agricultural villages began to produce crafts and commodities not just for their own consumption but for trade with other villages and with the urban centers. With increasing long-distance trade, regional and local marketplaces, as well as marketplaces, in the urban areas, began to emerge. Foodstuffs and other commodities produced by peasants, artisans, and craftsworkers were bought and sold in these marketplaces. Through these activities, rural villages became involved in widespread trading alliances and marketing activities. In early cities such as Ur, Memphis, Teotichuacán, and Tikal, markets were centrally located, offering both imported and domestically manufactured items as well as food. In addition, marketplaces arose throughout the countryside. A steady flow of goods developed from villages to towns to regional and national capitals.

The power of the rulers in agricultural states was often expressed in sumptuous palaces, monumental architecture, and a luxuriant lifestyle. The opulent lifestyle of the leaders could be contrasted with that of the other classes. It was in early states that slavery and the state-sanctioned use of force to enforce laws became common. The leaders of the state often sanctioned their positions through the manipulation of religious institutions and symbols. The division between the supernatural and social institutions that we make today had little meaning in these cultural settings (Trigger, 1993).

One major factor that contributed to the evolution of agricultural states was the development of a more sophisticated technology. To some extent, this represented modifications of existing technologies. For example, stone tools such as axes, hammers, knives, and scrapers were refined for more prolonged use. Increased knowledge of metallurgy enabled some agricultural peoples to create more durable tools. For example, copper, tin, and iron ores were smelted and cast into weapons, armor, ornaments, and other tools. Many technological innovations were dramatically expressed in myriad artwork and monumental construction, illustrated by such massive structures as the Pyramid of the Sun at Teotichuacán in Mexico, which rises more than 200 feet and covers some 650 square feet.

Agricultural states emerged in many parts of the world. Six of the earliest were in Mesopotamia (3500 B.C.); the Egyptian Old Kingdom (3500 B.C.); China (2500 B.C.); the Indus River valley, India (2500 B.C.);

and Mexico and Peru (300 B.C.). Other agricultural states flourished in many other areas, including Greece and Rome, tropical Africa, Southeast Asia, and, to a lesser extent, North America. These societies exhibited many of the signs of complexity, social stratification, and specialization that Childe used to define civilization (Connah, 1987; Service, 1975; Tainter, 1990; Trigger, 1993).

STUDYING COMPLEX SOCIETIES

Archaeologists draw on many sources of information to reconstruct the nature of early agricultural states. Variations in such features as settlement patterns and site characteristics, for example, can be associated with different forms of political organization or social stratification. Depending on a researcher's theoretical orientation and research design, a project may focus on certain aspects of a prehistoric state, perhaps on the link between the control of agricultural land and status differentiation or the expression of religious beliefs in art and architecture.

WRITTEN LANGUAGE

The advent of state-level organization gives archaeologists a source of information not available to researchers working on earlier periods: written sources (Daniels & Bright, 1995). As noted, Childe considered writing to be fundamental to the definition of civilization. However, surveying early states, archaeologist Bruce Trigger (1993) notes that there are no obvious differences in social, political, and economic organization between societies that had writing and those that did not. In the absence of writing, important events and records were remembered by special historians, at times with the use of memory aids such as the knotted strings utilized by the Inca of South America.

Writing systems may have originated in agricultural states in response to commercial and political needs, for example, as a means of keeping accounts for taxation. But as complex civilizations developed, writing systems became linked with historical, literary, and religious expression. Often their specific function varied in different cultural settings. Anthropologists have explored the results of literacy and its effects on society. Jack Goody (1987) has been

Egyptian hieroglyphics are one of the world's oldest forms of writing.
Source: © Copyright The British Museum.

doing research on this topic since the 1960s. One of his conclusions is that literacy gives great social and economic power to the official bureaucrats, lawyers, and priests of large-scale political states.

Written records may provide information on administration, ruling dynasties, political divisions, trade contacts, and locations of settlements. Because of the amount of information such sources provide, archaeologists working in some regions may gear their work toward locating and interpreting early libraries or archives. This is particularly true of research in Mesopotamia, where many records are preserved on clay tablets. Apart from the information preserved in the records themselves, repositories such as libraries indicate the degree of centralized authority and bureaucratic organization that was present.

WRITING SYSTEMS   Many agricultural states developed writing systems that enabled people to keep records, document their history, and produce religious texts. Early pictures, sometimes referred to as *pictographs,* may have been the precursors of early writing systems. **Ideographic writing systems**, one early form of writing, use simple pictures to communicate ideas: Each picture represents an idea. In actuality, this system involves neither language nor writing. Most likely it was independent of the language spoken by the people, and the meaning of the pictures can be read by anyone.

The Chinese writing system developed from an ideographic pictorial system. The Chinese characters, estimated to number between 70,000 and 125,000, have been conventionalized, but it is still easy to recognize their ideographic origin. In the Chinese system, the individual symbols are whole units that have meaning (one picture stands for one morpheme). There is no connection between sounds (phonemes) and the writing. This means that the Chinese who speak different dialects, such as Mandarin, Cantonese, or Hakka, can read the text without difficulty. Thus, Chinese writing has been one of the strongest cultural forces in unifying that civilization.

Other forms of writing emerged in different regions. **Hieroglyphic writing** evolved from an ideographic type of picture writing. Hieroglyphic writing simplifies a picture into a symbol, which has a direct relationship to the sound of a word. People in different parts of the world, like the Egyptians and Mayas, developed hieroglyphic systems independently of one another.

In another form of writing, called *syllabic writing,* the characters themselves express nothing but a sequence of sounds. These sounds make up the syllables of the language. Syllabic writing is much more efficient than ideographic or hieroglyphic writing because it uses a smaller number of characters. The ancient Semitic writing systems such as Arabic and Hebrew were syllabic systems. One modern language that still involves a mostly syllabic writing system is Japanese (although Japanese writing also includes Chinese characters).

Eventually, in the Mediterranean region, *alphabetic writing systems* evolved from syllabic forms. In alphabetic writing there is a sign for each sound (technically each phoneme) of the language rather than one for each word or for each syllable. This is the most efficient writing system because most languages have only twelve to sixty total sounds. Hence, an alphabet writing system can, with the fewest possible units, record every possible utterance in the language. The Greek writing system is the first to be considered fully alphabetic because it has a separate written symbol for each vowel and consonant. From the Greek system, the Romans adapted the Latin alphabet, which became the basis of the writing system of western Europe and, eventually, of English.

DEFINING STATES ARCHAEOLOGICALLY

Important clues about the scale and complexity of an ancient society can be gleaned by studying the size and distribution of settlements. The world's first cities developed from Neolithic villages. With the emergence of state societies, larger urban settlements, or *cities,* became focal points of government, trade, and religious life. The first states were linked to the control or dominance of particular territories, which often incorporated settlements of different size and function. Archaeologists can use this information to locate the centers of ancient states, infer political divisions, and define sites with specialized functions.

In Chapter 3 we saw how archaeologists gather information on past settlements through ground surveys, remote sensing techniques like aerial photography, and excavation. Drawing on these resources, an archaeologist can produce a map of early settlements. In addition to locating sites accurately, determining their age is of great importance; without this information, developments at a particular time cannot be evaluated.

CENTRAL PLACE THEORY   A concept archaeologists find useful in considering the implications of past settlement patterns is **central place theory,** developed by the German geographer Walter Christaller (1933). Surveying the contemporary distribution of towns in southern Germany in the 1930s, Christaller hypothesized that given uniform topography, resources, and opportunities, the distribution of sites in a region would be perfectly regular. Political and economic centers would be located an equal distance from one another; each in turn would be surrounded by a hierarchical arrangement of smaller settlements, all evenly spaced. The archaeological implications of the theory are clear: The largest archaeological site in a region might be assumed to be the capital, regional centers would be represented by slightly smaller sites, and village sites would be even smaller.

Central place theory is not without its limitations. Natural resources and topography are not uniform. In addition, the size of sites of the same level or function may vary. Nevertheless, the study of modern and ancient settlements confirms the basic premise of central place theory: Major centers are located some distance from one another and surrounded by sites in a hierarchically nested pattern. Mathematical models based on this premise have provided more sophisticated ways of modeling ancient states (Renfrew & Cooke, 1979).

EVIDENCE FOR ADMINISTRATION In addition to size, archaeologists use a host of other indicators to determine the position of sites in a settlement hierarchy. The primary center of the state can be expected to have more indications of administration, higher status, and more centralized authority than other settlements in the region. These features might be represented by monumental architecture, administrative buildings, archives, and storage facilities. In many so-

cieties people used seals to impress clay sealings with marks of ownership and authority (see the photo on this page). Such artifacts might be expected to be particularly prevalent in political, economic, or religious centers (Smith, 1990). Features such as these may suggest a site's role as a primary center even if it is smaller in size than other settlements.

The archaeological record also provides indications of administration outside of the primary center. There may be evidence of centralized administration such as that just noted, but to a lesser extent. Road systems and carefully maintained pathways might link settlements within a state. A feature that seems associated with many early states is the establishment of outposts in key locations along their margins (Algaze, 1993). Such outposts may have served to secure access to resources, to gain influence in peripheral areas, or to facilitate trade.

MONUMENTAL ARCHITECTURE

No archaeological discoveries captivate popular imagination more than the temples, palaces, and tombs of past civilizations. Aside from the obvious advantage of helping archaeologists locate sites, monumental architecture preserves information about the political organization, ritual beliefs, and lifeways of ancient people. As seen in Chapter 8, megalithic monuments such as Stonehenge were erected by prestate societies. However, the surplus of wealth, high-density populations, and organization harnessed by states made monumental architecture one of the most striking legacies of state societies. The great temples, pyramids, and palaces of ancient civilizations were the work of thousands of skilled artisans, laborers, and slaves, who quarried and transported huge stones and sculpted numerous artworks. Although some buildings had civil functions, the purpose of most monumental constructions in agricultural states was intertwined with religious conceptions and beliefs.

Monumental architecture served different functions in different cultural settings. For example, the 5,000-year-old ziggurats of the Sumerian state in Mesopotamia were both places of worship and centers of food distribution. The massive ziggurat at the Sumerian city of Warka (Uruk) had two flights of steps leading to a temple or shrine. The structure had extensive mosaic decorations and the earliest known use of columns in building construction. This

Clay sealings, such as these recovered at the site of Askut in the Sudan, provide clues about administration and state organization.

Source: Courtesy of Stuart Tyson Smith, University of California, Santa Barbara.

Sumerian ziggurats, such as the one shown here, served as places of worship and centers of food distribution.

temple, as well as other buildings, represents the product of hundreds of craftsworkers such as stonemasons and architects.

The pyramids, which are characteristic of the Egyptian Old Kingdom, are an example of a very different type of monumental construction. Based on beliefs of resurrection and life in an afterworld, the Egyptian pyramids, constructed as burial chambers for the pharaohs, contained many material items for use in the afterworld. The first pyramid was built under the direction of Pharaoh Djoser approximately 4,680 years ago. It is the smallest of the royal pyramids constructed at the capital at Memphis, near modern Cairo. The pyramid of Djoser was followed by more complex structures, culminating with the Great Pyramid at Giza, which is 481 feet tall and spans 13 acres. The Great Pyramid dates to the reign of Pharaoh Khufu, or Cheops, approximately 4,600 years ago. Next to the Great Pyramid, two smaller pyramids made of quarried limestone blocks were built by later pharaohs.

MONUMENTS IN THE AMERICAS Monument construction was also prevalent among civilizations in the Americas. For example, the Maya built many palaces, temples, and pyramid complexes. One of the Mayan centers, Tikal, which had a population of about 50,000 inhabitants, contained 300 large ceremonial and civic buildings dominated by large temple pyramids that were almost 200 feet high. At the

top of these structures were temples decorated with carvings made from a plaster or cement called *stucco*. These carvings represented Mayan religious beliefs in half-human–half-jaguar figures, snakes, eagles, alligators, and other spirit beings. The Maya believed in the *nagual,* an animal spirit that lived apart from humans but could at certain times merge with the human soul. Thus, many Mayan religious carvings expressed the concept of the *nagual.*

SPECIALIZATION

We have already discussed the growth of administration and bureaucracy and the intensification of agricultural practices in state-level societies. Similarly, the artwork and monument construction of the agricultural civilizations represents a dramatic shift in the scale of technology and the organization of production compared with that of small-scale societies. During the Neolithic, few people were able to devote themselves full time to nonagricultural pursuits: Potters, metalworkers, and weavers also farmed the land. In state societies, more people were able to concentrate on a variety of specialized tasks full time, supported by the food surplus made available by intensive agriculture.

Many of the agricultural civilizations produced brilliantly painted pottery, sculpture, and other artwork. Often enormous ovens, called *kilns,* were used to produce pottery bowls, flasks, and dishes, which were decorated with geometric designs and representations of animals and people. Elaborate watercolors painted on plaster walls called *frescoes,* as well as statues, intricate stone carvings, and beautiful murals, testify to the impressive development of the arts in agricultural civilizations. Sophisticated and elaborate metal objects also characterized many agricultural states.

In settlements of state societies, craft specialists were often concentrated in particular areas. Because each craft frequently had its own distinctive technology, these areas can be readily identified archaeologically. For example, in Mesoamerica, excavations have uncovered extensive areas associated with the flaking of *obsidian,* a volcanic glass. In other settings, special tools and work areas for weaving, sewing, leather working, potting, metalworking, and even beer brewing have been identified. In some societies, one's craft specialization and one's profession were linked to social position and status.

STATUS AND SOCIAL RANKING

Written accounts and illustrations, often depicting the wealth and power of rulers, underscore the disparities that existed between the rich and the poor in state societies. The kind of housing, clothing, and material goods people had access to was rigidly controlled. Many of these distinctions are recognizable archaeologically in palace complexes, exotic trade items, and greater than average concentrations of wealth. For example, Teotichuacán, which flourished in Mexico between A.D. 150 and 750, was a planned city with perhaps as many as 120,000 inhabitants at its peak (Millon, 1976; Millon et al., 1974). Neither the houses nor their furnishings were uniform; larger compounds with lavish interior decoration suggest a great deal of variation in wealth and status.

Important indications of status differences were also expressed in burials (Brown, 1971; Chapman et al., 1981). Through the study of skeletal remains, physical anthropologists are often able to determine the age, sex, and health of an individual. Surveying materials associated with the burial and its archaeological context, researchers gain insights into the deceased's social standing. For example, the amount of labor invested in the Egyptian pyramids indicates the importance of the pharaoh as well as the organizational power of the state. In some instances, servants were put to death and buried with a ruler.

Archaeologists often study *grave goods,* artifacts associated with a body, to evaluate status. It is logical to assume that the grave of an individual of higher status will have a larger number and more valuable array of goods than that of a commoner. However, it is important to recognize the extensive variation between and within cultural groups because of other variables, such as age, occupation, and gender.

TRADE AND EXCHANGE

Archaeological research indicates that trade has long been a feature of human societies. However, with growing social and political complexity, trade networks extended over ever-growing areas. Expanded mercantile exchange may be indicated by the appearance of standardized weights and measures as well as monetary systems, including coinage. Documentary records may also list trade contacts and the

The power of the rulers of early states was often expressed in art and architecture, here illustrated by a brass plaque depicting the Oba (king) of Benin (now in modern Nigeria) flanked by retainers.

cost and amounts of materials exchanged. More information comes from the trade materials themselves. Resources were not evenly distributed throughout the landscape, and items such as precious stones, metals, and amber were frequently traded over long distances. Raw materials might have been exchanged for finished products like pottery, hides, beads, and cloth. As some of these items have survived, they provide an archaeological record of past exchange (Adams, 1974; Earle & Ericson, 1977).

Certain kinds of raw materials, or *artifacts,* may be characteristic of a particular region. By plotting the distribution of these commodities, an archaeologist may be able to trace past trade networks. Typically, the quantity of a trade material at archaeological sites decreases as distance from the source increases, though interpretation of this information is less straightforward than one might think. Simply counting the instances of a particular artifact at sites fails to consider the site size, the variation in preservation, and the amount of archaeological work that has been undertaken (Renfrew & Bahn, 1996). To assess this trade successfully, archaeologists need to consider these variables.

Some trade materials can be readily recognized as exotic imports. At times, coins, sealings, or manufacturers' marks on such artifacts as pottery may even indicate specific sources. But the origin of some artifacts is less clear. Pottery styles can be copied, and stones of diverse origin can look quite similar. To identify an artifact's origins, archaeologists use a number of techniques of varying sophistication. The simplest method is to examine an object such as a stone or potsherd under low magnification for rocks, minerals, or other inclusions that can be linked to a particular location.

A much more sophisticated technique, referred to as **trace element analysis,** involves the study of the elements found in artifacts (Tite, 1972). The basic chemical configuration of certain kinds of stones is broadly similar—all obsidian, for example, contains silicon, oxygen, and calcium—but certain elements may be present in very small quantities. These *trace elements* may occur in quite different concentrations in materials from different sources. Their presence in artifacts can be plotted, providing a means of assessing distribution patterns.

THE ARCHAEOLOGY OF RELIGION

Allusion to the importance of religion in early states has pervaded the preceding discussion. Ritual structures such as temples and pyramids and the status afforded to religious leaders are rich testaments to religious beliefs. However, real understanding of the workings of ancient religions remains elusive. Aside from the obvious recourse to written texts, which when available offer a great deal of information, archaeologists must infer religious beliefs from material culture. A standard joke among archaeologists when confronting any artifact whose function is unknown is to describe it as a "ritual object," a reflex that serves to underscore the challenge of inferring the complex, nonmaterial aspects of past cultures.

Although the study of past religions is difficult, insights may be drawn from a number of different sources. Often religious beliefs are given a physical expression in places of worship. These locations may have artifacts, spatial arrangements, architectural features, and decorative elements dis-

tinct from other buildings at a site. In many cultures, worship involved offerings or animal and human sacrifices, traces of which may be preserved. Archaeological finds may also include representations of deities, ritual figures, or symbols that convey religious beliefs.

Some of the most fascinating reconstructions of ancient rituals have been undertaken by researchers working in South America. Here rich cultural traditions and excellent archaeological preservation combine to provide unique insights into past cultures. Surveying traces of the Moche, a small state in the valleys of coastal Peru that reached its zenith between about A.D. 100 and 600, archaeologists used detailed depictions of rituals on pottery and murals to interpret material recovered archaeologically (Donnan & Castillo, 1992).

THEORIES ABOUT STATE FORMATION

One of the major questions that anthropologists have attempted to answer is, Why did state societies emerge? Agricultural states arose independently in different world areas at different times and not at all in some regions. To explain state formation, researchers have posited an array of theories, which can be divided into two overarching perspectives: integrationist (or functionalist) theories and conflict theories (Cohen & Service, 1978; Haas, 1982; Lenski, 1966; Service, 1975).

Integrationist theories of state formation assume that society as a whole benefited from state organization, or in other words, the state was a positive, integrative response to conditions that a society faced. The benefits that certain individuals or groups in a society may have obtained were balanced by the key organizational or managerial functions they performed, which enabled the society to survive and prosper.

In contrast, **conflict theories** emphasize domination and exploitation. It is thought that state organization arose out of the ability of certain individuals or subgroups in a society to monopolize or control resources. State organization, therefore, has been advantageous only to the dominant elite in a society and, in general, very costly to subordinate groups like the peasantry (Fried, 1967).

INTEGRATIONIST PERSPECTIVES

An early integrationist view of state formation was proposed independently by Julian Steward (1955) and Karl Wittfogel (1957), who linked the development of the state to the construction and maintenance of irrigation projects needed in intensive agriculture. This view, sometimes known as the *hydraulic hypothesis,* suggests that the expansion of intensive agriculture created such problems as disputes among landowners and the need to build and operate irrigation canals, dikes, and other technology. To resolve these problems, labor had to be recruited, mobilized, fed, and organized. To fulfill all these functions, people developed centralized bureaucracies.

Archaeologists have challenged the hydraulic hypothesis on a number of points. One criticism is that many of the irrigation projects associated with agricultural civilizations were organized locally in communities without centralized governments. This appears to be the case in early Mesopotamia, Mesoamerica, South America, and China (Adams, 1966; Chang, 1986). For example, in Tiahuanaca, Peru, the archaeological research demonstrates that individuals constructed their own canals to irrigate their crops long before the emergence of any administrative hierarchy or state system.

Another problem with the hydraulic thesis is that in such areas as Egypt and India the state bureaucracy had developed prior to the initiation of large-scale irrigation. Therefore, instead of large-scale irrigation contributing to the rise of the state, it was actually a consequence of state development.

TRADE AND EXCHANGE   Other integrationist perspectives are illustrated by several theories that emphasize the role of trade in the emergence of the state (Rathje, 1971; Wright & Johnson, 1975). In these scenarios, the organizational requirements needed to produce trade items and to manage trade were a major driving force. Increased labor specialization associated with the production of trade items led to a concomitant rise in administrative specialization.

Although it seems clear that trade was important in many early states, to characterize its role as a primary cause in the rise of state society is probably simplistic. Increased understanding of prestate exchange systems and techniques such as trace element analysis have made the picture of early exchange far more complex than was previously

In areas such as the Nile Valley, the shaduf, a bucket on long weighted pole, was used to bring water to irrigation channels. The construction and control of water resources may have played an important role in the origin of early states.

thought (Sabloff & Lamberg-Karlovsky, 1975). Inter- and intraregional trade probably predates state formation in many instances, and increased trade may be a consequence of state society in some world areas. However, this is a theory that can be evaluated more thoroughly as techniques for attributing sources of trade items become more available.

WARFARE AND CIRCUMSCRIPTION   Anthropologist Robert Carneiro (1970) concentrates on the role of warfare in the formation of early states. Carneiro hypothesizes that population growth in a region with clearly defined boundaries, such as a narrow valley surrounded by high mountains, leads to land shortages, resulting in increased competition and warfare among villages. In this context efficient military organization is advantageous. Eventually one group becomes powerful enough to dominate. Members of this group become the state administrators, who rule over the less powerful groups. Thus, the state emerges in regions where land is in short supply and competition for scarce resources exists. Although Carneiro's focus on warfare might make this theory seem more like a conflict theory (see following discussion), it is integrationist in that the state ensures its members both land and security.

Carneiro (1970) suggests that population centers can be confined, or *circumscribed,* by factors other than geography. Circumscription can also occur when an expanding population is surrounded by other powerful societies. This "social circumscription" may prevent weaker groups from migrating into surrounding regions where they can enjoy greater autonomy.

SYSTEMS MODELS Other integrationist hypotheses are based on *systems model* approaches, which emphasize the requirements of agricultural states to organize large populations; to produce, extract, and exchange economic resources; to maintain military organizations; and to manage information (Flannery, 1972; Wright, 1977). Instead of emphasizing one variable such as irrigation or trade, these theorists attempt to understand the links among agriculture, technology, economy, and other specialized functions in an overall state system. In addition, systems theorists also consider the importance of centralized organization in long-distance trade and the value of a state bureaucracy in mobilizing the military to protect trade routes.

CONFLICT THEORIES

Conflict theories of state formation include various formulations, all of which stress the need to protect the interests of the dominant elite from other members of the society. Morton Fried (1967) developed a conflict model that focuses on population growth. According to Fried, as populations grew, vital resources such as land became increasingly scarce, causing various groups to compete for these resources. Ultimately, this competition led to the domination of a particular group, which then enjoyed privileged access to land and other economic resources. This dominant group constructed the state to maintain its privileged position. To accomplish this, the state, backed by the ruling elite, used force and repression against subordinate groups. Thus, in Fried's view, the state is coercive and utilizes force to perpetuate the economic and political inequalities in a society. This conflict model is closely related to the Marxist anthropological approaches.

Other researchers have underscored the evidence for stratification and domination by a ruling elite during the earliest phases of state formation. For example, Jonathan Haas (1982) uses data from various world areas to validate conflict theory. Though Haas concedes that the process of state formation included some integrative functions, he concludes that the elites dominated economic resources to the point that they could exert repressive and coercive power over the rest of the population. For example, Haas maintains that the large-scale monument construction of state societies required a ruling elite that coerced peasants to pay tribute and provide labor to the state. The ruling elite could use its control over economic resources to mobilize large-scale labor projects. Haas does not believe that chiefdoms could extract this labor and tribute from their populations.

Archaeologist Elizabeth Brumfiel (1983) approaches conflict from another direction. Drawing on evidence from the Aztec state, she hypothesizes that coercion and repression evolve from political rather than from the economic determinants proposed by Fried and Haas. Brumfiel focuses on the political networks and the elimination of competition utilized by various Aztec leaders to consolidate the authority of one ruling group. During one phase

The Inca fortress of Sacsahuaman, Peru. Conflict theorists point to the massive labor requirements used to construct monumental structures to underscore the exploitive nature of state societies.

of political competition, the rulers centralized their authority through organizational reforms that reduced the power of subordinate rulers and nobles. Brumfiel maintains that these manipulations are important preconditions for state formation and coercion.

CRITICISMS OF THE CONFLICT APPROACH Some theorists are critical of the different conflict models. Using various archaeological data, Elman Service (1975) and other researchers concede that inequality and conflict are basic aspects of state development, but they emphasize the enormous conflict-reducing capacities of state systems that coordinate and organize diverse groups of people. Service argues that the archaeological record does not indicate any class-based conflict that results in state formation. Following theorists such as Max Weber, the integrationists further argue that state systems become coercive or repressive only when they are weak. Service argues that strong centralized state systems provide benefits to all social groups and thereby gain political legitimacy and authority, which reduce the degree of repression and coercion needed to maintain order. In this sense, Service views state societies as differing only in degree from chiefdom societies.

Some writers have questioned the theoretical underpinning of conflict theories, which assumes that human ambition, greed, and exploitation are universal motivating factors (Tainter, 1990). If such characteristics are common to all societies, why didn't food surpluses, status differentiation, and class conflict appear in all societies? Hunting-and-gathering populations have social mechanisms that maintain egalitarian relationships and hinder individual ambition. If class conflict is a universal explanation of state formation, "How did the human species survive roughly 99 percent of its history without the state?" (p. 35). These factors suggest that the causal reasons for state formation are more complex than intragroup conflict.

PERSPECTIVES ON STATE FORMATION

The reasons for state formation are clearly complex, involving demographic, social, political, economic, environmental, and cultural factors. Certain theories seem to fit conditions in particular world areas but not others. Empirical evaluation of causal factors on the basis of archaeological information is challenging. Not surprisingly, many researchers have called for the combination of integrationist and conflict theories to reach a full understanding of the dynamics of state formation. Indeed, the theories surveyed are not, in truth, as simple as they have been presented, and many theorists recognize the importance of other causal factors. On the one hand, it is clear that the organizational and managerial capabilities of state society are worthwhile; on the other, it may be noted that the benefits of stratification are not as advantageous to all members of a population as some integrationist theories hypothesize. It is likely, as Ronald Cohen (1978) has suggested, that multiple roads to statehood exist.

STATES IN DIFFERENT WORLD AREAS

As in the case of agricultural origins, early research on the emergence of state-level societies and civilizations concentrated on a few world areas such as Southwest Asia and Egypt. More recent work, however, has provided insight into a variety of societies throughout the world. It is not possible to survey all of these, but the following discussion provides some idea of the diversity of the civilizations represented.

CIVILIZATIONS IN SOUTHWEST ASIA

Some of the earliest agricultural civilizations in the world emerged in Southwest Asia. Both the Neolithic revolution and the development of intensive agriculture in the area have been well documented, with evidence dating to about 6,000 years ago. Extensive civilizations arose at the base of the Zagros mountains in the Mesopotamian valley near the Tigris and Euphrates rivers. One of the earliest civilizations, known as the Sumerian Empire, contained twenty urban centers, such as Uruk, Ur, Ubaid, and Eridu, dating back to 2500 B.C. Some of these cities had populations of more than 10,000 people, covering areas of 6 square kilometers. The rulers of these cities tried to maintain political stability by building walled fortresses for protection from nomadic invaders.

By 2100 B.C., the city of Ur in southern Mesopotamia had become the center of an extensive empire that established a large-scale regional bureaucracy to collect taxes and tribute. The breadth of political control and organization of labor in the Ur Empire is manifested in the ziggurat, a large ceremonial center with a monumental structure made up of a series of platforms with a small temple at the top where the priests conducted rituals. In time, however, the costs of maintaining political domination over a vast region, coupled with large-scale population growth, overtaxed Ur's economic and physical resources. In 2000 B.C., the empire collapsed. Archaeologists have concluded that in the 1,000 years following the fall of Ur, the number of settlements in Mesopotamia declined by 40 percent (Tainter, 1990).

Other major empires in northern Mesopotamia included Babylon (2000–323 B.C.) and Assyria (1920–1780 B.C.). These empires established widespread trade routes throughout the Middle East. They maintained political control over extensive rural regions, requiring peasants to provide surplus food and labor for monument construction. Babylon's ruler, Hammurabi, developed the first written legal codes to maintain a standardized system of political rules for the Babylonian Empire.

Other agricultural peoples of Southwest Asia were the Phoenicians, the Arameans, and the Hebrews, who settled the kingdom of Israel in about 1000 B.C. In addition, in the country now known as Iran, the Persian civilization rose to conquer much of the Middle East in the sixth century B.C.

The agricultural civilizations of the Middle East influenced one another through the rapid spread of technology and culture. Contacts among the civilizations were enhanced by the movement of trading ships in the Mediterranean. In addition, long-distance trade by land was conducted throughout the region. For example, via extensive caravan routes the Bedouins carried goods from port cities in the Arabian peninsula across the desert to cities such as Damascus, Jerusalem, and Cairo. Because of these economic and cultural connections, all these civilizations had broad knowledge of metallurgy, highly developed writing and coinage systems, and an extensive network of roads, as well as sophisticated ecclesiastical religions with full-time priests.

AGRICULTURAL CIVILIZATIONS IN AFRICA

When people think of African civilization they likely think of Egypt. The pyramids, tombs, and temples of ancient Egypt have captured our imagination. The treasures of the pharaohs, such as those of Tutankhamun (fourteenth century B.C.) are spectacular by any standard. State-level society in Egypt can be traced back to the unification of Upper and Lower Egypt in about 3100 B.C. This empire, known as the Old Kingdom, maintained a highly centralized bureaucracy headed by monarchs, called pharaohs, who were believed to possess supernatural authority. Urban areas such as Memphis and Thebes, with populations of more than 100,000, sprang up next to the Nile River, where agricultural production exceeded that of Mesopotamia. In contrast to the Sumerian cities, urban centers such as Thebes and Memphis were cut off from surrounding nomadic tribes by desert regions; thus, they did not have to construct walled fortresses to maintain security. Various Egyptian states continued to exist until the Persians conquered the region in 525 B.C.

NUBIA Farther south, the fertile flood plain of the Nile, which was so important to the people of ancient Egypt, was also invaluable to people living to the south in an area traditionally known as Nubia, which today is part of the country of Sudan. Evidence of one of Nubia's earliest civilizations is provided by the site of Kerma, located 500 kilometers north of Khartoum. Kerma may have been the capital of the Nubian state of Kush that is referred to in Egyptian texts of the Middle Kingdom. The town's ruins include a spectacular monumental mud-brick structure called a *deffufa*. Large *tumuli* (mounds), containing hundreds of human sacrificial victims, and associated settlement areas are also indicative of the town's importance. Although some Egyptian influences are evident at Kerma, it was clearly an indigenous state, the origins of which can be traced back to the second or third millennium B.C.

A number of other civilizations are associated with Nubia, among the best known being Meroe, which flourished between 500 B.C. and A.D. 300. Meroe is located south and east of Kerma, above the fifth cataract of the Nile. Unique characteristics of Meroitic civilization include temples to the lion-god Apedemek, a distinct system of kingship, and a unique form of writing. Used from about the second century B.C. to the fourth century A.D., the Meroitic script remains undeciphered today. Meroe reached its peak after 300 B.C., when trade connections were made with Egypt and the classical Graeco-Roman world. The civilization declined rapidly during the third century A.D.

THE EAST AFRICAN COAST   Coastal East Africa was linked with other areas by an extensive sea trade. Settlements had developed in coastal Kenya and Tanzania by the first century A.D. Arabic and classical Greek references indicate a long period of contact with peoples to the north and east. The arrival of Persian Gulf settlers may date back to the ninth century A.D., with Indian colonists arriving five centuries later. The remains of some of these settlements are quite impressive. Gedi, on the Kenyan coast, covers approximately 45 acres and probably flourished between the thirteenth and the sixteenth centuries A.D. The ruins of the town include impressive stone-walled structures, mosques, and Islamic tombs. Examples of other coastal settlements can be seen at Kilwa and Lamu. These are only a few examples, however: An estimated 173 towns on the coast have stone ruins (Connah, 1987). Studies of these communities have traditionally focused on external trade, but more recent work is stressing the importance of indigenous populations.

GREAT ZIMBABWE   One of the best-known states of southeastern Africa is known as Great Zimbabwe (1250–1450). The central city, also known as Great Zimbabwe, was situated near the Zambezi River and contained more than 200 large stone buildings in its center. A temple or ritual enclosure within the city is the largest prehistoric structure found in sub-Saharan Africa. It consists of a great circular building surrounded by walls 24 feet high. These imposing remains were surrounded by less permanent structures made of timber and clay, which archaeologists believe may have housed 18,000 people.

Archaeologist Graham Connah (1987) has described Great Zimbabwe as "the best-known and most ill-used archaeological site in Africa." Early ethnocentric European theories suggested that the site was a copy of the Queen of Sheba's palace, perhaps built by a group of lost white settlers. Poor archaeological work destroyed much of the information that could have been obtained by excavation. More recent work indicates that Great Zimbabwe can best be understood as a continuation of indigenous African cultural traditions. The origins of Great Zimbabwe begin with twelfth-century A.D. Iron Age sites, and the settlement itself gained prominence between the thirteenth and fifteenth centuries. Archaeological data suggest continuities with the Bantu-speaking Shona people, who still occupy the area.

Great Zimbabwe was a center of an extensive empire. Sites with similar wall construction, ceremonial structures, and artifacts are found throughout the modern nation of Zimbabwe and South Africa. The most important economic activities included cultivation of cereal crops, cattle raising, mining of gold, and long-distance trade along the East African coast. The artistic achievements of this civilization can be seen in magnificent soapstone sculptures of crocodiles, birds, and other animals.

WEST AFRICA   West Africa includes a large area of great environmental and cultural diversity. Many sections have not been well studied by archaeologists. Generally, the region seems to have developed larger and more nucleated settlements than the eastern part of the continent. Increasing trade contacts and political organization in the Senegambia are evidenced by the appearance of *tumuli* (large burial mounds with exotic grave goods) and megaliths. Examples of the latter are the sites of Kerbach and Wasu, which have circles of carefully dressed stones. Unfortunately, many of these sites remain largely unexamined, and no associated living areas have been excavated.

The Yoruba people of southwestern Nigeria developed large urban centers in both the savannas and the rainforest. One forest city, known as Ife, contained a large palace complex and many other monuments with stone and terra-cotta sculptures (Connah, 1987). Ife was a major political and ceremonial center where artisans produced bronze castings of religious figures. The artwork suggests a high degree of technological sophistication and craft specialization.

Much more is known about the later West African states of ancient Ghana and Mali, which flourished in the western Sudan. Both of these states are known from Arabic accounts: Ghana is mentioned in Arabic accounts from the eighth to tenth centuries A.D. The origins of Mali may date back almost as far, but Mali reached its apogee much later. It is described by Ibn Battuta in A.D. 1352. The wealth of ancient Mali was well known, and Malian cities such as Timbuktu were familiar to European writers of the period. Unfortunately, although the capitals of ancient Mali and Ghana have tentatively been located, excavation has not yet provided a clear picture of their origins.

A great deal of information has appeared about African states in the past few decades. However, much more work needs to be undertaken. Evidence

CRITICAL PERSPECTIVES

CONTACTS BETWEEN TWO WORLDS?

The structural similarities of Egyptian pyramids to those of Native American civilizations have fostered theories about contact between the Near East and the Americas. For example, one idea promoted in the sixteenth century was that the indigenous peoples of the Americas were the descendants of the so-called Lost Tribes of Israel. Another theory proposed that the ancient continents of Atlantis and Mu connected the continents of America and Europe, enabling people to migrate between the two.

In a speculative book entitled *American Genesis* (1981), Jeffrey Goodman proposes that all populations throughout the world are related to an original population located in California, which he asserts was the original Garden of Eden. Goodman contends that from this location, humans inhabited the Americas and crossed the continents of Atlantis and Mu to enter Europe and Asia. Zoologist Barry Fell (1980) has proposed that America was settled by colonists from Europe and North Africa sometime between 3000 and 1000 B.C. In an even more radical vein, popular writer Erich Von Daniken (1970) has argued that extraterrestrial beings colonized the Americas and left behind the ideas for the development of civilization.

Archaeologists and physical anthropologists who have analyzed the question of early contacts between continents are skeptical of these theories. As we saw in Chapter 7, the consensus regarding the peopling of the Americas is that migrations occurred across the Bering Strait from Asia. All of the dental evidence and skeletal remains indicate that Native Americans have the same physical characteristics as Asians. There have been no archaeological discoveries connecting ancient Israelites or any other societies with those of the Americas. In addition, geologists have not found any evidence of any lost continents such as Atlantis and Mu. Certainly, archaeologists have not identified any evidence of spaceships.

The similar pyramid construction in the Near East and the

indicates that trade was an important factor in many areas, but lack of archaeological work in associated settlements and surrounding areas makes it difficult to evaluate what resources may have been important. Many parts of Africa have not been studied at all. For example, very little is known about the culture that produced the spectacular tenth-century A.D. metalworking at Igbo Ukwu, Nigeria. Other regions, such as central Africa, are virtually unknown.

EARLY ASIAN CIVILIZATIONS

Major agricultural states emerged in Asia following the Neolithic period. One developed in the Indus River valley in southern Asia. Two of Asia's largest cities, Harappa and Mohenjo-Daro, developed on the Indus River in what is today the country of Pakistan. Mohenjo-Daro developed about 2,500 years

ago and had a population of nearly 35,000. The city was well planned and laid out in a grid, with single-room apartments, multiroom houses, courtyards, and the ancient world's most sophisticated sewer and plumbing system. In the center of the city stood an enormous mound constructed of mud brick, which was protected by fortifications. Close to this ceremonial and political center were a large public bathhouse and a bathing pool that may have been used for ritual purposes (Fagan, 1989).

CHINA Independent urban centers also developed in China. Beginning about 2,500 years ago, during the Shang dynasty, intensive agricultural production culminated in the emergence of the earliest major urban areas of China, which were situated near the Yellow River. One ancient Shang city, known as Yin, which had large structures such as palaces and plat-

Americas, in addition to other evidence, has led to some interesting hypotheses regarding transoceanic contacts. A number of serious scholars of *diffusionism* have offered hypotheses that are still being evaluated by archaeologists. Anthropologist George Carter (1988) has suggested that plants and animals spread from Asia to America before the age of Columbus. Evidence such as pottery design, artwork, and plants has led some archaeologists to hypothesize pre-Columbian diffusion across the Pacific (Jett, 1978).

Most archaeologists have not collected enough convincing data to confirm these hypotheses. Although similarities exist between the Near East and Native American pyramids, they appear to be based on the universal and practical means of monument construction: There are a finite number of ways to construct a high tower with stone blocks. For this reason, children from widely different cultural settings often construct pyramids when playing with blocks.

Archaeologist Kenneth Feder (1999) reviewed many of the myths, mysteries, and frauds that have been perpetrated about the past. He thinks that many people are prone to believe these speculations rather than accept the scientifically verified evidence. Because the past cannot be experienced directly, people are drawn to speculative accounts of what occurred. This does not mean that anthropologists have absolutely disproved their findings about such phenomena as connections between the two worlds. Anthropologists, however, must rely on the precise evaluation of hypotheses to substantiate their explanations of whether there were any contacts between these areas.

Points to Ponder

1. What ideas have you heard regarding such connections?

2. What types of evidence would convince you of connections between civilizations? How would you go about assessing information about these ideas?

3. What would it take to convince you that extraterrestrial creatures brought civilization to earth?

4. What does an extraterrestrial origin imply about the abilities of humans to develop their own civilizations?

---

form altars, appears to have been the center of this civilization (Chang, 1975, 1976, 1986). These urban centers were divided into residential, craft production, and royal neighborhoods, each with appropriate structures. The cities had walled fortresses, designed to keep out nomadic invaders such as Mongol pastoralists from regions farther north. From these early cities, China developed an increasingly large agricultural state incorporating millions of people.

SOUTHEAST ASIA   A number of agricultural states also emerged in Southeast Asia. Wet-rice cultivation and extensive irrigation projects provided the basis for large state societies such as the Pagan dynasty of Burma (1044–1287), the Ayudhyan dynasty of Thailand (1350–1760), the Khmer Empire of Cambodia (802–1431), and the Majapahit kingdom of Indonesia (1350–1425).

The Khmer civilization was typical of the Southeast Asian empires. It had a large urban center that contained more than 20,000 temples, including the enormous temple of Angkor Wat. The Khmer rulers, who were considered semidivine, had a bureaucratic staff of 300,000 priests supported by the tribute of millions of peasants (Sardesai, 1989).

EMPIRES OF THE AMERICAS

Following the emergence of an agricultural system based on the production of maize, beans, and squash, a series of indigenous complex societies developed in Mesoamerica and South America (see the box "Contacts between Two Worlds?"). In lowland Mesoamerica, a sedentary farming civilization known as the Olmec developed at about 1500 B.C. The Olmec civilization is regarded as the "mother civilization" of Mesoamerica, because many of the features represented provided a template for later societies. Olmec achievements include large-scale monument construction, hieroglyphic writing, and a calendar system. These cultural accomplishments indicate that the Olmec society was highly organized; yet whether it represents state- or chiefdom-level organization is uncertain (Coe, 1977; Service, 1978).

Another major civilization, known as the Maya, arose in lowland Mesoamerica around 300 B.C. After several centuries, the Maya developed a diversified agricultural system that was a combination of slash-and-burn horticulture, intensive horticulture based on hillside terracing, and large-scale irrigation agriculture. The Mayan population has been estimated at 3 million (Coe, Snow, & Benson, 1986). From large ceremonial centers such as Tikal in northern Guatemala and Palenque in Mexico, the Maya maintained extensive trade networks with peoples throughout the region. In these centers an elite class emerged that traced its royal lineages back to various deities. Archaeologists agree that the Mayan centers, with hieroglyphic writing, knowledge of astronomy, calendars, priests, and large monuments, were states.

In the central highland area of Mexico, known as the Valley of Mexico, a major city emerged that became the center of an agricultural empire of perhaps 1 million inhabitants. Most archaeologists believe that this empire eventually incorporated both Mayan lowland and highland ceremonial centers. Known as Teotihuacán, it is estimated to have had a population of more than 150,000 by about A.D. 200 (Sanders & Price, 1968). By A.D. 600, it was the sixth-largest city in the world. Its boundaries extended over an area of about 8 square miles. Teotichuacán was divided into quarters and district neighborhoods, with as many as 2,200 apartment houses and 400 crafts workshops. It had more than seventy-five temples, including the Pyramid of the Sun and the Pyramid of the Moon. The Pyramid of the Sun is the largest monument in pre-Columbian America.

The inhabitants of the Valley of Mexico developed a highly effective pattern of cultivation known as *chinampas*. The valley contained many lakes and swamps that were used as gardens. In *chinampas* cultivation, farmers dug up the muddy bottom from lakes and swamps and planted maize, beans, and squash in this rich soil. This system of cultivation was extremely productive and allowed farmers to

plant and harvest several times a year to support dense populations.

The decline of the Teotihuacán civilization, which occurred about A.D. 700, is still being investigated by archaeologists. The evidence from artifacts suggests that the flow of goods into the city from outlying regions was abruptly reduced. The city and its monuments were systematically burned. The population dropped within fifty years to one-fourth its former size (Tainter, 1990). This decline was followed by a period of political fragmentation that enabled other groups, such as the Toltec and eventually the Aztec, to dominate the region politically. The Aztec settled in the city of Tenochtitlan, located at the current site of Mexico City. By 1400, they had established political control over an empire of 25 million people.

ANDEAN CIVILIZATIONS

The Andean civilizations of Peru, including Moche and Huari (200 B.C.–A.D. 600) in the north, Tiahuanaco (A.D. 200–1100) in the south near Lake Titicaca, and the Incan Empire (1476–1534), developed intensive agriculture based on elaborate systems of irrigation and canals. The Andean peoples, like the Mesoamericans, relied on mixed patterns of agriculture, involving slash-and-burn cultivation in some areas and more intensive agriculture in others, to support large urban centers. The major food crops were maize, potatoes, and a plant called quinoa, the seeds of which are ground for flour. The most important domesticated animals were the llama, which provided both meat and transportation, and the alpaca, which produced wool for textile production.

As with Teotihuacán, Andean cities became the centers of regional states. The empire of Huari dominated almost the entire central Andean region and developed locally based urban centers to control outlying provinces. In addition, vast trading links produced interconnections with many regions. Tiahuanaco controlled a large rural area characterized by massive construction and irrigation projects that required large amounts of state-recruited labor. These empires collapsed by A.D. 1100 (Tainter, 1990). Later, the Inca settled in the urban capital of Cuzco and consolidated imperial domain over all of modern-day Peru and adjacent areas. Supported by a well-organized militia and an efficient state organization dominated by a divine emperor, the Inca ruled over a population of 6 million.

The Pyramid of the Moon looms over the ancient city of Teotihuacán in central Mexico. The city, which may have had a population of 150,000 by A.D.600, included residential complexes and craft centers, as well as dozens of temples and religious structures.

THE COLLAPSE OF STATE SOCIETIES

Perhaps no aspect of agricultural states is more intriguing than the question of why they ceased to function. As discussed in the chapter opening, no agricultural states exist today: Lost ruins, palaces hidden in tropical forests, and temples buried beneath shifting sand captivate the attention of archaeologists and the public alike. These images grip our attention all the more because of the lusterless prospect that the downfall of these ancient societies offers insight into the limitations of our own civilization. As Joseph Tainter (1990: 2) notes: "Whether or not collapse was the most outstanding event of ancient history, few would care for it to become the most significant event of the present era."

In looking at their demise, it is important to note that many early agricultural states were exceedingly successful. They flourished in many different world areas for hundreds of years. The ancient Egyptian Old Kingdom, which spanned a period of almost 1,000 years, is particularly notable, but many lasted longer than the 200-odd years the United States has been in existence. The apparent success of these states makes the reasons for their collapse all the more enigmatic. What accounts for the loss of centralized authority; the decline in stratification and social differentiation; the interruption of trade; and the fragmentation of large political units that seem to document the end of many different civilizations? These features extend beyond the end of specific governments or political systems; rather, they seem to reflect the breakdown of entire cultural systems.

REASONS FOR COLLAPSE

In examining the downfall of complex societies, writers have posited many different theories. Among the earliest were notions that collapse was somehow an innate, inevitable aspect of society, to be likened to the aging of a biological organism. Plato, for example, wrote that "since all created things must decay, even a social order . . . cannot last forever, but will decline" (quoted in Tainter, 1990: 74). Although romantically appealing, such interpretations lack explanatory value and cannot be evaluated by empirical observation.

More recent scholars have sought a more precise understanding of the factors that contributed to collapse. Many theories have focused on the depletion of key resources as a result of human mismanagement or climatic change. In an agricultural state, conditions that interfered with or destroyed the society's ability to produce agriculture surplus would have had serious consequences. If the society was unable to overcome this depletion, collapse would result. Reasons such as this have been posited as contributing to the collapse of areas often included in discussions of the origins of complex societies, such

CRITICAL PERSPECTIVES

THE DOWNFALL OF THE MOCHE

By 200 B.C. a number of small state-level societies had emerged in northwestern South America along the coast and hinterland of present-day Peru. The best known of these is that of the Moche, which was centered in the Moche and nearby Chicama valleys. At its peak the state controlled satellite communities in a number of neighboring valleys—eventually extending its influence hundreds of miles to the north and south (Donnan & McClelland, 1979).

The archaeological record provides a rich testament to the wealth and power of the Moche. Ceremonial complexes, such as those near the modern city of Trujillo, include adobe-brick temple platforms, pyramids, and associated room complexes that may have served as palaces. In 1987, the riches found in an undisturbed tomb of a Moche warrior-priest near the modern village of Sipán

led some to dub the discovery "a King Tut of the New World." Although the Moche possessed no writing system, the interpretation of the Sipán tomb and other Moche sites is aided by exquisite depictions of Moche life on pottery vessels.

The Moche kingdom flourished for almost 800 years. However, by 600 A.D., the classic Moche ceremonial centers were abandoned, and power shifted to other areas. Archaeological excavations and recent studies of photographs taken by the space shuttle *Challenger* have provided insights into what may have caused this collapse (Moseley & Richardson, 1992). Excavations during the 1970s provided part of the answer to the Moche mystery. Prior to that time Moche ceremonial structures were known but no one had found any settlements, leading archaeologists to believe that the popula-

tions that had built the monumental structures had lived elsewhere. Deep excavations at Moche sites during 1972, however, revealed residences, civic buildings, and high-status cemeteries buried under almost 30 feet of sand—evidence that the Moche centers had been destroyed at their peak of prosperity.

Interpretations of the archaeological findings have been aided by recent insights into the environmental conditions that may have ravaged the Moche sites and buried them with sand. Between A.D. 500 and 600, the Moche region experienced devastating rainfall and large-scale flooding characteristic of a climatic phenomenon today known as *El Niño*. During normal climatic conditions a cold stream of water known as the Humboldt or Peru Current flows through the Pacific Ocean along the west coast of South

as Mesopotamia (Adams, 1981), Egypt (Butzer, 1984), and Mesoamerica (Haas, 1982). The role of environmental degradation in the collapse of one society is examined in more detail in the box "The Downfall of the Moche."

In some cases, researchers have suggested that resource depletion may be the result of sudden catastrophic events involving earthquakes, volcanic eruptions, or floods, which have an impact on agricultural lands as well as other resources. One of the most well-known theories of this kind links the destruction of Minoan civilization in the Mediterranean to the eruption of the Santorini volcano on the island of Thera (Marinatos, 1939).

Other theories have suggested that conditions within societies have led to collapse. Many of these stress the tension or conflict resulting from social stratification. For example, mismanagement, excessive taxes, demands for food and labor, or other forms of exploitation by a ruling class are seen as instigating revolts or uprisings by the disaffected peasant class. Without the support of the peasants the political system cannot function and the state collapses (Guha, 1981; Lowe, 1985; Yoffe, 1979).

Alternatively, some researchers have viewed collapse as the result of the societies' failure to respond to changing conditions. For example, Elman Service (1960: 97) argued: "The more specialized and adapt-

America. This cold water limits rainfall along the coast, producing the New World's driest desert in northern Chile and Peru. During an El Niño event, which may last 18 months, this pattern suddenly changes, with disastrous consequences. Warm water flows along the coast and the normally arid coastland is beset with torrential rains and flooding, while severe droughts plague the usually wetter highlands of Peru and Bolivia. Archaeological evidence indicates that flood waters possibly from one or more El Niño events destroyed Moche settlements and stripped as much as 15 feet from some sites. Although the Moche survived the flooding, by A.D. 600 sand dunes engulfed irrigation canals, fields, and architecture.

Recent studies of photographs taken by the space shuttle *Challenger* in 1983 and earlier satellite photos have helped explain the dune encroachment (Moseley & Richardson, 1992). The high-altitude photographs of the Peruvian coast link the formation of new beach ridges and dune systems to earthquakes and El Niño. The *Challenger* photographs indicate that a new beach ridge formed between 1970 and 1975. The ridge's appearance was linked to the Rio Santa earthquake of 1970, which caused massive landslides that dumped huge amounts of earth into the dry river valleys. The torrential rains of the El Niño event of 1972–1973 carried this debris into the ocean, where strong currents deposited it in a new beach ridge. The ridge, in turn, provided the sand for the dunes that swept inland.

The available evidence seems to support the theory that El Niño events, possibly exacerbated by earthquakes, may have contributed to the downfall of the Moche state. Additional evidence suggests that there may have been periods of extreme drought prior to A.D. 600. The combination of these environmental disasters beyond human control could have seriously undermined the state's ability to produce the agricultural surplus it needed to survive. In the absence of any means of overcoming these problems, the downfall of the Moche state may have been inevitable.

Points to Ponder

1. What other evidence could be used to evaluate this interpretation of the collapse of the Moche state?

2. Even if excellent evidence for environmental disasters is uncovered, can other reasons for the collapse of the Moche be ruled out?

3. Is it likely that it will be possible to create general models that will explain the collapse of societies in all cultural and environmental settings?

4. Do you think that study of ancient states offers any insights into problems facing modern societies?

ed a form in a given evolutionary stage, the smaller its potential for passing to the next stage." Underlying this interpretation is the assumption that successful adaptation to a particular environmental or cultural setting renders a society inflexible and unable to adapt to changing conditions. In this setting less complex societies with greater flexibility overthrew older states.

Anthropologist Joseph Tainter notes in a survey that most of the models that have been presented focus on specific case studies, not on the understanding of collapse as a general phenomenon. He notes that most researchers assume that the decline in complexity associated with collapse is a catastrophe: "An end to the artistic and literary features of civilization, and the umbrella of service and protection that an administration provides, are seen as fearful events, truly paradise lost" (Tainter, 1990: 197). Tainter argues that in reality collapse instead represents a logical choice in the face of declining returns. When people's investment in complexity fails to produce benefits, they opt for disintegration.

Theories concerning the collapse of states can be evaluated in light of ethnographic, historical, and archaeological evidence. Upon surveying information from different states, it appears that reasons for collapse are exceedingly complex; the specific manifestation varies in individual settings—

just as the specific features that define states differ. Adequate appraisal is dependent on the existence of a great deal of information about the society under study, including its technological capabilities, information on population, agricultural yields, and climatic conditions in the region. The difficulty involved in assessing competing models is perhaps best illustrated by the fact that very different theories have often been used to explain the decline of the same society.

 SUMMARY _____

During the Neolithic, people in many regions of the world developed techniques such as irrigation, plow cultivation, and the use of fertilizers, which allowed fields to be cultivated year after year. These intensive agricultural practices produced food surpluses, which contributed to further changes in human subsistence and political organization. Approximately 5,500 years ago in some world areas, these developments led to the appearance of complex societies frequently called civilizations—a term that V. Gordon Childe used to indicate the presence of a complex set of traits, including population concentration in cities, agricultural surpluses, monumental architecture, craft specialization, stratified social organization, and institutionalized political authority.

Today, anthropologists generally focus on the role and function of specific features and are less concerned with overarching definitions. One aspect of early societies that interests anthropologists is the appearance of an institutionalized form of government, the state, run by full-time officials. Early states were characterized by a great deal of social inequality. The ruler's position was based on the control of agricultural surpluses and was sanctioned by religious institutions and symbols.

Archaeologists can draw on many sources of information to reconstruct the nature of early agricultural states. They may be able to use written records, a source of information that researchers working on earlier periods did not have. Many different kinds of archaeological information, including settlement patterns, site features, and artifacts, are also available.

A question of interest to many anthropologists has been the origin of political complexity and the rise of agricultural states. The many different theories that have been proposed can be divided into two major positions: the integrationist and the conflict perspectives. Integrationist theorists view the early state as having functioned to provide benefits to all members of a society. The conflict theorists believe that the state was created to enforce status divisions and primarily benefited the ruling class. Reasons for state formation probably varied in different cultural settings and incorporated elements of both integrationist and conflict theories.

Archaeologists have been conducting research in different areas of the world, including the Americas, Africa, and Southwest, South, Southeast, and East Asia, in order to test their models of state formation. This research needs to continue in order to offer insights that can be gleaned from a fully global perspective.

 QUESTIONS TO THINK ABOUT _____

1. Examine the criteria that were used by V. Gordon Childe to define civilization. What are the problems with his definition?

2. What is meant by *central place theory?* Using specific examples, discuss how this concept is useful for understanding the nature of ancient agricultural states.

3. Why is the presence of monumental architecture considered to be a good indication of the presence of a stratified society? According to Jonathan Haas, what is the role played by monumental architecture in the formation of the state?

4. What techniques can archaeologists use to recognize evidence of trade and exchange in ancient

civilizations? What role did long-distance exchange play in the emergence of social status and social complexity?

5. Using specific examples, explain the differences between integrationist and conflict theories of the origins of state formation. Which of these perspectives do you feel provides a more realistic interpretation, and why?

6. Define *specialization* as it applies to the evolution of ancient states. When does specialization first appear, and how can it be recognized archaeologically? Discuss the difference between specialization as seen in state societies and specialization in Neolithic societies.

 KEY TERMS

agricultural states

central place theory

chiefdom

conflict theories

hieroglyphic writing

ideographic writing system

integrationist theories

specialization

state

trace element analysis

tribes

 SUGGESTED READINGS

CONNAH, GRAHAM. 1987. *African Civilizations: Precolonial Cities and States in Tropical Africa: An Archaeological Perspective*. Cambridge: Cambridge University Press. A survey of complex societies in sub-Saharan Africa by one of the leading Africanists, this volume provides a survey of the limited archaeological information available about this large, diverse world area.

DAVIES, W. V. 1987. *Egyptian Hieroglyphs*. Berkeley: University of California Press. A basic introduction to one of the world's oldest writing systems, this volume provides a brief, readable introduction to the concepts of hieroglyphic writing, its development, its cultural setting, and its translation by archaeologists. This book is one of a series of volumes by the University of California Press surveying early writing systems.

SASSON, JACK M., ed. 1994. *Civilizations of the Ancient Near East*. New York: Charles Scribners and Sons. This authoritative compendium, including chapters by 189 contributors, surveys many of the key discoveries and interpretations that have refined our understanding of some of the world's earliest civilizations.

TAINTER, JOSEPH A. 1990. *The Collapse of Complex Societies*. New York: Cambridge University Press. A seminal attempt to provide a general model of collapse in complex societies. Although Tainter aimed at addressing his own theoretical position, the volume provides an excellent survey and critique of previous work on the subject.

TRIGGER, BRUCE. 1993. *Early Civilizations: Ancient Egypt in Context*. Cairo: American University in Cairo Press. A very readable survey of early states, broader in scope than the title suggests.

Chapter 10

PRACTICING
ANTHROPOLOGY

THE PRECEDING CHAPTERS ILLUSTRATE some of the research that physical anthropologists and archaeologists are engaged in. These studies are far-ranging and address a diversity of questions concerning the emergence of humankind and past cultures. These issues appeal to the human intellect and speak to fundamental questions about our origins, and to both the diversity and commonalties of human populations throughout the world. While many of these issues are of more philosophical than practical import, some of the research and information is very relevant to immediate concerns of the present. As seen in Chapter 1, one of the most important developments in the field of anthropology is that of applied anthropology, the use of anthropological data to offer solutions to problems faced by modern society. On one hand, it is the application of physical anthropological and archaeological methods and techniques to address concerns of the present. On the other, it deals with the relevance, use, and presentation of archaeological sites, human remains, and information about the past to modern society. This chapter introduces some of these applications and the issues and concerns faced by physical anthropologists and archaeologists in presenting information about the past.

THE ROLES OF THE APPLIED ANTHROPOLOGIST

The popular, if not accurate, images of anthropologists vary from the adventurous explorer seeking out lost treasure to the absentminded academic working away in the dusty halls of a university or museum. These perspectives do not provide a valid picture of the modern physical anthropologist or archaeologist. Anthropologists are increasingly engaged in a variety of activities that have direct relevance to the modern world. Rather than being confined to the halls of the university, an increasing number of anthropologists have become practitioners of anthropology. Indeed, according to a 1990 survey, 50 percent of those individuals with doctorates in anthropology develop careers outside the academic area (Givens & Fillmore, 1990). Similarly, a 1994 survey indicated that only 35 percent of the membership of the Society for American Archaeology, the primary national organization for archaeologists working in North America, was made up of researchers employed in academic positions, while approximately 8 percent were employed in museums (Zeder 1997: 47–48). Of the remainder of the membership, the largest portion, 23 percent, were employed by federal, state, or local governments, followed by 18 percent who worked in private businesses. Comparable trends can be seen in the membership of other archaeological organizations. Many physical anthropologists work with or are employed by public agencies such as the United States Agency for International Development (USAID) and the National Institutes of Health (NIH). Some also work for private firms.

Distinguishing **applied anthropology** from other anthropological pursuits in many respects presents a false dichotomy. Methodical and theoretical concerns are shared by all; the difference lies in perceptions of the practitioners' objectives, an arbitrary division based on the practicality of the intended outcomes. As Bronislaw Malinowski observed more than a half-century ago: "Unfortunately, there is still a strong but erroneous opinion in some circles that practical anthropology is fundamentally different from theoretical or academic anthropology. The truth is that science begins with application. . . . What is application in science and when does 'theory' become practical? When it first allows us a definite grip on empirical reality" (Malinowski 1945).

The work of many anthropologists can be seen as applied in some sense. In an overview of applied anthropology, Erve Chambers (1985) classified the different roles of applied anthropologists. While he was primarily concerned with the applied aspects of cultural anthropology, his observations are equally relevant to the work of physical anthropologists and archaeologists. One role noted by Chambers is that of *representative,* in which the anthropologist becomes the spokesperson for the particular group being studied. Anthropologists at times have represented Native American communities in negotiations with state and federal authorities, mining companies, and development organizations. Anthropologists can also be seen as *facilitators.* In this capacity, anthropologists actively help bring about change or development in the community being researched. For example, they may take a proactive, participatory role in economic or social change to improve medical care, education, or public facilities. An alternative position is the *informant* role, in which the applied anthropologist transfers cultural knowledge obtained from anthropological research to the government or other agency that wants to promote change in a particular area. The U.S. government has employed anthropologists as on-site researchers to provide data on how local-level service clients and delivery agencies respond to government policy. Informally, many archaeologists and anthropologists become involved in local activities and educational programs that present anthropological findings to the public.

Yet another role of applied anthropologists is that of *analyst.* Rather than being just a provider of data, the practicing anthropologist sometimes becomes engaged in the actual formulation of policy. In archaeology in particular, this has become an important area with the passage of the National Historic Preservation Act in 1966, the Native American Graves Repatriation Act of 1990 and other, related legislation. These laws have afforded increased protection to some archaeological resources and mandated the consideration of archaeological resources in planning development. Archaeologists have increasingly found employment in federal, state, or local governments reviewing proposals for development and construction projects that impact cultural resources and archaeological sites. Another role Chambers identified is that of *mediator,* which involves the anthropologist as an intermediary among different interest groups that are participating in a development project. This may include private developers, government officials, and the people who will be affected by the project. As mediator, the anthropologist must try to reconcile differences among these groups, facilitating compromises that ideally will benefit all parties involved in the project. The following discussions highlight some of the applied work that physical anthropologists and archaeologists are engaged in.

PHYSICAL ANTHROPOLOGY

As seen in the preceding chapters, physical anthropologists deal with the biological aspects of humans and human ancestors in the past and the present. Much of the basic information gathered consists of the measurement, observation, and explanation of various physical characteristics. *Anthropometry,* for example, concerns the measurement of human body parts, and *osteometry* is the measurement of skeletal elements. This information is basic to the interpretation of fossil hominids as well as human remains recovered from archaeological sites. However, some of this information has immediate relevance to the present. Such information may be used in combination with engineering data to design ergonomically efficient airplane cockpits, work environments, or equipment. Such data may also provide an important aid to police in investigations of murders or the identification of disaster victims. Physical anthropological study of the causes of diseases, when combined with knowledge of cultural anthropology, offers important insight into perceptions of medical treatment in different cultural settings. Some of these examples of practicing anthropologists are considered in this section.

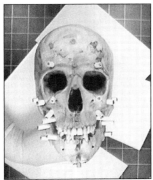 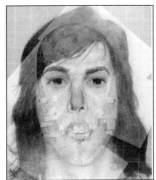  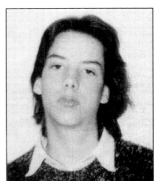

A specialized area of forensic anthropology deals with the reconstruction of facial features. The photographs illustrate (from left) the victim's skull, a reconstruction of the face, a sketch based on the reconstruction, and a photograph of how the victim actually appeared in life.

Source: Courtesy of Gene O'Donnell, FBI.

FORENSIC ANTHROPOLOGY

A fragmentary skeleton is accidentally found in a desolate part of the desert. Through a series of twists and turns an enterprising detective pieces together clues to a twenty-year-old murder and brings a fugitive to justice. Such a scenario is the stuff of mystery novels, but real-life criminal investigations often do depend on the identification of fragmentary skeletal remains. **Forensic anthropology** can be defined as the application of physical anthropological data to law. Physical anthropologists in this area of specialization are often called to assist police when unidentified humans remains are found. Whereas medical doctors focus on the soft tissues, forensic anthropologists study the hard tissues—the skeletal remains (Işcan and Kennedy 1989). Analysis of such material would begin by reconstructing the skeleton and joining together the often fragmentary and broken remains. Missing pieces might be reconstructed or estimated. The materials are then carefully measured and compared to anthropological data. Such research can yield the sex, approximate age, height, and physical characteristics of an individual.

The skeleton also provides a record of medical problems, illnesses, and the overall health of a person. The bones may preserve information about a person's health at the time of death, as well as the living conditions and health problems the individual faced during his or her lifetime. For example, broken bones, although healed, still leave a trace on the skeleton. Arthritis, certain infections, dietary stress, and nutrition may also be in evidence. This kind of information may provide insight into living conditions in the distant human past, as, for example, when considering the consequences of domestication (see Chapter 8), but it also provides details that may be very helpful to the police in identifying victims. Unidentified skeletal remains from a white female, 5'4" to 5'6", 40 to 45 years of age, with a healed fracture of the left leg, and traces of arthritis in the hands would dramatically reduce the number of potential fits with reported missing persons files.

A specialized area within forensic anthropology deals specifically with the reconstruction of faces (Prag & Neave, 1997). Using average skin depth, muscle patterns, and knowledge about the skeleton, the researcher can create an image of what a person looked like when alive. This interdisciplinary work draws on information from anatomy, facial surgery, pathology, dentistry, and the skills of an artist, as well as physical anthropology. Reconstruction of a face based solely on information provided by the skull may be done using a computer or sketched by an artist, but researchers also rely on a detailed model of a skull, which they then cover with clay. Muscles of clay are sculpted over the skull, which are then covered with clay representing the overlying tissues. The thickness of the skin covering the skull is based on average thickness at different points of the skull, estimated for individuals of different ages, sexes, body builds, and ethnic groups. A final model is prepared using plaster of Paris, which is colored and given hair. While the final products may not be exact portraits, their resemblance to the living individuals has proven remarkable.

Forensic anthropology may also offer important clues about the circumstances of a person's death and the treatment of the body after death (Haglund & Sorg 1997). Damage or trauma to the bones may provide a primary indicator of the cause of death. For example, bullet wounds, stabbings, and blunt force trauma may be identified in skeletal remains. Careful study of the skeleton may also indicate if an individual was killed where the body was found or at another location and then transported to the site. Forensic anthropologists may be able to determine whether the body was disturbed or transported after burial. Such information may be extremely important in determining the cause of death. As in the case of archaeological and paleontological investigations, the context of the findings is very important. Hence, physical anthropologists with archaeological training can help ensure that all of the remains are recovered.

Because the cause of death may be central to a murder investigation and trial, the forensic anthropologist is often called upon to testify as an expert witness. In such cases the forensic anthropologist impartially presents his or her findings. These may prove or disprove the identity or cause of death of the victim: The ultimate concern of the forensic anthropologist is not the outcome of the trial, but the evidence provided by the skeletal remains.

The amount of information extracted from skeletal remains can be surprising. Many illustrations from actual criminal cases can be recounted (Rathbun & Buikstra 1984; Stewart 1979). For example, fractures of the hyoid or the thyroid, a small bone and ossified cartilage of the throat may indicate strangulation. The location and kind of breaks may offer clues to the type of weapon used, as well as the position of the attacker relative to the victim. In the vein of a Sherlock Holmes novel, it may actually be possible to determine that a fatal blow was struck from behind by a right-handed assailant.

Forensic anthropologists have also played important roles in the identification of victims of natural disasters, airplane crashes, war, and genocide (Snow et al., 1989; Snow & Bihorriet, 1992; Stewart, 1970; Stover, 1981, 1992). Many of the methods and techniques used by modern forensic anthropologists

Excavation of the burial site of civilians who were killed by government troops at El Morote, El Salvador, during the country's civil war in the 1980s. Forensic anthropologists often play an important role in the identification of victims of natural disasters, airplane crashes, wars, and genocide.

were needed during and after World War II to assist with the identification and repatriation of the remains of soldiers killed in battle. This remained an important role for forensic anthropologists during the Korean and Vietnam wars (Stewart 1979: 11–12). In these cases the physical remains recovered are matched against the life histories provided by medical and dental records. In some instances the positive identification may be dependent on relatively minor variation in bony structures. The role of physical anthropologists in locating and identifying American soldiers killed or missing in action in Vietnam continues to this day. Forensic anthropologists and archaeologists have assisted in the documentation of human rights abuses and recovery of victims from mass graves in Argentina, Brazil, Croatia, El Salvador, Haiti, Iraq, Rwanda, and other world areas.

MEDICAL ANTHROPOLOGY

Another subfield of anthropology, **medical anthropology,** represents the intersection of cultural anthropology with physical anthropology. Medical anthropologists may study disease, medicine, curing, and mental illness in cross-cultural perspective. Some focus specifically on **epidemiology,** which examines the spread and distribution of diseases and how this is influenced by cultural patterns. For example, these anthropologists may be able to determine whether coronary (heart) disease or cancer is related to particular cultural or social dietary habits, such as the consumption of foods high in sodium or saturated fats. They also study cultural perceptions of illness and their treatment. These studies can often help health providers to design more effective means for delivering health care and formulating health care policies.

An illustration of medical anthropology is the work of Louis Golomb (1985), who conducted ethnological research on the curing practices in Thailand. Golomb did research on Buddhist and Muslim medical practitioners who rely on native spiritualistic beliefs to diagnose and cure diseases. These practices are based on earlier Hindu, magical, and animistic beliefs that had been syncretized with Buddhist and Muslim traditions. Practitioners draw on astrology, faith healing, massage, folk psychotherapy, exorcism, herbs, and charms and amulets to treat patients. The most traditional practitioners are curer-magicians, or *shamans,* who diagnose and treat every illness as an instance of spirit possession or spirit attack. Other practitioners are more skeptical of the supernatural causation of illness and diagnose health problems in reference to natural or organic causes. They frequently use herbal medicines to treat illnesses.

Golomb discovered that although Western-based scientific forms of medicine may be available, many Thais still rely on traditional practitioners. He found that even urban-educated elite, including those who had studied in the United States and other Western countries, adhered to both supernatural and scientific views. Golomb referred to this as *therapeutic pluralism.* He observed that patients do not rely on any single therapeutic approach but rather use a combination of therapies that include elements of ritual, magic, and modern scientific medications. Parasites or germs are rarely seen as the only explanations of disease; a sick person may go to a clinic to receive medication to relieve symptoms but may then seek out a traditional curer for a more complete treatment. Golomb emphasized that the multiplicity of alternative therapies encourages people to play an active role in preserving their health.

Medical anthropologists do studies to help improve basic health care delivery in countries around the world.

In Thailand, as in many other countries undergoing modernization, modern medical facilities have been established based on the scientific treatment of disease. Golomb found that personnel in these facilities are critical of traditional medical practices. Nevertheless, he discovered that although the people in the villages often respect the modern doctor's ability to diagnose diseases and prescribe medications to relieve symptoms, in most cases they don't accept the scientific explanation of the disease. In addition, villagers feel that modern medical methods are brusque and impersonal, because doctors do not offer any psychological or spiritual consolation. Doctors also do not make house calls and rarely spend much time with patients. This impersonality in the doctor-patient relationship is also due to social-status differences based on wealth, education, and power. Golomb found that many public health personnel expected deference from their rural clientele. For these reasons, many people preferred to rely on traditional curers.

Through his study of traditional medical techniques and beliefs, Golomb isolated some of the strengths and weaknesses of modern medical treatment in Thailand. His work contributed to a better understanding of how to deliver health care services to rural and urban Thais. For example, the Thai Ministry of Public Health began to experiment with ways of coordinating the efforts of modern and traditional medical practitioners. Village midwives and traditional herbalists were called on to dispense modern medications and pass out information about nutrition and hygiene. Some Thai hospitals have established training sessions for traditional practitioners to learn modern medical techniques. Golomb's studies in medical anthropology offer a model for practical applications in the health field for other developing societies.

INTERVENTIONS IN SUBSTANCE ABUSE

Another area of research and policy formulation for applied medical anthropologists is substance abuse, many of the causes of which may be explained by social and cultural factors. For example, Michael Agar (1973, 1974) did an in-depth study of heroin addicts based on the addicts' description of U.S. society and its therapeutic agencies in particular. His research involved taking on the role of a patient himself so that he could participate in some of the problems that exist

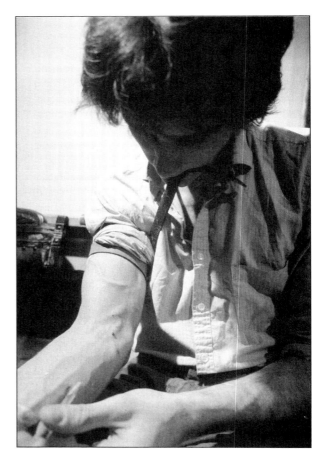

Some applied anthropologists have been doing studies of drug addiction to assist agencies in the prevention of drug abuse.

between patients and staff. From that perspective he was better able to understand the "junkie" worldview.

Through his research, Agar isolated problems in the treatment of heroin addiction. He found that when the drug methadone was administered by public health officials as a substitute for heroin, many heroin addicts not in treatment became addicted to methadone, which was sold on the streets by patients. This street methadone would often be combined with wine and pills to gain a "high." In some cases, street methadone began to rival heroin as the preferred drug, being less expensive than heroin, widely available, and in a form that could be taken orally rather than injected. By providing this information, Agar helped health officials monitor their programs more effectively.

In a more recent study, Philippe Bourgois spent three and a half years investigating the use of crack cocaine in Spanish Harlem in New York City. In his award-winning ethnography, *In Search of Respect: Selling Crack in El Barrio* (1996), Bourgois noted that policymakers and drug-enforcement officials minimize the influences of poverty and low status in dealing with crack addiction. Through his investigation of cultural norms and socioeconomic conditions in Spanish Harlem, Bourgois demonstrated that crack dealers are struggling to earn money and status in the pursuit of the American Dream. Despite the fact that many crack dealers have work experience, they find that many of the potential jobs in construction and factory work are reserved for non-Hispanics. In addition, unpleasant experiences in the job world lead many to perceive crack dealing as the most realistic route toward upward mobility. Most of the inner-city youths who deal crack are high school dropouts who do not regard entry-level, minimum-wage jobs as steps to better opportunities. In addition, they perceive the underground economy as an alternative to becoming subservient to the larger society. Crack dealing offers a sense of autonomy, position, and rapid, short-term mobility.

Bourgois compared the use of crack to the millenarian movements such as the Ghost Dance of Native Americans or the cargo cults of Melanesia. As he observed:

> *Substance abuse in general, and crack in particular, offer the equivalent of a millenarian metamorphosis. Instantaneously, users are transformed from being unemployed, depressed high school dropouts, despised by the world—and secretly convinced that their failure is due to their own inherent stupidity, "racial laziness," and disorganization—into being a mass of heart-palpitating pleasure, followed only minutes later by a jaw-gnashing crash and wide-awake alertness that provides their life with concrete purpose: get more crack—fast! (1989: 11)*

Bourgois's depictions of the culture and economy of crack dealers and users provided useful policy suggestions. He concluded that most accounts of crack addiction deflect attention away from the economic and social conditions of the inner city, and that by focusing on the increases of violence and terror associated with crack, U.S. society is absolved from responsibility for inner-city problems. He suggested that rather than use this "blame-the-victim" approach, officials and policymakers need to revise their attitudes and help develop programs that resolve the conditions that encourage crack use.

APPLIED ARCHAEOLOGY

One of the problems that humanity faces is how to safeguard the cultural heritage preserved in the archaeological record. Although archaeology may address questions of general interest to all of humanity, it is also important in promoting national heritage, cultural identity, and ethnic pride. Museums the world over offer displays documenting a diversity of local populations, regional histories, important events, and cultural traditions. The number of specialized museums, focusing on particular peoples, regions, or historic periods have become increasingly important. Archaeologists must be concerned with the preservation of archaeological sites and the recovery of information from sites threatened with destruction, as well as the interpretation and presentation of their findings to the more general public.

Preservation of the past is a challenge to archaeologists, government officials, and the concerned public alike, as archaeological sites are being destroyed at an alarming rate. Archaeological materials naturally decay in the ground and sites are constantly destroyed through geological processes, erosion, and animal burrowing (see Chapter 2). Yet while natural processes contribute to the disappearance of archaeological sites, by far the greatest threat to the archaeological record is human activity. Construction projects such as dams, roads, buildings, and pipelines all disturb the ground and can destroy archaeological sites in the process. In many instances, archaeologists work only a few feet ahead of construction equipment, trying to salvage any information they can before a site disappears forever.

Some archaeological sites are intentionally destroyed by collectors searching for artifacts that have value in the antiquities market, such as arrowheads and pottery. Statues from ancient Egypt, Mayan terracotta figurines, and Native American pottery may be worth thousands of dollars on the antiques market. To fulfill the demands, archaeological sites in many world areas are looted by pothunters, who dig to retrieve artifacts for collectors, ignoring the traces of ancient housing, burials, and cooking hearths.

TABLE 10.1 Major Federal Legislation for the Protection of Archaeological Resources in the United States

Antiquities Act of 1906	Protects sites on federal lands
Historic Sites Act of 1935	Provides authority for designating National Historic Landmarks and for archaeological survey before destruction by development programs
National Historic Preservation Act of 1966 (amended 1976 and 1980)	Strengthens protection of sites via National Register and in integrates state and local agencies into national program for site preservation
National Environmental Policy Act of 1969	Requires all federal agencies to specify impact of development programs on cultural resources
Archaeological Resources Protection Act of 1979	Provides criminal and civil penalties for looting or damaging sites on public and Native American lands
Convention of Cultural Property of 1982	Authorizes U.S. participation in 1970 UNESCO convention to prevent illegal international trade in cultural property
Cultural Property Act of 1983	Provides sanctions against U.S. import or export of illicit antiquities
Federal Abandoned Shipwreck Act of 1988	Removes sunken ships of archaeological interest from marine salvage jurisdiction; provides for protection under state jurisdiction
Federal Reburial and Repatriation Act of 1990	Specifies return of Native American remains and cultural property to Native American groups by U.S. museums

Source: From *Discovering Our Past: A Brief Introduction to Archaeology* by Wendy Ashmore and Robert J. Sharer. Copyright ©1988 by Mayfield Publishing Company. Reprinted by permission of the publisher.

Removed from their context, with no record of where they came from, such artifacts are of limited value to archaeologists. The rate of destruction of North American archaeological sites is such that some researchers have estimated that 98 percent of sites predating the year 2000 will be destroyed by the middle of the twenty-first century (Herscher, 1989; Knudson, 1989).

The rate at which archaeological sites are being destroyed is particularly distressing because the archaeological record is an irreplaceable, nonrenewable resource. That is, after sites are destroyed, they are gone forever, along with the unique information about the past that they contained. In many parts of the world, recognition of this fact has led to legislation aimed at protecting archaeological sites (Table 10.1). The rational for this legislation is that the past has value to the present and, hence, should be protected and interpreted for the benefit of the public.

PRESERVING THE PAST

Recognition of the value and nonrenewable nature of archaeological resources is the first step in a planning process. Archaeological resources can then be systematically identified and evaluated. Steps can be taken to preserve them by limiting development or designing projects in a way that will preserve the resource. For example, the projected path of a new road might be moved to avoid an archaeological site, or a building might be planned so that the foundations do not extend into a historic burial ground (see the photos of the African burial ground). Alternatively, if a site must be destroyed, effective planning can ensure that information about the site is recovered by archaeologists prior to its destruction.

One of the most spectacular examples of salvage archaeology arose as a result of the construction of a dam across the Nile River at Aswan, Egypt, in the 1960s. The project offered many benefits, including water for irrigation and the generation of electricity. However, the rising water behind the dam threatened thousands of archaeological sites that had lain undisturbed and safely buried by desert sand for thousands of years. The Egyptian government appealed to the international community and archaeologists from around the world to mount projects to locate and excavate the threatened sites.

Among the sites that were to be flooded by the dam was the temple of Pharaoh Rameses II at Abu Simbel, a huge monument consisting of four colos-

Located just blocks from Wall Street in New York City, an eighteenth-century African burial ground was accidentally uncovered during construction of a federal office building in 1991. The 427 burials excavated at the site are testament to the enslaved Africans that made up the second largest slave population in colonial America. As many as 20,000 individuals may have been buried at the site. Following discovery, local community protests over the treatment and interpretation of the remains led to a delay in construction, modification of the construction plan, and the increased involvement of African American researchers in the analysis of the finds.

Source: Courtesy of the General Services Administration.

sal figures carved from a cliff face on the banks of the Nile River. With help from the United Nations Educational, Scientific, and Cultural Organization (UNESCO), the Egyptian government was able to cut the monument into more than a thousand pieces, some weighing as much as 33 tons, and reassemble them above the floodwaters. Today the temple of Rameses can be seen completely restored only a few hundred feet from its original location. Numerous other archaeological sites threatened by the flooding of the Nile were partly salvaged or recorded. Unfortunately, countless other sites could not be located or even recorded before they were flooded.

The first legislation in the United States designed to protect historic sites was the Antiquities Act of 1906, which safeguarded archaeological sites on federal lands (see Table 10.1). Other, more recent legislation, such as the **National Historic Preservation Act** passed in 1966, has extended protection to sites threatened by projects that are funded or regulated by the government. The Federal **Abandoned Shipwreck Act** of 1988 gives states jurisdiction over shipwreck sights. This legislation has had a dramatic impact on the number of archaeologists in the United States and has created a new area of specializa-

tion, generally referred to as **cultural resource management (CRM).** Whereas most archaeologists had traditionally found employment teaching or working in museums, many are now working as applied archaeologists, evaluating, salvaging, and protecting archaeological resources that are threatened with destruction. Applied archaeologists conduct surveys before construction begins to determine if any sites will be affected. Government agencies such as the Forest Service have developed comprehensive programs to discover, record, protect, and interpret archaeological resources on their lands (Johnson & Schene, 1987).

Unfortunately, current legislation in the United States leaves many archaeological resources unprotected. In many countries, excavated artifacts, even those located on privately owned land, become the property of the government. This is not the case in the United States. One example of the limitations of the existing legislation is provided by the case of Slack Farm, located near Uniontown, Kentucky (Arden, 1989). Archaeologists had long known that an undisturbed Native American site of the Late Mississippian period was located on the property. Dating roughly to between 1450 and 1650, the site was particularly important because it was the only surviving Mississippian site from the period of first contact with Europeans. The Slack family, who had owned the land for many years, protected the site and prevented people from digging (Arden, 1989). When the property was sold in 1988, however, conditions changed. Anthropologist Brian Fagan described the results:

> *Ten pot hunters from Kentucky, Indiana, and Illinois paid the new owner of the land $10,000 for the right to "excavate" the site. They rented a tractor and began bulldozing their way through the village midden to reach graves. They pushed heaps of bones aside and dug through dwellings and potsherds, hearths, and stone tools associated with them. Along the way, they left detritus of their own—empty pop-top beer and soda cans—scattered on the ground alongside Late Mississippian pottery fragments. Today Slack Farm looks like a battlefield—a morass of crude shovel holes and gaping trenches. Broken human bones litter the ground, and fractured artifacts crunch underfoot. (1989; 15)*

The looting at the site was eventually stopped by the Kentucky State Police, using a state law that prohibits the desecration of human graves. Archaeologists went to the site attempting to salvage what information was left, but there is no way of knowing how many artifacts were removed. The record of America's prehistoric past was irrevocably damaged.

Regrettably, the events at Slack Farm are not unique. Many states lack adequate legislation protecting archaeological sites on private land. For example, Arkansas had no laws protecting unmarked burial sites until 1991. As a result, Native American burial grounds were systematically mined for artifacts. In fact, one article written about the problem was titled "The Looting of Arkansas" (Harrington, 1991). Although Arkansas now has legislation prohibiting the unauthorized excavation of burial grounds, the professional archaeologists of the Arkansas Archaeological Survey face the impossible job of trying to locate and monitor all of the state's archaeological sites. This problem is not unique. Even on federal lands, the protection of sites is dependent on a relatively small number of park rangers and personnel to police large areas. Even in national parks such as Mesa Verde or Yellowstone archaeological sites are sometimes vandalized or looted for artifacts. Much of the success that there has been in

protecting sites is largely due to the active involvement of amateur archaeologists and concerned citizens who bring archaeological remains to the attention of professionals.

The preservation of the past needs to be everyone's concern. Unfortunately, however well intended the legislation and efforts to provide protection for archaeological sites may be, they are rarely integrated into comprehensive management plans. For example, a particular county or city area might have a variety of sites and resources of historic significance identified using a variety of different criteria and presented in different lists and directories. These might include National Historic sites, designated though the National Historic Preservation office; state files of archaeological sites; data held by avocational archaeological organizations and clubs; and a variety of locations of historical importance identified

Although many countries have legislation protecting their archaeological resources, enforcing the laws is often difficult. Here a farmer in Mali displays artifacts looted from an archaeological site. Removed from their archaeological context, such finds have limited value to archaeologists.

by county or city historical societies. Other sites of potential historical significance might be identified though documentary research or oral traditions. Ideally all of these sources of information should be integrated and used to plan development. Such comprehensive approaches to cultural resource management plans are rare rather than the norm.

Important strides have been taken in planning and coordinating efforts to identify and manage archaeological resources. Government agencies, including the National Park Service, the military, and various state agencies, have initiated plans to systematically identify and report sites on their properties. There have also been notable efforts to compile information at the county, state, and district levels. Such efforts are faced with imposing logistical concerns. For example, by the mid-1990s, over 180,000 historic and prehistoric archaeological sites had been identified in the American Southeast (including the states of Alabama, Arkansas, Florida, Georgia, Kentucky, Louisiana, Mississippi, North Carolina, South Carolina, and Tennessee). In addition, an estimated 10,000 new sites are discovered each year (Anderson & Horak, 1995). A map of these resources reveals a great deal of variation in their concentration. On one hand, this diversity reflects the actual distribution of sites; on the other, the areas where archaeological research has and has not been undertaken. Incorporating the thousands of new site reports into the database requires substantial commitment of staff resources. What information should be recorded for each site? What computer resources are needed? The volume of information is difficult to process with available staff, and massive backlogs of reports waiting to be incorporated into the files often exist. Nevertheless, this kind of holistic perspective is needed to ensure effective site management and the compliance of developers with laws protecting archaeological sites. It also provides a holistic view of past land use that is of great use to archaeologists.

THE STUDY OF GARBAGE

The majority of archaeological research deals with the interpretation of past societies. Whether the focus is on the Stone Age inhabitants of Australia or the archaeology of nineteenth-century mining communities in the American West, the people being examined lived at a time somewhat removed from the present. There are, however, some archaeolo-

gists who are concerned with the study of the refuse of modern society and the application of archaeological methods and techniques to concerns of the present—and the future. The focus for these researchers is not the interpretation of past societies but the immediate application of archaeological methods and techniques to the modern world. The topics examined range from the use of archaeological data in marketing strategies to the best methods for marking nuclear waste sites. Archaeologists, who routinely examine artifacts thousands of years old can, for example, provide important perspectives of the suitability of different materials and burial strategies that can be used to bury nuclear waste (Kaplan & Mendel, 1982).

One of the more interesting examples of this kind of applied archaeology is the Garbage Project, or *Le Projet du Garbage,* that grew out of an archaeological method and theory class at the University of Arizona in 1971 (Rathje, 1992; Rathje & Ritenbaugh, 1984). Archaeologists William L. Rathje and Fred Gorman were so intrigued by the results of the student projects that they established the Garbage Project the following year, and the project is still ongoing. The researchers gather trash from households with the help of the City of Tucson Sanitation foremen who tag the waste with identification numbers that allow the trash bags to be identified with specific census tracts within the city. The trash bags are not identified with particular households, and personal items and photographs are not examined. Over the years research has been broadened to the study of trash from other communities, including Milwaukee, Marin County, and Mexico City, and also to the excavation of modern landfills in Chicago, San Francisco, and Phoenix using archaeological methods.

The Garbage Project has provided a surprising amount of information on a diversity of topics. On one hand, the study provides data that are extremely useful in monitoring trash disposal programs. As Rathje observed, study of waste allows the effective evaluation of current conditions, the anticipation of changing directions in waste disposal, and, therefore, more effective planning and policy making (1984: 10). Reviewing data on the project, Rathje noted a number of areas in which this archaeological research has dispelled some common notions about trash disposal and landfills. Despite common perceptions, items such as fast-food packaging, polystyrene foam, and disposal di-

Archaeological study of modern garbage has provided important insights into waste management procedures, marketing, food wastage, and recycling.
Source: Courtesy of C. R. DeCorse.

apers do not account for a substantial percentage of landfills. Rathje observed:

> Of the 14 tons of garbage from nine municipal land-fills that the Garbage Project has excavated and sort-ed in the past five years, there was less than a hundred pounds of fast-food packaging—that is, con-tainers or wrappers for hamburgers, pizzas, chicken, fish and convenience-store sandwiches, as well as the accessories most of us deplore, such as cups, lids, straws, sauce containers, and so on. (Rathje, 1992: 115)

Hence, fast-food packaging makes up less than one half of 1 percent of the weight of landfills. The percentage by volume is even lower. Rathje further noted that despite the burgeoning of materials made from plastic, the amount of plastic in landfills has re-mained fairly constant since the 1960s, or even de-

creased. The reason for this, he believes, is that while more things are made of plastic, many objects are now made with *less* plastic. A plastic milk bottle that weighed 120 grams in the 1960s, today weighs 65 grams.

Rathje found that the real culprit in landfills re-mains plain old paper. A year's subscription to the *New York Times* is roughly equivalent to the volume of 18,660 crushed aluminum cans or 14,969 flattened Big Mac containers. While some predicted that com-puters would bring about a paperless office, the pho-tocopier and millions of personal printers ensure that millions of pounds of paper are discarded each year: "Where the creation of paper waste is concerned, technology is proving to be not so much a contra-ceptive as a fertility drug" (Rathje, 1992: 116). He also observed that despite popular perception, much of the paper in landfills is *not* biodegrading. Because of the limited amount of moisture, air, and biologi-cal activity in the middle of a landfill, much of the na-tion's newsprint is being preserved for posterity.

The Garbage Project has also provided informa-tion on a diversity of issues connected with food waste, marketing and the disposal of hazardous ma-terials. In these studies, archaeology has provided a unique perspective. Much of the available data on such topics has typically been provided by ques-tionnaires, interviews, and data collection methods that rely on the cooperation of informants. The prob-lem is that informants often present biased respons-es, consistently providing lower estimates of the alcohol, snack food, or hazardous waste that they dispose of than is actually the case. Archaeology, on the other hand, presents a fairly impartial material record. While there are sampling problems in ar-chaeological data—some material may be sent down the garbage disposal and not preserved in a land-fill—the material record can provide a fairly unam-biguous record of some activities.

The Garbage Project has examined the discard of food and food wastage for the U.S. Department of Agriculture; the recycling of paper and aluminum cans for the Environmental Protection Agency; stud-ies of candy and snack food consumption for den-tal associations and manufacturers; and the impact of a new liquor store on alcohol consumption in a Los Angeles neighborhood. In the latter case, researchers conducted both interviews and garbage analysis be-fore and after the liquor store opened. The interview data suggested no change in consumption patterns before and after the store's opening. Study of the trash, however, showed a marked increase in the discard of beer, wine, and liquor cans and bottles. Studies such as these have wide applications both for marketing and policymakers.

WHO OWNS THE PAST?

A critical issue for modern archaeologists and physical anthropologists is **cultural patrimony,** that is, who owns the human remains, artifacts, and associated cultural materials that are recovered in the course of research projects. Are they the property of the scientists who collected or excavated them? The descendants of the peoples discovered archaeologically? The owners of the land on which the materials were recovered? Or the public as a whole? Resolution of this issue has at times been contentious, and the position taken by anthropologists has not always been the best. Prior to the twentieth century, laws governing the deposition of antiquities were nonexistent or unclear at best and the owner often became the person, institution, or country with the most money or the strongest political clout. Colonial governments amassed tremendous collections from their territories throughout the world. The spoils of war belonged to the victors. Such a position remained the norm until after the turn of the century. Rights of conquest were only outlawed by the Hague Convention of 1907 (Fagan, 1992; Shaw, 1986).

Prior to the twentieth century there was also little or no legislation governing human remains. Researchers appropriated excavated skeletal material, medical samples, and even cadavers of the recently deceased (Blakely & Harrington, 1997). Native American remains from archaeological sites were displayed to the public, despite the fact that some of the descendant communities found such display inappropriate or sacrilegious. Scientific value was the underlying rational for ownership, though until the latter half of the twentieth century there was little discussion of this issue. As in the case of antiquities, value and ownership were vested in the politically stronger, whether a colonial government or the politically enfranchised within a country. Such remains

had scientific value and this was viewed as more important than the interests of other groups or cultural values.

Ironically, such views would seem to fly in the face of some of the basic tenets of modern anthropology, which underscore sensitivity and openness to other cultural perspectives and beliefs. In fact, archaeological resources and human remains at times do provide unique, irreplaceable information that cannot be obtained through any other source. Archaeologists and physical anthropologists have provided information extremely important in documenting the past of Native Americans and indigenous peoples throughout the world, at times serving to underscore their ties to the land and revealing cultural practices forgotten from memory. But what is the cost of such information if the treatment and methods of obtaining the artifacts and remains represented are abhorrent to the populations whose history is represented? Researchers of the present cannot afford to ignore the views and concerns of the focus of their research.

Recognition of the validity of different concerns and perspectives of cultural patrimony has not made resolution of debate easier. Artifacts now in museums were, in some instances, obtained hundreds of years ago in ways that were consistent with the moral and legal norms of that time (see the box "The Elgin Marbles"). Many antiquities have legitimately changed ownership numerous times. Not infrequently information about the original origins may be unclear, and there are differences of opinion or uncertainly about the cultural associations of some artifacts or cultural remains. These issues aside, there remain fundamentally different perspectives about the role of the descendant population.

NATIVE AMERICAN GRAVES PROTECTION AND REPATRIATION ACT

Perhaps the most important legislation impacting the treatment and protection of archaeological and physical anthropological resources in the United States is the **Native American Graves Protection and Repatriation Act (NAGPRA),** passed on November 16, 1990 (McKeown, 1998). This legislation is the most comprehensive of a series of recent laws dealing with the deposition of Native American burials and cultural properties. NAGPRA and related legislation requires that federal institutions con-

CRITICAL PERSPECTIVES

THE ELGIN MARBLES

The story of the Parthenon sculptures—or Elgin marbles as they came to be called—is a twisted tale of the nineteenth-century quest for antiquities, international politics, and the complexities of cultural heritage. The Parthenon, perched on a hilltop overlooking Athens, is a striking symbol of both ancient Greece and the modern Greek nation-state. It was built by the Greek ruler Pericles to commemorate the Greek victory over the Persians at Plataea in 479 B.C. A temple to Athena, the patron goddess of Athens, the Parthenon was deemed by Pericles to be one of the most striking edifices in the city. The Parthenon is clearly the most striking of the buildings in the Acropolis, the cluster of classical structures that cover Mount Athena. It is regarded by some to be one of the world's most perfect buildings. It was distinguished by a full surrounding colonnade, and the exterior walls were decorated with a processional frieze. The pediments, or peaked eaves, in the east and the west also had strikingly detailed sculptures.

The structure has endured for centuries, and it has come to embody classical civilization to the world. In recent years, the Parthenon has been the focus of several restoration efforts that have stabilized the structure, removed more recent additions, and replaced some of the fallen masonry.

The Parthenon still overlooks Athens and it hopefully will for years to come. But while the Parthenon is an architectural trea-

Two young horsemen join a procession of sculpted figures on the Parthenon frieze. The marbles were taken from Greece in the early nineteenth century and are now on display in the British Museum, London.

Source: © Copyright The British Museum.

sure, today only traces of the magnificent art that once adorned it remain. To view its friezes and sculptures, you have to go to the British Museum in London, where they are beautifully displayed in a specially designed room. To understand why statuary of such clear cultural significance to Greece is to be found in England, one has to go back to the early nineteenth century and the exploits of Thomas Bruce, the 7th Earl of Elgin (Jackson, 1992). By the early nineteenth century, Britain was in the midst of a classical revival. The country's well-to-do traveled to Europe to visit the historic ruins of ancient Greece and Rome. The wealthier purchased statuary and antiquities for their estates. Patterns and illustrations from classical Greek and Roman motifs were reproduced and incorporated into architectural features, jewelry, and ceramic designs. Within this

setting, Elgin, a Scottish nobleman, set out to obtain sketches and casts of classical sculptures that might be used at his estate, then being built near the Firth of Forth.

In 1799, Elgin was appointed British ambassador to the Ottoman Empire, which extended over much of the eastern Mediterranean from western Europe to Egypt. By the late eighteenth century, the Ottomans had ruled Greece for 350 years. A major military power and one-time masters of the Mediterranean, the Ottomans have been viewed by historians in a variety of ways, but one thing is certain: They were not overly concerned with the glories of ancient Greece. During their rule the Parthenon was used as a mosque, then as an ammunition dump; also, various Turkish structures were built on the site. Much of the north colonnade was destroyed in the Venetian bombard-ment of 1687. Some of the Parthenon marble was ground to make lime, and bits of statuary were broken off (Jackson, 1992). In 1800, one of the world's most perfect buildings was in a sorry state.

Elgin initially proposed that the British government finance a survey of the art of the Acropolis as a resource for British art. When this initiative was turned down, Elgin made his own plans and contracted laborers. Initially his workers were to make copies of the Parthenon sculptures. In 1801, however, Britain defeated Napoleon's forces in Egypt, saving the Ottoman Empire. Coincidentally, Elgin soon obtained a permit from the Ottomans not only to copy and make molds of the Parthenon art but also to "take away any pieces of stone with old inscriptions or sculptures thereon" (Jackson, 1992: 137). During 1801 and 1802, scaffolding was erected and hundreds of laborers went to work on the Parthenon with blocks and tackles and marble saws. Some sculptures broke or crashed to the ground. Twenty-two ships conveyed the marbles, loaded in hundreds of crates, back to England.

The marbles hardly proved to be good fortune for Elgin. The expense of obtaining them ruined his credit and he discarded the idea of installing them at his estate. Totaling all of his expenses, Elgin estimated that he had spent over £60,000. To recoup his costs, he began negotiations for sale of the marbles to the government for display in the British Museum. After long parliamentary debate they were sold for £35,000 in 1816.

More than half of this amount went to clear Elgin's debt.

Elgin's treatment of the Parthenon's marbles had its contemporary critics. Among the most vocal opponents was none other than the Romantic poet and celebrator of Greek art and culture Lord Byron, who immortalized the story of the Elgin marbles in the poems *Childe Harold's Pilgrimage* and *Curse of Minerva.* Disgusted by what he saw as the desecration of the Parthenon, Byron asked by what right Elgin had removed these treasures of national cultural significance. Greece gained independence from Turkey in 1830, and the Parthenon became integrally tied to the new nation's identity. The first restoration efforts began soon after independence. In the years since their installation in the British Museum, the ownership of the Parthenon's marbles and demands for their return have periodically been raised by Greeks and Britons alike, but to no avail. In the 1980s, then Greek Minister of Culture Melina Mercouri charged the British with vandalism and argued that the continued possession of the marbles by the British Museum was provocative. Although garnering substantial international support from ministers of culture from around the world, these efforts also proved unsuccessful.

In his defense of his actions, Elgin pointed to the poor condition of the Parthenon and the ill treatment that the sculptures had received. If left in place they would surely have continued to deteriorate. Why not remove them and have them cared for and appreciated by those who could afford to

preserve them? Despite criticisms, Elgin believed he was saving the sculptures from the ravages of time and neglect. Time has proven Elgin at least somewhat correct. The Parthenon continues to present a complex and continuous preservation problem. Time has ravaged the remains of the sculptures that were not removed, and deterioration of the monument accelerated rather than diminished throughout the twentieth century. Stonework and architectural details have been eaten away by erosion, pollution, and acid rain, as well as by early and poorly conceived restoration efforts. As recently as 1971, a UNESCO report stated that the building itself was so weakened that it was in danger of collapse. Recent supporters of retaining the marbles argue that the marbles were obtained honestly with the permission of the government then in power. Other ancient Greek treasures such as the Venus de Milo (currently on display in the Louvre in Paris) have also been removed from the country. Are these to be returned as well? For the time being it seems that the marbles remain in London.

Points to Ponder

1. Do you feel Elgin was right or wrong to remove the marbles?

2. Should the Elgin marbles be returned to Greece? On what basis did you make your decision?

3. The conservation of the Parthenon and the preservation of the sculptures are valid concerns. How can these concerns be reconciled with the question of ownership?

"Split levels should work fine.......Where do you want to put the Jacuzzis?"

Archaeologists often find themselves balancing concern for preservation of the past and development projects aimed at meeting the needs of modern society.

sult with the lineal descendants of Native American groups and Native Hawaiians prior to the initial excavation of Native American human remains and associated artifacts on federal or tribal lands. Under this legislation federal agencies and institutions receiving federal funding are also required to **repatriate**—or return—humans remains and cultural items in their collections *at the request* of the descendant populations of the relevant Native American group. NAGPRA also dictates criminal penalties for trade in Native American human remains and cultural properties.

The impact of NAGPRA has been profound, not only on the way in which many archaeological projects are conducted but also on the way in which museums and institutions inventory, curate, and manage their collections. The law has, at times, placed very different worldviews in opposition. For many Native Americans the past is intricately connected to the present, and the natural world—animals, rocks, and trees, as well as cultural objects—may have spiritual

meaning (Naranjo, 1995). This perspective is fundamentally different from that of most museums where both human remains and cultural artifacts are treated as nonliving entities and the continuing spiritual links with the present are, at least at times, unrecognized. Museums are traditionally concerned with the collection and exhibition of objects. Reburial or repatriation of collections is the antithesis of their mission. As one scholar noted: ". . . no museum curator will gladly and happily relinquish anything which he has enjoyed having in his museum, of which he is proud, which he has developed an affection for, and which is one of the principal attractions of his museum" (Shaw, 1986: 46).

NAGPRA and repatriation also raise pragmatic concerns. Return of objects or remains is dependent on complete and accurate inventory of all of a museum's holdings. Yet often museums have amassed collections over many decades, and the full extent of their collections may not be in a readily assessable form. While records may be accurate, the number of human remains or Native American artifacts from a particular tribal group may not be known, not because of poor records but because such information was not a priority.

A case in point is the collection of the Peabody Museum of Archaeology and Ethnology at Harvard University. Founded in 1866, the Peabody has a massive collection from all over the world, including substantial Native American and ancient Mesoamerican holdings. In the 1970s and 1980s, before the passage of NAGPRA, the museum repatriated several burials, collections, and objects at the request of various constituencies. NAGPRA spurred the museum to complete a detailed inventory. They found that the estimated 7,000 human remains grew to an inventory of about 10,000, while the amount of archaeological objects grew from 800,000 to 8 million (Isaac, 1995). Following NAGPRA guidelines, the Peabody sent out summaries of collections to the 756 recognized tribal groups in the United States. Determining the cultural affiliations, the relevant descendant communities, and the need for repatriation of all of these items is a daunting task. Many museums have undertaken major inventories, revamped storage facilities, and hired additional staff specifically to deal with the issue of repatriation. Impending repatriation of collections and human remains has also spurred many institutions and researchers to reexamine old collections. Such study is necessary to ensure that the presumed age and cultural attribution of individual remains are correct. Of course, all of these concerns have serious budgetary considerations.

While NAGPRA has produced conflicts, it has also provided an avenue for new cooperation between Native Americans and researchers. Native claims will in some instances necessitate additional research on poorly documented groups. Indeed, anthropological or archaeological research may be critical to assessing the association and ownership of cultural materials and human remains. On other hand, anthropologists are given the opportunity to share their discoveries with those populations for whom the knowledge is most relevant.

 SUMMARY

Applied anthropology is a specialization within anthropology that has offered new opportunities for anthropologists to provide practical solutions to issues and concerns of the modern world. There are applied aspects in each of four subdisciplines of anthropology. These "practicing anthropologists" serve as consultants to public and private agencies and help solve local and global problems. Applied anthropologists cooperate with government officials and others in establishing policies and development projects, addressing a diversity of concerns ranging from health care systems to the management of cultural resources. This chapter specifically considers some of the applied components of physical anthropology and archaeology.

Many aspects of physical anthropology have immediate relevance to the present. For example, data concerning the measurement or study of the human skeleton or musculature may be used in combination with engineering data to design ergonomically efficient airplane cockpits, work environments, or equipment. Such data may also provide an important aid to police in murder investigations or the identification of disaster victims. This particular area of expertise dealing with the use of physical anthropological data in a legal context is called forensic anthropology. Physical anthropological understanding of the causes of diseases, when combined with knowledge of cultural anthropology, offers important insight into perceptions of medical treatment in different cultural settings.

The archaeological record is a nonrenewable resources. A relatively recent development in the field of archaeology is cultural resource management, that is, the identification, protection, and management of archaeological resources. Legislation passed by state and federal authorities in the United States requires the preservation of both prehistoric and historic materials. Applied archaeologists are involved in identifying important sites that may be endangered by development. They conduct surveys and excavations to preserve data that are important to understanding the cultural heritage of the United States. Cultural resource management offers new career opportunities for archaeologists in government agencies, universities, and consulting firms.

A major concern of archaeologists and physical anthropologists alike is the ownership of the human remains and archaeological materials that they recover. Prior to the twentieth century, researchers paid scant attention to the rights or concerns of descendant populations. Laws governing the deposition of antiquities were unclear or nonexistent. Researchers of the present, however, cannot afford to ignore the views and concerns of those who are the focus of their research. Recent legislation in the United States has dealt with the protection and deposition of Native American burials and cultural properties. Among the most important of these is the Native American Graves Protection and Repatriation Act (NAGPRA). This legislation requires that federal institutions consult with the lineal descendants of Native American groups prior to the excavation of Native American human remains and associated artifacts on federal or tribal lands, and requires the repatriation, or return of humans remains and cultural items at the request of the descendent Native American populations. NAGPRA has had profound impact not only on the way in which archaeological projects are conducted, but also on the way in which museums and institutions manage their collections. While NAGPRA has produced conflicts, it has also provided an avenue for new cooperation between Native Americans and researchers.

QUESTIONS TO THINK ABOUT

1. What is *applied anthropology?* Discuss some of the different roles as they apply to present-day anthropological studies.

2. Forensic anthropology may offer information of critical importance in a police investigation. Consider the different types of information that can, and cannot, be provided.

3. What are some of the issues and concerns that a medical anthropologist might be concerned with?

4. What is *cultural resource management?* What resources are being managed, and how are they evaluated?

 KEY TERMS

Abandoned Shipwreck Act
applied anthropology
cultural patrimony
cultural resource management
epidemiology

forensic anthropology
medical anthropology
National Historic Preservation Act

Native American Graves Protection and Repatriation Act
repatriation

 SUGGESTED READINGS

BLAKELY, ROBERT L. and HARRINGTON, JUDITH M. 1997. *Bones in the Basement: Postmortem Racism in Nineteenth-Century Medical Training*. Washington, DC: Smithsonian Institution Press. A study of a nineteenth-century collection of some 9,800 dissected and amputated human bones excavated from the basement of the Medical College of Georgia in Augusta. The volume deals with the analysis of the discoveries, as well as the social implications of the medical practices represented.

BURNS, KAREN RAMSEY. 1999. *Forensic Anthropology Training Manual*. Upper Saddle River, NJ: Prentice Hall. A practical guide to the osteometric techniques used to make age, sex, and racial inferences on the basis of skeletal remains. Although this volume is intended to complement a hands-on osteology course and is explicitly not a self-help manual, it provides a useful, accessible discussion of the techniques used by physical anthropologists to analyze skeletal remains.

CHAMBERS, ERVE. 1985. *Applied Anthropology: A Practical Guide*. Englewood Cliffs, NJ: Prentice Hall. A thorough introduction to the field of applied anthropology, with many examples of how anthropologists resolve social problems.

FAGAN, BRIAN. 1992. *The Rape of the Nile*. Providence, RI: Moyer-Bell. An extremely readable volume tracing the growth and development of the European quest for antiquities of ancient Egypt.

JOHNSON, RONALD W., and MICHAEL G. SCHENE. 1987. *Cultural Resources Management*. Malabar, FL: Robert E. Krieger. A compilation of essays focusing on the management of archaeological resources and the preservation of standing historic monuments. The work serves as a useful introduction to cultural resource management and the diverse problems confronting archaeologists in their efforts to preserve the past.

RATHBUN, T.A. and J. E. BUIKSTRA. 1984. *Human Identification: Case Studies in Forensic Anthropology*. Springfield, IL: Charles C. Thomas. This book provides detailed illustrations of some of the techniques and methods used by forensic anthropologists.

SCHMIDT, PETER R. and RODERICK J. MCINTOSH, eds. 1996. *Plundering Africa's Past*. Bloomington, IN: Indiana University Press. A compilation of articles examining the reasons for the rapid destruction of archaeological resources in Africa and the efforts that have been made to stop it.

Glossary

Abandoned Shipwreck Act of 1988 This act and related legislation define state ownership of abandoned shipwrecks, and thus provide critical protection to underwater archaeological resources.

acclimatization The physiological process of being or becoming accustomed to a new physical environment.

Acheulian A Lower Paleolithic stone tool tradition distinguished from that of the Oldowan by the bifacial flaking of stone tools and a greater range of tools, such as the hand ax. Associated with *Homo erectus.*

adaptation The evolutionary adjustment of an organism to a certain environmental setting or, in genetic terms, the shift in allele frequencies as a result of changing environmental conditions.

adaptive radiation The relatively rapid evolution of a species in a new environmental niche.

aerial photography Photographs of archaeological sites and the landscape taken from the air. Helpful to archaeologists in mapping and locating sites.

agricultural states States in which the power of the ruling elite was vested in the control of agricultural surpluses. This type of control typifies early states.

alleles The alternate forms of a gene.

amino acids The chemical compounds that form protein molecules.

analogy Similarities in organisms that have no genetic relatedness.

anatomically modern *Homo sapiens* The most recent form of humans, distinguished by the anatomical characteristics typical of living humans.

anthropology The systematic study of humankind, including the major subfields of archaeology, physical anthropology, cultural anthropology, linguistics, and applied anthropology.

antiquaries Collectors whose interest lies in the object itself, not in where the fossils might have come from or what the artifact and associated materials might tell about the people that produced them. Collectors of this kind characterized the early history of archaeology.

apartheid A political, legal, and social system developed in South Africa in which the rights of different population groups were based on racial criteria.

applied anthropology The subdiscipline of anthropology that focuses on the use or application of anthropological data, techniques, and methods to solve problems in the modern world.

arboreal Living in trees.

archaeological sites Places of past human activity that are preserved in the ground.

archaeology The subfield of anthropology that focuses on study of the lifestyles, history, and evolution of human societies through examination of the artifacts, traces, and physical remains they have left behind.

Archaic A post-Pleistocene hunting-and-gathering adaptation in the Americas characterized by tools suitable for broad-spectrum collecting and more intensive exploitation of localized resources. Corresponds in terms of general trends with the Mesolithic of Europe.

archaic *Homo sapiens* The earliest forms of *Homo sapiens,* dating back more than 200,000 years. Characterized by some anatomical features found in earlier hominid species, such as *Homo erectus,* but not in anatomically modern humans.

artifacts Any object made or modified by humans, including everything from chipped stone tools and pottery to plastic soft-drink bottles and computers.

artificial selection The process in which people select and direct the breeding of certain plant and animal species to produce characteristics that better meet human needs or that are desirable to humans.

balanced polymorphism   The mixture of homozygous and heterozygous genes of a population in a state of equilibrium.

band   The least complex and, most likely, the oldest form of political system; characterized by small groups, mostly consisting of family or extended family members.

baton method   A method of percussion flaking of tools that involves the use of bone or antler. This provides more precise control than is possible with a hammer stone.

beliefs   Specific cultural conventions concerning true or false assumptions shared by a particular group.

bifacial flaking   The stone tool manufacturing technique whereby flakes or chips are removed from both sides of a stone to produce a sharp edge.

bipedalism   The ability to walk erect on two legs.

brachiation   An arm-over-arm suspensory locomotion used by some primates to move in trees.

broad-spectrum collecting   The exploitation of varied food resources in local environments.

catastrophism   A theory that suggests that many species have disappeared since the time of creation because of major catastrophes such as floods, earthquakes, and other major geological disasters.

central place theory   The theory in geography and archaeology that, given uniform topography, resources, and opportunities, the distribution of sites within a region should be perfectly regular. Political and economic centers should be located an equal distance from one another, and each in turn would be surrounded by a hierarchical arrangement of smaller settlements, all evenly spaced. Developed by the German geographer Walter Christaller in the 1930s.

chief   A person who owns, manages, and controls the basic productive factors of the economy and has privileged access to strategic and luxury goods.

chiefdom   A political system with a formalized and centralized leadership, headed by a chief. Chiefdom political systems are more complex than tribal societies in that they are formalized and centralized. Despite their size and complexity, though, chiefdoms are still fundamentally organized by kinship principles.

chromosomes   Structures within the cell nucleus that contain the hereditary information.

clinal distribution   Plotting the varying distribution of physical traits in various populations on maps by clines, or zones.

clines   The zones on a map used to plot physical traits of populations.

Clovis culture   A Paleo-Indian tool tradition in the Americas characterized by distinctive spear points and associated with the hunting of late Pleistocene megafauna.

Clovis-first hypothesis   The theory that maintains that the Clovis culture represents the initial human settlement of the Americas.

composite tools   Tools made from several components, such as harpoons or spears.

conflict theories   Theories that argue that state-level organization is beneficial only to the ruling elite and is generally costly to subordinate groups such as the peasantry.

context   The specific location in the ground of an artifact or fossil and all associated materials.

continental drift   The separation of continents that occurred over millions of years as a result of the geological process of plate tectonics.

continuous variation   A phenomenon whereby variation in a particular trait or characteristic cannot be divided into discrete, readily definable groups but varies continuously from one end of the spectrum to the other.

correlation   The simultaneous occurrence of two variables.

cosmologies   Culturally specific ideas that present the universe as an orderly system, including answers to basic questions about the place of humankind in the universe.

cultivation   The systematic planting and harvesting of plants to support the subsistence activities of a population.

cultural anthropology   The subdiscipline of anthropology that focuses on the study of contemporary societies. Also called *ethnology*.

cultural patrimony   The ownership of cultural properties, such as human remains, artifacts, monuments, sacred sites, and associated cultural materials.

cultural relativism   The view that cultural traditions differ from one another because of their unique historical circumstances, and that they should be viewed and evaluated from that perspective.

cultural resource management   The attempt to protect and conserve artifacts and archaeological resources for the future.

culture   A shared way of life that includes material products, values, beliefs, and norms that are transmitted within a particular society from generation to generation.

datum point   A reference point in an archaeological excavation, often some permanent feature or marker, from which all measurements of contour, level, and location are taken.

deductive method A scientific research method that begins with a general theory, develops a specific hypothesis, and tests it.

dendrochronology A numerical dating technique based on the varying pattern of environmental conditions preserved in the annual growth rings of trees.

dentition The number, form, and arrangement of teeth.

deoxyribonucleic acid (DNA) A chain of chemicals contained in each chromosome that produces physical traits that are transmitted to the offspring during reproduction.

diffusionism The spread of cultural traits from one society to another.

diurnal Active during the day.

division of labor The specialization of economic roles and activities within a society.

domestication The systematic, artificial selection of traits in plants or animals to make them more useful to human beings.

dominance hierarchy The relative social status or ranking order found in some primate social groups.

dominant The form of a gene that is expressed in a heterozygous pair.

drives Basic, inborn, biological urges that motivate human behavior.

ecological niche The specific environmental setting to which an organism is adapted.

economy The social relationships that organize the production, distribution, and exchange of goods and services.

egalitarian A type of social structure that emphasizes equality among people with different statuses.

enculturation The process of social interaction through which people learn their culture.

environmental niche A locale that contains various plants, animals, and other ecological conditions to which a species must adjust.

epidemiology The study of disease patterns and their causes.

ethnicity Cultural differences among populations based on attributes such as language, religion, lifestyle, and cultural ideas regarding common descent or specific territory.

ethnoarchaeology A subdiscipline within archaeology that concentrates on the collection and interpretation of ethnographic data to interpret archaeological materials.

ethnocentrism The practice of judging another society by the values and standards of one's own.

ethnography A description of a society written by an anthropologist who conducted field research in that society.

ethnology The subfield of anthropology that focuses on the study of different contemporary cultures throughout the world. Also called *cultural anthropology*.

evolution Process of change within the genetic makeup of a species over time.

experimental studies Studies involving the replication of tools or activities to infer how ancient tools may have been made and used.

faunal correlation The dating of fossils through the comparison of similar fossils from better dated sequences.

faunal succession The principle of faunal succession, literally "animal" succession, recognizes that life forms change through time. First noted by the English geologist William Smith.

features Nonmovable artifacts or traces of past human activity, such as an ancient fire hearth, a pit dug in the ground, or a wall.

fission-track dating A numerical dating method based on the decay of an unstable isotope of uranium. Used to date volcanic rocks.

flotation A specialized recovery technique in archaeology in which material from an excavation is placed in water to separate soil from organic remains, such as plants, seeds, and charcoal.

foramen magnum The opening in the base of the skull through which the spinal cord passes.

forensic anthropology A subdiscipline within physical anthropology that deals with the identification of human skeletal remains for legal purposes.

fossil localities Places where fossils are found. These may be locations where predators dropped animals they killed, places where creatures were naturally covered by sediments, or sites where early humans or primates actually lived.

fossils The preserved remains, impressions, or traces of living creatures from past ages. They form when an organism dies and is buried by soft mud, sand, or silt.

founder effect A type of genetic drift resulting from the randomly determined genetic complement present in the founders of an isolated population.

gametes Sex cells (such as egg and sperm in humans). They contain only half of the chromosomes found in ordinary body, or somatic, cells.

gender Specific behavioral traits attached to each sex by a society and defined by culture.

gene A discrete unit of hereditary information that determines specific physical characteristics or traits of an organism.

gene flow The exchange of genes between populations as a result of interbreeding.

gene pool The total collection of all the alleles within a particular population.

genetic drift Change in allele frequencies within a population as a result of random processes of selection.

genetics The study of genes, the units of heredity.

genotype The specific genetic constitution of an organism.

gestation period The length of time the young spend in the mother's womb.

gradualistic theory, or **gradualism** A theory suggesting that the rate of evolutionary change is relatively constant and occurs gradually over a long time.

Hardy-Weinberg theory of genetic equilibrium An idealized mathematical model that sets hypothetical condtions under which no evolution is taking place. Developed independently by G. H. Hardy and W. Weinberg, the model is used to evaluate evolutionary processes operating on a population.

heredity The genetic means through which traits within populations of plants and animals are passed from one generation to another.

heterozygous Having two different alleles in a gene pair.

hieroglyphic writing *hiero* meaning "sacred," and *gyphein* meaning "carving." An early pictographic writing system in which symbols denoted the ideas and concepts.

historical linguistics The comparison and classification of different languages to explore the historical links among them.

holistic A broad comprehensive approach to the study of humankind drawing from all of the subfields of anthropology and integrating both biological and cultural phenomena.

home bases Archaeological sites that, according to some theoretical interpretations, were locations where early hominids gathered for food sharing, child care, and social interaction.

hominids Members of the family Homindea in the order Primates, which includes modern humans and their direct ancestors who share bipedal locomotion.

hominoids Members of the primate superfamily Hominoidea, which includes all apes and hominids.

homology Traits that have a common genetic origin but may differ in form and function.

homozygous Having the same alleles in a gene pair.

hunter-gatherer A society that depends on hunting and gathering gathering of wild plants and animals for subsistence.

hybrid An organism produced by parents with different features, that is, a heterozygous organism.

hypothesis A testable proposition concerning the relationship among different variables within the collected data.

ideographic writing system An early form of writing in which simple pictures are used to communicate ideas, an individual picture expressing each idea. In actuality, this system involves neither language nor writing.

ideology Cultural symbols and beliefs that reflect and support the interests of specific groups within a society.

inductive method A method of investigation in which a scientist first makes observations and collects data and then formulates hypotheses.

instincts Fixed, complex, genetically based, unlearned behaviors that promote the survival of a species.

integrationist theories A variety of theories that argue that state organization is, on the whole, advantageous and beneficial to all members of a society.

intelligence The capacity to process and evaluate information for problem solving.

knowledge The storage and recall of learned information.

knuckle walking A quadrupedal form of locomotion using the hind limbs and the knuckles of the front limbs.

lactase deficiency The inability to digest lactose, the sugar found in milk.

language A system of symbols with standard meanings through which members of a society communicate.

law of supraposition States that in any succession of rock layers, the lowest rocks were deposited first and the upper rocks have been in place for progressively shorter periods. This assumption forms the basis of stratigraphic dating.

Levalloisian technique A refined type of percussion flaking used during the Middle Paleolithic.

Linguistic anthropology The subdiscipline of anthropology dealing with the study of language.

linguistics The study of language.

Lower Paleolithic The earliest stage of the Stone Age characterized by the Oldowan and Acheulian stone tool industries, roughly spanning the period between 2.4 million and 200,000 years ago and including the tools produced by pre-*Homo sapiens* hominids.

material culture The physical traces, including artifacts, features, and archaeological sites, that a society produces.

medical anthropology The study of disease, health care systems, and theories of disease and curing in different societies.

megaliths Large stone complexes such as those found at Stonehenge that were used for burial chambers or astronomical observations. Such structures are the principal defining characteristic of the Megalithic, a Neolithic tradition found in western Europe and Britain.

meiosis The process by which gametes, which contain only half the number of chromosomes present in the original cell, are formed.

Mesolithic The Middle Stone Age of the Old World. A post-Pleistocene hunting-and-gathering adaptation characterized by tools suitable for broad-spectrum collecting and more intensive exploitation of localized resources. Corresponds in terms of general trends with the Archaic of the Americas.

microliths Small flakes of stone probably used for harpoon barbs and specialized cutting tools during the Mesolithic.

Middle Paleolithic A stage within the Paleolithic characterized by innovations in tool technology, particularly the Levalloisian technique. Dated to approximately 200,000 to 40,000 years ago and associated with archaic *Homo sapiens.*

middens Ancient dumps or trash heaps.

migration rate The rate at which people move into and out of a specific territory.

mitosis The process by which somatic cells divide to produce new cells with the full number of chromosomes present in the original cell.

molecular dating A system for dating the divergence among different species by comparing amino acid sequences or DNA material.

Mousterian A Middle Paleolithic stone tool tradition associated with Neandertals in Europe.

multiregional evolutionary model The view that *Homo sapiens* evolved from *Homo erectus* concurrently in different regions of the world.

multivariate analysis A complex form of analysis that examines the distributions and interrelations among multiple variables or characteristics as, for example, patterns of disease, blood groups, and demographics in human populations.

mutation A change in the genotype of an individual through the alteration of the chromosomes or DNA.

myth Assumed knowledge about the universe, the natural and supernatural worlds, and humanity's place within these worlds.

National Historic Preservation Act Passed in 1966, this act and subsequent legislation provide critical support for the protection of archaeological sites via the National Register and integrates state and local agencies into a national program for site preservation.

Native American Graves Protection and Repatriation Act Passed in 1990, this law and related legislation require all federal agencies to consult with Native American and Native Hawaiian groups concerning existing collections and the intentional excavation of Native American burials and associated cultural remains and repatriate such materials if requested.

natural selection The evolutionary process resulting from the natural environment acting on traits within a species.

Neolithic Literally the "New Stone Age," it was first characterized by polished stone tools. Later it came to refer to the shift from food gathering to food production, as well as a suite of other characteristics and cultural features such as sedentary village life, animal husbandry, and pottery production. Though the term is still used, this package of characteristics is clearly not associated in many cultural settings.

niche A locale that contains various plants, animals, and other ecological phenomena to which a species must adjust.

nocturnal Active during the night.

nonmaterial culture Intangible products of human society, including values, beliefs, and knowledge.

norms Shared rules that define how people are supposed to behave under certain circumstances.

numerical dating Dating technique that provides an assessment of the age of a fossil, artifact, or geological feature in years before present. Also sometimes called *absolute* or *chronometric dating.*

Oldowan The oldest stone tool industry, dating back 2.4 million years in Africa, characterized by simple hammerstones, choppers, and cutting tools.

omnivorous Possessing a diverse, generalized diet consisting of plants, fruits, nuts, seeds, insects, and animals.

opposable thumb A highly flexible thumb that can touch the tips of the fingers; it is characteristic of the human primate.

paleoanthropology The study of human evolution and the behavior of early human ancestors through the analysis of fossil remains, the ancient environment and associated artifacts.

paleoecology *paleo,* meaning "old," and *ecology,* meaning "study of environment." The area of research focusing on the reconstruction and interpretation of ancient environments.

paleoethnobotany The field of study that specializes in the interpretation of the relationship between plants and ancient human populations.

Paleolithic The period of time characterized by the use of stone tools. Often divided on the basis of innovations in tool technology. *See* Lower, Middle, and Upper Paleolithic.

palynology The study of pollen grains, the minute male reproductive part of plants. It may provide a means of reconstructing past environments and of relative dating.

participant observation The method by which the ethnologist learns the culture of the group being studied by participating in the group's daily activities.

pastoralists Groups whose subsistence activities center on the care of domesticated animals.

percussion flaking The production of tools by striking a stone with a hammerstone or other strong object to remove chips or flakes.

phenotype The external, observable physical characteristics of an organism that result from the interaction of the organisms genetic make-up (the genotype) and its distinctive life history.

plate tectonics The gradual movement of plates on the earth's surface as a result of geological processes, one consequence of which is the movement of continents or continental drift.

polymorphism A trait that exhibits variation within a species.

polytypic A species exhibiting physical variation in different regional populations.

population A group of organisms that can potentially interbreed and occupy a given territory at the same time.

postorbital constriction A feature of *Homo erectus* in which the front portion of the skull is narrow and the forehead is low.

potassium-argon dating A numerical dating method based on the decay of an unstable isotope of potassium into the inert gas, argon. It is used by paleoanthropologists to date volcanic rocks.

pre-Clovis hypothesis The hypothesis that the initial settlement of the Americas took place prior to the arrival of the populations represented by Clovis spear points and associated artifacts.

primates A diverse order of mammals, including humans, monkeys, and apes, that share similar characteristics.

primatology The study of primates.

proton magnetometer A sensor that can detect differences in the soil's magnetic field caused by buried features and artifacts.

punctuated equilibrium The theory of evolution that species remain relatively stable for long periods, with major changes (punctuations) and new species arising very

rapidly as a result of mutations or changes in selective pressures.

quadrupeds Species that use all four limbs for locomotion.

qualitative data Nonstatistical information that tends to be the most important aspect of ethnological research.

quantitative data Data that can be expressed as numbers, including census materials, dietary information, and income and household-composition data.

races Divisions within a species based on identifiable hereditary traits.

racism Beliefs and practices that advocate the superiority of certain races and the inferiority of others.

radioactive decay A process in which radioisotopes, unstable atoms of certain elements, break down or change.

radiocarbon dating A numerical dating technique based on the decay of the unstable isotope carbon-14. Can be used to date organic material such as charcoal, wood fragments, or skeletal material as much as 80,000 years old.

random sample A nonsubjective, representative sample of people, artifacts, or attributes.

recessive Designating a gene that is unexpressed when occurring in a heterozygous pair with a dominant form.

relative dating A variety of dating methods that can be used to establish the age of a fossil, artifact, or geological feature relative to another.

repatriation The return of human remains or cultural property to a descendant community or interest group. *See* Native American Graves Protection and Repatriation Act.

replacement model The paleoanthropological theory that *Homo sapiens* evolved in one world area and then expanded, replacing regional populations of earlier hominids.

research design A proposal in which the objectives of a project are set out and the strategy for recovering the relevant data is outlined.

residue studies The examination of tools for minute traces, or residues, of the materials they were used to cut or work.

resistivity The measurement of variation in the electrical current passing between electrodes placed in the ground that indicates differences in the soil's moisture content, which in turn reflects buried ditches, foundations, or walls that retain moisture to varying degrees.

rituals Repetitive behaviors that communicate sacred symbols to members of society.

sagittal crest A bony ridge along the top of the skull.

scientific method   A method used to investigate the natural and social world that relies on critical thinking, logical reasoning, and skeptical thought.

seriation   A relative dating method based on the assumption that any particular artifact, attribute, or style will appear, gradually increase in popularity until it reaches a peak, and then progressively decrease.

sex   Biological and anatomical differences between males and females.

sexual dimorphism   The presence of different characteristics in males and females of the same species.

shamans   Part-time religious specialists who are believed to be linked with supernatural beings and powers.

silica gloss   A distinctive residue left on stone blades used by people to harvest plants.

sites   *See* archaeological sites.

social grooming   A social activity of many primates that involves removing ticks, fleas, dirt, and debris from one another. It may serve as an important means of maintaining sociality.

social learning   A type of learning in which an organism observes another organism's response to a stimulus and then adds that response to its own collection of behaviors.

social stratification   Inequality of statuses within a society.

social structure   The sum of the patterns of relationships within a society.

society   A group of people who reside within a specific territory and share a common culture.

sociobiology   The systematic study of the biological basis of social behavior, including the development of culture.

sociolinguistics   The subdiscipline within linguistics that deals with the use of language in various social and cultural settings to explore the links between language and social behavior.

somatic cells   Body cells; unlike gametes (sex cells), they have the full number of chromosomes.

specialization   The organization of production whereby certain individuals concentrate, or specialize, in certain activities or types of production, such as pottery making, stone carving, or metalworking.

speciation   The development of new species.

species   Groups of organisms with similar physical characteristics that can potentially interbreed successfully.

state   A form of political system with centralized, institutionalized bureaucracy or government. In contrast to bands, tribes, and chiefdoms, states systems are not principally based on kinship.

status   A recognized position that a person occupies within society.

stone caches   Stockpiles of stone tools and unprocessed stone that have been deposited in a particular area by early hominids.

strata   A group of equivalent statuses based on ranked divisions within a society.

stratigraphic dating   A form of relative dating by assessing whether a layer of rock is recent or old.

structural linguistics   An area of research that investigates the structure of language patterns as they presently exist.

subsistence patterns   The means by which people obtain their food supply.

survey   An examination of a particular area, region, or country to locate archaeological sites or fossil localities.

taphonomy   The study of the variety of natural and behavioral processes that led to the formation of a fossil locality. This may include traces of the activities of early human ancestors, as well as natural agencies such as erosion, decay, and animal activities.

taxonomy   The science of classification, which provides scientists a convenient way of referring to, and comparing, living and extinct organisms.

technology   All the human techniques and methods of reaching a specific goal in subsistence or in modifying or controlling the natural environment.

terrestrial   Living on land.

theories   General, overarching statements that explain hypothesis about natural or social phenomena.

trace element analysis   The study of elements in artifacts that may provide a distinctive signature of the artifacts' origin.

transitional forms   The hominid fossils that are either advanced *Homo erectus* or early *Homo sapiens*.

tribe   Political organization based on kinship, age, or gender divisions but without formal, centralized political institutions. The political organization of tribes is more complex than that of bands but not as complex as that of chiefdoms and states.

uniformitarianism   The geological view that the earth's geological features are the result of gradual, natural processes that can still be observed.

unilineal evolution   The belief, widespread during the nineteenth century, that societies were evolving in a single direction toward complexity and industrial progress.

Upper Paleolithic   The late Stone Age, dating back to about 40,000 years ago.

use-wear Studies of the damage or traces of use present on tools. Such traces may provide indications of how a tool was used.

unifacial flaking The method of stone tool manufacture whereby flakes or chips are removed from one side of the stone to produce a sharp edge.

variable Specific characteristics or phenomena that vary from case to case, such as temperature, age, size, color, and the like.

vegiculture The propagation of plants by selectively dividing and replanting living plants.

worldview An integrated system of beliefs and cosmologies about natural and supernatural realities.

References

Abu-Lughod, Lila. 1987. *Veiled Sentiments: Honor and Poetry in a Bedouin Society*. Berkeley: University of California Press.

ADAMS, ROBERT McC. 1966. *The Evolution of Urban Society*. Chicago: Aldine.

————. 1974. "Anthropological perspectives on ancient trade." *Current Anthropology,* 15(3), 239–258.

————. 1981. *Heartland of Cities*. Chicago: Aldine.

AGAR, MICHAEL. 1973. *Ripping and Running.* New York: Academic Press.

————. 1980. *Cognition and Ethnography.* Minneapolis, MN: Burgess.

AIKENS, C. MELVIN, & HIGUCHI, TAKAYASU. 1981. *Prehistory of Japan.* New York: Academic Press.

ALGAZE, GUILLERMO. 1993. "Expansionary dynamics of some early pristine states." *American Anthropologist,* 95(2), 304–333.

ALVAREZ, L.W., ET AL. 1980. "Extraterrestrial cause for the Cretaceous-Tertiary extinction." *Science,* 208, 1095-1108.

ANDERSON, DAVID G. & HORAK, VIRGINIA (Eds.). 1995. *Archaeological Site File Management: A Southeastern Perspective. Readings in Archaeological Resource Protection Series No.3.* Atlanta, GA: Interagency Archaeological Service Division.

ANDRESEN, JOHN M., BYRD, BRIAN F., ELSON, MARK D., McGUIRE, RANDALL H., MENDOZA, RUBEN G., STASKI, EDWARD, & WHITE, J. PETER. 1981. "The deer hunters: Star Carr reconsidered." *World Archaeology,* 13(1), 31–46.

ANDREWS, PETER, & MARTIN, LAWRENCE. 1987. "Cladistic relationships of extant and fossil hominoid primates." *Journal of Human Evolution,* 16, 101–118.

ARDEN, HARVEY. 1989. "Who owns our past?" *National Geographic,* 75(3), 378.

ARDREY, ROBERT. 1961. *African Genesis.* New York: Dell.

AUEL, JEAN M. 1981. *The Clan of the Cave Bear.* New York: Bantam.

BAILEY, R.C., HEAD, G., JENIKE, M., OWN, B., RECHTMAN, T., & ZECHENTER, E. 1989. "Hunting and gathering in tropical rain forests: Is it possible?" *American Anthropologist,* 91(1), 59–82.

BARASH, DAVID P. 1987. *The Hare and the Tortoise: Culture, Biology, and Human Nature.* New York: Penguin.

BASS, GEORGE FLETCHER. 1963. "Underwater archaeology: Key to history's warehouse." *National Geographic,* 124(1), 138-156.

————. 1973. *Archaeology Beneath the Sea.* New York,: Harper & Row.

BEALS, K.L. 1972. "Head form and climatic stress." *American Journal of Physical Anthropology,* 37, 85–92.

BEARD, K.C., TEAFORD, M.F., & WALKER, A. 1986. "New wrist bones of *Proconsul africanus* and *P. nyanzae* from Rusinga Island, Kenya." *Folia Primatology,* 47, 97–118.

BEAUDRY, MARY C. (Ed.). 1988. *Documentary Archaeology in the New World.* Cambridge: Cambridge University Press.

BENTON, M.J. 1986. "More than one event in the late Triassic mass extinction." *Nature,* 321, 857–861.

BINFORD, LEWIS. 1968. "Post-Pleistocene adaptations." In Lewis R. Binford & Sally Binford (Eds.), *New Perspectives in Archaeology* (pp. 313–341). New York: Academic Press.

————. 1985. "Human ancestors: Changing views of their behavior." *Journal of Anthropological Archaeology,* 1, 5–31.

BINFORD, LEWIS R., & BINFORD, SALLY. 1966. "A preliminary analysis of functional variability in the Mousterian of Levallois facies." *American Anthropologist,* 68(2), 238–295.

BINFORD, LEWIS R., & HO, CHUAN KUN. 1985. "Taphonomy at a distance: Zhoukoudian, the cave home of Beijing man?" *Current Anthropology,* 26, 413–442.

BIRDSELL, JOSEPH B. 1981. *Human Evolution* (3rd ed.). Boston: Houghton Mifflin.

BISHOP, J.A., COOK, L.M., & MUGGLETON, J. 1978. "The response of two species of moths to industrialization in northwest England: II. Relative fitness of moths and population size." *Philosophical Transactions of the Royal Society of London,* B 281, 517–542.

BLAKELY, ROBERT L., & HARRINGTON, JUDITH M. 1997. *Bones in the Basement: Postmortem Racism in Nineteenth-Century Medical Training.* Washington, DC: Smithsonian Institution Press.

BLINDERMAN, C. 1986. *The Piltdown Inquest.* Buffalo, NY: Prometheus Books.

BOAZ, NOEL T., & ALMQUIST, ALAN J. 1997. *Biological Anthropology: A Synthetic Approach to Human Evolution.* Upper Saddle River, NJ: Prentice Hall.

BOGIN, B.A. 1978. "Seasonal pattern in the rate of growth in height of children living in Guatemala." *American Journal of Physical Anthropology,* 49, 205–210.

BORDES, FRANCOIS. 1968. *The Old Stone Age*. New York: McGraw-Hill.

BOSERUP, ESTER. 1965. *The Conditions of Agricultural Growth: The Economics of Agrarian Change Under Population Pressure*. Chicago: Aldine.

BOURGOIS, PHILIPPE. 1989. "Crack in Spanish Harlem: Culture and economy in the inner city." *Anthropology Today*, 5(4), 6–11.

———. 1996. *In Search of Respect: Selling Crack in El Barrio*. Cambridge: Cambridge University Press.

BOWER, BRUCE. 1990. "Biographies etched in bone." *Science News*, 138, 106–108.

———. 1991. "Fossil finds expand early hominid anatomy." *Science News*, 139, 182.

BRACE, C.L. 1964. "The fate of the 'Classic' Neanderthals: A consideration of hominid catastrophism." *Current Anthropology*, 5, 3–43.

———. 1967. *The Stages of Human Evolution: Human and Cultural Origins*. Englewood Cliffs, NJ: Prentice Hall.

———. 1989. "Medieval thinking and the paradigms of paleoanthropology." *American Anthropologist*, 91(2), 442–446.

BRACE, C.L., & MONTAGU, M.F.A. 1965. *Man's Evolution: An Introduction to Physical Evolution*. New York: Macmillan.

BRAIDWOOD, ROBERT J. 1960. "The agricultural revolution." *Scientific American*, 203, 130–141.

BROWN, JAMES A. (ED.). 1971. "Approaches to the social dimensions of mortuary practices." *Memoirs of the Society for American Archaeology* No. 25.

BROWN, MICHAEL H. 1990. *The Search for Eve*. New York: HarperCollins.

BROWN, T. M., & ROSE, K. D. 1987. "Patterns of dental evolution in early Eocene Anaptomorphine Commomyodael from the Bighorn Basin, Wyoming." *Journal of Paleontology*, 61, 1-62.

BRUES, A.M. 1977. *People and Races*. New York: Macmillan.

BRUMFIEL, ELIZABETH. 1983. "Aztec state making: Ecology, structure, and the origin of the state." *American Anthropologist*, 85(2), 261–284.

———. 1991. "Weaving and cooking: Women's production in Aztec Mexico." In Joan M. Gero & Margaret Conkey (Eds.), *Engendering Archaeology: Women and Prehistory* (pp. 224-251). Cambridge: Blackwell.

BUTZER, KARL W. 1984. "Long-term Nile flood variation and political discontinuities in Pharonic Egypt." In J. Desmond Clark & Steven A. Brandt (Eds.), *From Hunters to Farmers: The Causes and Consequences of Food Production in Egypt* (pp. 102–112). Berkeley: University of California Press.

BYARD, P.J. 1981. "Quantitative genetics of human skin color." *Yearbook of Physical Anthropology*, 24, 123–137.

CAMPBELL, BERNARD G. 1987. *Humankind Emerging* (5th ed.). Glenview, IL: Scott, Foresman.

CANN, R.L., BROWN, W.M., & WILSON, A.C. 1987. "Mitochondrial DNA and human evolution." *Nature*, 325, 31–36.

CARLISLE, R. (ED.). 1988. *Americans Before Columbus: Perspectives on the Archaeology of the First Americans*. Pittsburgh: University of Pittsburgh Press.

CARLISLE, R.C., & ADAVASIO, J.M. 1982. *Collected Papers on the Archaeology of Meadowcroft Rockshelter and the Cross Creek Drainage*. Pittsburgh: University of Pittsburgh, Department of Anthropology.

CARNEIRO, ROBERT. 1970. "A theory of the origin of the state." *Science*, 169, 733–738.

CARTER, GEORGE F. 1988. "Cultural historical diffusion." In Peter J. Hugill & Bruce D. Dickson (Eds.), *The Transfer and Transformation of Ideas and Material Culture*. College Station: Texas A&M University Press.

CHAGNON, NAPOLEON A. 1974. *Studying the Yanomanö*. New York: Holt, Rinehart & Winston.

———. 1997. *Yanomanö* (5th ed.). New York: Harcourt Brace.

CHAMBERS, ERVE. 1985. *Applied Anthropology: A Practical Guide*. Englewood Cliffs, NJ: Prentice Hall.

CHANG, KWANG-CHIH. 1970. "The beginnings of agriculture in the Far East." *Antiquity*, 44(175), 175–185.

———. 1975. "From archaeology to history: The Neolithic foundations of Chinese civilization." In Chang Chun-shu (Ed.), *The Making of China: Main Themes in Premodern Chinese History* (pp. 38–45). Englewood Cliffs, NJ: Prentice Hall.

———. 1976. *Early Chinese Civilization: Anthropological Perspectives*. Cambridge, MA: Harvard University Press.

———. 1986. *The Archaeology of Ancient China* (4th ed.). New Haven, CT: Yale University Press.

CHAPMAN, ROBERT, KINNES, IAN, & RANDSBORG, KLAVS (Eds.). 1981. *The Archaeology of Death*. Cambridge: Cambridge University Press.

CHASE, P.G., & DIBBLE, H.L. 1987. "Middle Paleolithic symbolism: A review of current evidence and interpretations." *Journal of Anthropological Archaeology*, 6, 263–293.

CHILDE, V. GORDON. 1936. *Man Makes Himself*. London: Watts.

———. 1950. "The Urban Revolution" *Town Planning Review*, 21, 3–17.

———. 1952. *New Light on the Most Ancient East*. London: Routledge & Kegan Paul.

CHRISTALLER, WALTER. 1933. *Die Zentralen Orte in Suddeuyschland. Jena*, Germany: Karl Zeiss.

CIOCHON, R.L., & CORRUCCINI, R. (Eds.). 1983. *New Interpretations of Ape and Human Ancestry*. New York: Plenum Press.

CIOCHON, R.L., OLSEN, J., & JAMES, J. 1990. *Other Origins: The Search for the Giant Ape in Human Prehistory*. New York: Bantam Books.

CLARK, GRAHAME, & PIGGOTT, STUART. 1965. *Prehistoric Societies*. New York: Knopf.

CLARK, J. DESMOND. 1970. *The Prehistory of Africa*. New York: Praeger.

CLARK, J. G. D. 1979. *Mesolithic Prelude*. Edinburgh: Edinburgh University Press.

CLARK, J. DESMOND, & BRANDT, STEVEN A. 1984. *From Hunters to Farmers: The Causes and Consequences of Food Production in Africa*. Berkeley: University of California Press.

CLARK, J.D., & HARRIS, J.W.K. 1985. "Fire and its roles in early hominid lifeways." *African Archaeological Review*, 3, 3–27.

CLEMENS, W.A. 1974. "Purgatorius, an early paromomyid primate mammalia." *Science*, 184, 903–906.

COE, MICHAEL. 1977. *Mexico* (2nd ed.). New York: Praeger.

Coe, Michael, Snow, Dean, & Benson, Elizabeth. 1986. *Atlas of Ancient America*. New York: Facts on File.

Cohen, Mark Nathan. 1977. *The Food Crisis in Prehistory*. New Haven, CT: Yale University Press.

Cohen, Mark Nathan, & Armelagos, George J. 1984. *Paleopathology at the Origins of Agriculture*. New York: Academic Press.

Cohen, Ronald. 1978. "Introduction." In R. Cohen & E. Service (Eds.), *Origins of the State: The Anthropology of Political Evolution*. Philadelphia: Institute for the Study of Human Issues.

Cohen, Ronald, & Service, Elman (Eds.). 1978. *Origins of the State: The Anthropology of Political Evolution*. Philadelphia: Institute for the Study of Human Issues.

Connah, Graham. 1987. *African Civilizations: Precolonial Cities and States in Tropical Africa: An Archaeological Perspective*. Cambridge: Cambridge University Press.

Conroy, Glenn C. 1990. *Primate Evolution*. New York: W.W. Norton.

Conroy, Glenn C., & Vanier, Michael. 1990. "Endocranial features of *Australopithecus africanus* revealed by 2- and 3-D computer tomography." *Science*, 247, 839–841.

Cowan, C. Wesley, & Watson, Patty Jo T. B. (Eds.). 1992. *The Origins of Agriculture: An International Perspective*. Washington, DC: Smithsonian Institution Press.

Crawford, G.W. 1992. "Prehistoric plant domestication in East Asia." In C. Wesley Cowan & Patty Jo Watson, *The Origins of Agriculture: An International Perspective* (pp. 7–38). Washington, DC: Smithsonian Institution Press.

Dagosto, M. 1988. "Implications of postcranial evidence for the origin of euprimates." *Journal of Human Evolution*, 17, 35–56.

Dahlberg, Frances (Ed.). 1981. *Woman the Gatherer*. New Haven, CT: Yale University Press.

Daniel, Glyn. 1981. *A Short History of Archaeology*. New York: Thames & Hudson.

Daniels, Peter T., & Bright, William (Eds.). 1995. *The World's Writing Systems*. New York: Oxford University Press.

Darwin, Charles. 1979. *The Illustrated Origin of Species*. New York: Hill & Wang.

Day, Michael H. 1986. *Guide to Fossil Man* (4th ed.). Chicago: University of Chicago Press.

Deacon, H. J. 1992. "Southern Africa and modern human origins." *Philosophical Transactions of the Royal Society of London*, B 337,177-183.

DeCorse, Christopher. 1992. "Culture contact, change, and continuity on the Gold Coast, 1400–1900 A.D." *African Archaeological Review*, 10, 163–196.

———. 1998. "Culture contact and change in West Africa. In James G. Cusick (Ed.), *Studies in Culture Contact: Interaction, Culture Change, and Archaeology* Occasional Paper 25 (pp. 358-377). Bloomington, IN: Center for Archaeological Investigations.

Deetz, James. 1996. *In Small Things Forgotten: An Archaeology of Early American Life*. New York: Anchor Books.

de Lumley, H. 1969. "A Paleolithic camp at Nice." *Scientific American*, 220, 42–59.

Dennell, Robin W. 1992. "The origins of crop agriculture in Europe." In C. Wesley Cowan & Patty Jo Watson (Eds.), *The Origins of Agriculture: An International Perspective* (pp. 71–100). Washington, DC: Smithsonian Institution Press.

Diamond, Jared. 1987. "The worst mistake in the history of the human race." *Discover*, May, pp. 64–66.

Dillehay, Thomas D. 1989. *Monte Verde: A Late Pleistocene Settlement in Chile* (Vol. 1). Washington, DC: Smithsonian Institution Press.

Dillehay, Thomas D., & Meltzer, David J. 1991. *The First Americans: Search and Research*. Boca Raton, FL: CRC Press.

Donnan, Christopher, & Castillo, Luis Jaime. 1992. "Finding the tomb of a Moche priestess." *Archaeology*, 45(6), 38–42.

Donnan, Christopher B., & McClelland, Donna. 1979. *The Burial Theme in Moche Iconography*. Washington, DC: Dunbarton Oaks.

Dorn, R., et al. 1986. "Cation ratio and accelerator radiocarbon dating of rock varnish on archaeological artifacts and land forms in the Mojave Desert." *Science*, 213, 830–833.

Dorn, R., & Whitley, D. 1988. "Cation-ratio dating of petroglyphs using PIXE." *Nuclear Instruments and Methods in Physics Research*, 35, 410–414.

Dubois, E. 1894. "*Pithecanthropus erectus*, transitional form between man and the apes." *Scientific Transactions of the Royal Dublin Society*, 6, 1–18.

Earle, T., & Ericson, J. (Eds.). 1977. *Exchange Systems in Prehistory*. New York: Academic Press.

Ebert, J.I. 1984. "Remote sensing application in archaeology." In Michael Schiffer (Ed.), *Advances in Archaeological Method and Theory* (Vol. 7, pp. 293–362). New York: Academic Press.

Evans, Robert K., & Rasson, Judith A. 1984. "Ex Balkanis lux? Recent developments in Neolithic and Chalcolithic research in southeastern Europe." *American Antiquity*, 49, 713–741.

Fagan, Brian. 1988. "Black Day at Slack Farm." *Archaeology*, 41(4), 15-16, 73.

———. 1992. *Rape of the Nile*. Providence, RI: Moyer-Bell.

———. 1998. *People of the Earth: An Introduction to World Prehistory* (9th ed.). New York: HarperCollins.

Falk, D. 1983. "A reconsideration of the endocast of *Proconsul africanus*: Implications for primate brain evolution." In R.L. Ciochon & R. Corruccini (Eds.), *New Interpretations of Ape and Human Ancestry*. New York: Plenum Press.

Feder, Kenneth L. 1999. *Frauds, Myths, and Mysteries: Science and Pseudoscience in Archaeology* (3rd ed.). Mountain View, CA: Mayfield.

Fedigan, Linda M. 1983. "Dominance and reproductive success in primates." *Yearbook of Physical Anthropology*, 26, 91–129.

———. 1986. "The changing role of women in models of human evolution." *Annual Review of Anthropology*, 15, 25–66.

Fell, Barry. 1980. *Saga America*. New York: Times Mirror.

Flannery, Kent. 1965. "The ecology of early food production in Mesopotamia." *Science*, 147, 1247–1256.

———. 1972. "The cultural evolution of civilizations." *Annual Review of Ecology and Systematics*, 4, 399–426.

———. 1973. "The origins of agriculture." *Annual Review of Anthropology*, 2, 271–310.

———. 1985. "Los origenes de la agricultura en Mexico: Las teorias y la evidencia." In T. Rojas-Rabiela & W.T. Sanders (Eds.), *Historia de la Agricultura: Epoca Prehispanica-Siglo* (XVI, pp. 237–265). Coleccion Biblioteca del INAH. Instituto Nacional de Antropologia e Historia.

FLEAGLE, JOHN G. 1983. "Locomotor adaptations of Oligocene and Miocene hominoids and their phyletic implications." In R. Ciochon & R. Corruccini (Eds.), *New Interpretations of Ape and Human Ancestry* (pp. 301–324). New York: Plenum Press.

———. 1988. *Primate Adaptations and Evolution.* New York: Academic Press.

FLEAGLE, J.G., BOWN, T.M., OBRADOVICH, J.O., & SIMONS, E.L. 1986. "How old are the Fayum primates?" In J.G. Else & P.C. Lee (Eds.), *Primate Evolution* (pp. 133–142). Cambridge: Cambridge University Press.

FLEAGLE, J.G., & KAY, R.F. 1987. "The phyletic position of the Parapithecidae." *Journal of Human Evolution,* 16, 483–531.

FORD, RICHARD I. 1985. "Prehistoric food production in North America." University of Michigan, Museum of Anthropology, Paper, 75.

FOSSEY, DIAN. 1983. *Gorillas in the Mist.* Boston: Houghton Mifflin.

FRAYER, DAVID W., WOLPOSS, MILFORD H., THORNE, ALAN G., SMITH, FRED H., & POPE, GEOFFREY G. 1993. "Theories of modern human origins: The paleontological test." *American Anthropologist,* 95(1): 14–50.

FRIED, MORTON. 1967. *The Evolution of Political Society: An Essay in Political Anthropology.* New York: Random House.

FRISANCHO, A.R. 1979. *Human Adaptation: A Functional Interpretation.* St. Louis, MO: C.V. Mosby.

FUTUYMA, D.J. 1995. *Science on Trial: The Case for Evolution.* New York: Pantheon Books.

GARDNER, HOWARD. 1983. *Frames of Mind: The Theory of Multiple Intelligences.* New York: Basic Books.

GARN, STANLEY. 1971. *Human Races* (3rd ed.). Springfield, IL: Charles C. Thomas.

GILKEY, LANGDON. 1986. "The creationism issue: A theologian's view." In Robert W. Hanson (Ed.), *Science and Creation: Geological, Theological, and Educational Perspectives* (pp. 174-188). New York: Macmillan.

GINGERICH, P.D. 1986. "Plesiadapis and the delineation of the order Primates." In B. Wood, L. Martin, & P. Andrews (Eds.), *Major Topics in Primates and Human Evolution* (pp. 32–46). Cambridge: Cambridge University Press.

———. 1990. "African dawn for primates." *Nature,* 346, 411.

GIVENS, DAVID, & FILLMORE, RANDOLPH. 1990. "AAA pilot survey of nonacademic departments: Where the MA's are." *Anthropology Newsletter,* 31(5).

GLADKIH, M.I., KORNEITZ, H.L., & SOFFER, O. 1984. "Mammoth-bone dwelling on the Russian plain." *Scientific American,* 25(5), 164–175.

GOLDING, WILLIAM G. 1981. *The Inheritors.* New York: Harcourt Brace Jovanovich.

GOLOMB, LOUIS. 1985. *An Anthropology of Curing in Multiethnic Thailand.* Urbana: University of Illinois Press.

GOODALL, JANE VAN LAWICK. 1971. *In the Shadow of Man.* New York: Dell.

———. 1986. *The Chimpanzees of Gombe.* Cambridge, MA: Harvard University Press.

———. 1990. *Through a Window: Thirty Years Observing the Chimpanzees of Gombe.* Boston: Houghton Mifflin.

GOODMAN, ALAN H., MARTIN, DEBRA L., & ARMELAGOS, GEORGE J. 1984. "Indications of stress from bone and teeth." In Mark Nathan Cohen & George J. Armelagos (Eds.), *Paleopathology at the Origins of Agriculture* (pp. 13–49). New York: Academic Press.

GOODMAN, JEFFREY. 1981. *American Genesis.* New York: Berkley.

GOODMAN, M., BABA, M.L., & DARGA, L.L. 1983. "The bearings of molecular data on the cladogenesis and times of divergence of hominoid lineages." In R.L. Ciochon & R. Corruccini (Eds.), *New Interpretations of Ape and Human Ancestry* (pp. 67–86). New York: Plenum Press.

GOODMAN, M., & LASKER, G.W. 1975. "Molecular evidence as to man's place in nature." In R.H. Tuttle (Ed.), *Primate Functional Morphology and Evolution* (pp. 70–101). The Hague: Mouton.

GOODY, JACK. 1987. *The Interface Between the Written and the Oral.* Cambridge: Cambridge University Press.

GORMAN, CHESTER A. 1969. "Hoabhinian: A pebble-tool complex with early plant associations in Southeast Asia." *Science,* 163, 671–673.

———. 1977. "A priori models and Thai prehistory: A reconsideration of the beginning of agriculture in Southeast Asia." In Charles A. Reed (Ed.), *Origins of Agriculture* (pp. 321–355). The Hague: Mouton.

GOSSETT, THOMAS F. 1963. *Race, the History of an Idea in America.* Dallas, TX: Southern Methodist University Press.

GOULD, R.A. 1980. *Living Archaeology.* Cambridge: Cambridge University Press.

GOULD, S.J. 1977. *Ever Since Darwin.* New York: W.W. Norton.

GOULD, S.J., & ELDREDGE, NILES. 1972. "Punctuated equilibrium: The tempo and mode of evolution reconsidered." *Paleobiology,* 3, 115–151.

GOULD, S.J., & LEWONTIN, R. C. 1979. "The spandrels of San Marco and the Panglossian paradigm: A critique of the adaptionist programme." *Proceedings of the Royal Society of London,* B 205, 581-598.

GREENBERG, JOSEPH H. 1986. "The settlement of the Americas." *Current Anthropology,* 27, 477–497.

GREENE, JEREMY. 1990. *Maritime Archaeology: A Technical Handbook.* London: Academic Press.

GUHA, ASHOK S. 1981. *An Evolutionary View of Economic Growth.* Oxford: Clarendon Press.

HAAS, JONATHAN. 1982. *The Evolution of the Prehistoric State.* New York: Columbia University Press.

HABGOOD, PHILLIP. 1985. "The origin of the Australian Aborigines." In Phillip Tobias (Ed.), Hominid evolution: *Past, present, and future.* New York: Alan R. Liss.

HAGLUND, K., & SORG, M.H. 1997. *Forensic Taphonomy: The Postmortem Fate of Human Remains.* Boca Raton, FL: CRC Press.

HALL, MARTIN. 1988. "Archaeology under apartheid." *Archaeology,* 41(6), 62–64.

HALVERSON, JOHN. 1987. "Art for art's sake in the Paleolithic." *Current Anthropology,* 28, 63–89.

HAMILTON, D.L., & WOODWARD, ROBYN. 1984. "A sunken 17th century city: Port Royal, Jamaica." *Archaeology,* 37(1), 38–45.

HARLAN, J.R. 1971. "Agricultural origins: Centers and noncenters." *Science,* 174, 468–474.

———. 1992. "Indigenous African agriculture." In C. Wesley Cowan & Patty Jo Watson (Eds.), *The Origins of Agriculture: An International Perspective* (pp. 59–70). Washington, DC: Smithsonian Institution Press.

HARLOW, HARRY F., & HARLOW, MARGARET K. 1961. "A study of animal affection." *Natural History,* 70, 48–55.

HARRINGTON, SPENCER P.M. 1991. "The looting of Arkansas." *Archaeology,* 44(3), 22–30.

———. 1992. "Shoring up the Temple of Athena." *Archaeology,* 45(1), 30-43.

HASSAN, FEKRI A. 1981. *Demographic Archaeology.* New York: Academic Press.

HAWKINS, GERALD S. 1965. *Stonehenge Decoded.* New York: Doubleday.

HAYNES, C.V., JR. 1991. "Geoarchaeological and paleohydrological evidence for a Clovis-age drought in North America and its bearing on extinction." *Quartinary Research,* 35, 438–450.

HENRY, DONALD O. 1984. "Preagricultural sedentism: The Natufian example." In Douglas T. Price & James A. Brown (Eds.), *Prehistoric Hunter-Gatherers: The Emergence of Cultural Complexity* (pp. 365–384). New York: Academic Press.

HERRNSTEIN, RICHARD J., & MURRAY, CHARLES. 1994. *The Bell Curve: Intelligence and Class Structure in American Life.* New York: Free Press.

HERSCHER, ELLEN. 1989. "A future in ruins." *Archaeology,* 42(1). 67-70.

HOLLOWAY, R.L. 1985. "The poor brain of *Homo sapiens neanderthalensis:* See what you please." In E. Delson (Ed.), *Ancestors: The Hard Evidence* (pp. 319–324). New York: Alan R. Liss.

HOPKINS, KEITH. 1982. "Aspects of the paleogeography of Beringia during the late Pleistocene." In D. Hopkins, J. Matthews, C. Schweger, & S. Young (Eds.), *The Paleoecology of Beringia* (pp. 3–28). New York: Academic Press.

HOWELLS, W.W. 1976. "Explaining modern man: Evolutionists versus migrationists." *Journal of Human Evolution,* 5, 477–496.

HOYLE, FRED. 1977. *On Stonehenge.* San Francisco: W.H. Freeman.

IRWIN, G. 1993. *Prehistoric Exploration and Colonization of the Pacific.* Cambridge: Cambridge University Press.

ISAAC, BARBARA. 1995. "An epimethean view of the future at the Peabody Museum of Archaeology and Ethnology at Harvard University." *Federal Archaeology,* Offprint Series, Fall/Winter, 18-22.

ISAAC, G.L. 1978. "The Food-Sharing Behavior of Protohuman Hominids." *Scientific American,* 238(4), 90–108.

———. 1984. "The archaeology of human origins: Studies of the Lower Pleistocene in East Africa 1971–1981." In Fred Wendorf & Angela E. Close (Eds.), *Advances in World Archaeology* (Vol. 3, pp. 1–87). New York: Academic Press.

ISÇAN, M.Y.S., & KENNEDY, K.A.R. 1989. *Reconstruction of Life from the Skeleton.* New York: Alan R. Liss.

JACKSON, DONALD DALE. 1992. "How Lord Elgin first won—and lost—his marbles." *Smithsonian,* 23(9), 135-146.

JENNINGS, JESSE D. 1989. *Prehistory of North America.* Mountain View, CA: Mayfield.

JENSEN, ARTHUR. 1980. *Bias in Mental Testing.* New York: Free Press.

JETT, STEPHEN C. 1978. "Precolumbian transoceanic contacts." In Jesse D. Jennings (Ed.), *Ancient North Americans.* San Francisco: W.H. Freeman.

JOHANSON, DONALD, & EDEY, MAITLAND. 1981. *Lucy: The Beginnings of Humankind.* New York: Simon & Schuster.

JOHANSON, DONALD C., & SHREEVE, JAMES. 1989. *Lucy's Child: The Discovery of a Human Ancestor.* New York: Avon Books.

JOHANSON, DONALD, & WHITE, TIMOTHY. 1979. "A systematic assessment of early African hominids." *Science,* 203, 321–330.

JOHANSON, D.C., WHITE, T.D., & COPPENS, Y. 1978. "A new species of the genus *Australopithecus* (Primates: Hominidae) from the Pliocene of eastern Africa." *Kirtlandia,* 28, 1–14.

JOHANSON, DONALD C., ET AL. 1982. "Pliocene hominid fossils from Hadar, Ethiopia. *American Journal of Physical Anthropology,* 57(4), 1–719.

JOHNSON, ALLEN, & EARLE, TIMOTHY. 1987. *The Evolution of Human Societies: From Foraging Group to Agrarian State.* Stanford, CA: Stanford University Press.

JOHNSON, RONALD W., & SCHENE, MICHAEL G. 1987. *Cultural Resources Management.* Malabar, FL: Robert E. Krieger.

JOLLY, ALISON. 1985. *The Evolution of Primate Behavior* (2nd ed.). New York: Macmillan.

KAPLAN, MAUREEN F., & MENDEL, JOHN E. 1982. "Ancient glass and the disposal of nuclear waste." *Archaeology,* 35(4), 22-29.

KAY, R.F., PLAUKIN, M., WRIGHT, P.C., GLANDER, K., & ALBRECHT, G.H. 1988. "Behavioral and size correlates of canine dimorphism in platyrrhine primates." *American Journal of Physical Anthropology,* 88, 385–397.

KEEGAN, WILLIAM F. (Ed.). 1987. *Emergent Horticultural Economies of the Eastern Woodlands Occasional Publications No. 7.* Carbondale: Southern Illinois University, Center for Archaeological Investigations.

KEELEY, L.H., & TOTH, N. 1981. "Microwear polishes on early stone tools from Koobi Fora, Kenya." *Nature,* 293, 464–465.

KENNEDY, G.E. 1984. "The emergence of *Homo sapiens*: The postcranial evidence." *Man,* 19, 94–110.

KENYON, KATHLEEN M. 1972. "Ancient Jericho." In *Old World Archaeology: Foundations of Civilization, Readings from Scientific American.* San Francisco: W.H. Freeman.

KETTLEWELL, H.B.D. 1957. "Industrial Melanism in Moss and Its Contribution to Our Knowledge of Evolution." *Proceedings of the Royal Institute of Great Britain,* 36, 1–14.

KINSEY, WARREN G. (Ed.). 1987. *The Evolution of Human Behavior: Primate Models.* Albany: State University of New York Press.

KITCHER, PHILLIP. 1982. *Abusing Science: The Case Against Creationism.* Cambridge, MA: MIT Press.

KLEIN, RICHARD G.1992. "The archaeology of modern human origins." *Evolutionary Anthropology,* 1(1), 5–14.

KNUDSON, RUTHANN. 1989. "North America's threatened heritage." *Archaeology,* 42(1), 71-75.

KRINGS, M., STONE, A., SCHMITZ, R.W., KRAINITSKI, H., STONEKING, M. AND PAABO, S. 1997. "Neandertal DNA Sequences and the Origins of Modern Humans." *Cell,* 90 19–30.

Kruckman, L. 1987. "The role of remote sensing in ethnohistorical research." *Journal of Field Archaeology,* 14, 343–351.

LAHR, MARTA MIRAZON, & FOLEY, ROBERT. 1994. "Multiple dispersals and modern human origins." *Evolutionary Anthropology* 3(2), 48-60.

LAMBERG-KARLOVSKY, C.C. 1989. *Archaeological Thought in America.* Cambridge: Cambridge University Press.

LARSEN, CLARK SPENCER. 1995. "Biological changes in human populations with agriculture." *Annual Review of Anthropology,* 24, 185-213.

LAVILLE, H., RIGUAD, J., & SACKETT, J. 1980. *Rock Shelters of the Perigord*. New York: Academic Press.

LEACH, EDMUND. 1988. "Noah's second son." *Anthropology Today*, 4(4), 2-5.

LEAKEY, L.S.B. 1959. "A new fossil skull from Olduvai." *Nature*, 201, 967–970.

———. 1961. "Exploring 1,750,000 years into man's past." *National Geographic*, 120(4), 564–589.

LEAKEY, M.D. 1971. *Olduvai Gorge* (Vol. 3). Cambridge: Cambridge University Press.

LEAKEY, MEAVE G., ET AL. 1995. "New four-milion-year-old hominid species from Kanapoi and Allia Bay, Kenya." *Nature*, 376, 565–571.

LEE, RICHARD B. 1969. "!Kung Bushman subsistence: An input-output analysis." In A.P. Vayda (Ed.), *Environment and Cultural Behavior: Ecological Studies in Cultural Anthropology*. Garden City, NY: Natural History Press.

LEE, RICHARD B., & DEVORE, IRVEN (Eds.). 1968. *Man the Hunter*. Chicago, Aldine.

LEGROS CLARK, W.E. 1962. *The Antecedents of Man*. Edinburgh: Edinburgh University Press.

LENSKI, GERHARD E. 1966. *Power and Privilege: A Theory of Social Stratification*. New York: McGraw-Hill.

LEONE, MARK P., & POTTER, PARKER B. 1988. *The Recovery of Meaning: Historical Archaeology in the Eastern United States*. Washington, DC: Smithsonian Institution Press.

LEVINTON, JEFFREY. 1988. *Genetics, Paleontology, and Macroevolution*. New York: Cambridge University Press.

LEWONTIN, R. 1972. "The apportionment of human diversity." In Theodore Dobzhansky & William C. Steere (Eds.), *Evolutionary Biology* (Vol. 6, pp. 381–398). New York: Plenum Press.

LI, WEN-HSIUNG, & TANIMURA, MASAKO. 1987. "The molecular clock runs more slowly in man than in apes and monkeys." *Nature*, 326, 93-96.

LIVINGSTON, FRANK B. 1971. "Malaria and human polymorphisms." *Annual Review of Genetics*, 5, 33–64.

LOEHLIN, J.C., LINDZEY, G., & SPUHLER, J.N. 1975. *Race Differences and Intelligence*. San Francisco: W.H. Freeman.

LORENZ, KONRAD. 1966. *On Aggression*. New York: Harcourt, Brace & World.

LOVEJOY, OWEN C. 1981. "The origin of man." *Science*, 211, 341–350.

———. 1984. "The natural detective." *Natural History*, 93(10), 24–28.

———. 1988. "Evolution of human walking." *Scientific American*, 259(5), 118–125.

LOVEJOY, OWEN C., & MEINDL, R.S. 1972. "Eukaryote mutation and the protein clock." *Yearbook of Physical Anthropology*, 16, 18–30.

LOWE, JOHN W.G. 1985. *The Dynamics of Apocalypse: A Systems Simulation of the Classic Maya Collapse*. Albuquerque: University of New Mexico Press.

MACNEISH, RICHARD. 1970. *The Prehistory of the Tehuacan Valley*. Austin: University of Texas Press.

MALINOWSKI, BRONISLAW. 1945. In P. Kaberry (Ed.), *The Dynamics of Culture Change: An Inquiry into Race Relations in Africa*. New Haven, CT: Yale University Press.

MARINATOS, SPIRIDON. 1939. "The volcanic destruction of Minoan Crete." *Antiquity*, 13, 425–439.

McCLUNG DE TAPIA, EMILY. 1992. "The origins of agriculture in Mesoamerica and Central America." In C. Wesley Cowan & Patty Jo Watson (Eds.), *The Origins of Agriculture: An International Perspective* (pp. 143–172). Washington, DC: Smithsonian Institution Press.

McHENRY, HENRY M. 1982. "The pattern of human evolution studies on bipedalism, mastication, and encephalization." *Annual Review of Anthropology*, 11, 151–173.

———. 1988. "New estimates of body weight in early hominids and their significance to encephalization and megadentia in robust australopithecines." In F.E. Grine (Ed.), *Evolutionary History of the Robust Australopithicines* (pp. 133–148). Hawthorne, NJ: Aldine.

McKEOWN, C. TIMOTHY. 1998. Ethical and legal issues, Complying with NAGPRA. In Rebecca A Buck, Amanda Murphy, & Jennifer Schansberg (Eds.) *The New Museums Registration Methods*. American Association of Museums.

MELLAART, JAMES. 1975. *The Earliest Civilizations of the Near East*. London: Thames and Hudson.

MELLARS, P.A. 1988. "The origin and dispersal of modern humans." *Current Anthropology*, 29, 186–188.

———. 1989. "Major issues in the emergence of modern humans." *Current Anthropology*, 30(3), 349–385.

MELTZER, D.J. 1993. "Pleistocene peopling of the Americas." *Evolutionary Anthropology*, 1(5), 157–169.

MILLER, N.F. 1992. "The origins of plant cultivation in the Near East." In C. Wesley Cowan & Patty Jo Watson (Eds.), *The Origins of Agriculture: An International Perspective* (pp. 39–58). Washington, DC: Smithsonian Institution Press.

MILLON, RENÉ. 1976. "Social relations in ancient Teotihuacan." In Eric R. Wolf (Ed.), *The Valley of Mexico: Studies in Prehispanic Ecology and Society*. Albuquerque: University of New Mexico Press.

MILLON, RENÉ, DREWITT, R. BRUCE, & COWGILL, GEORGE. 1974. *Urbanization at Teotihuacan, Mexico*. Austin: University of Texas Press.

MOLNAR, STEPHEN. 1992. *Human Variation: Races, Types, and Ethnic Groups* (3rd ed.). Englewood Cliffs, NJ: Prentice Hall.

MONTAGU, ASHLEY (Ed.). 1968. *Man and Aggression*. London: Oxford University Press.

MORRIS, DESMOND. 1967. *The Naked Ape*. New York: McGraw-Hill.

MOSELEY, MICHAEL E., & RICHARDSON, JAMES B. 1992. "Doomed by natural disaster." *Archaeology*, 45(6), 44–45.

MOTULSKY, ARNO. 1971. "Metabolic polymorphisms and the role of infectious diseases in human evolution." In Laura Newell Morris (Ed.), *Human Populations, Genetic Variation, and Evolution*. San Francisco: Chandler.

MOUNTAIN, JOANNA L. 1998. "Molecular evolution and modern human origins." *Evolutionary Anthropology*, 7(1), 21–38.

MURDOCK, GEORGE PETER. 1968. "The current status of the world's hunting and gathering peoples." In Richard Lee & Irven DeVore (Eds.), *Man the Hunter* (pp. 13–20). Chicago: Aldine.

MYERS, DAVID G. 1989. *Psychology* (2nd ed.). New York: Worth.

NARANJO, TESSIE. 1995. "Thoughts on two worldviews." *Federal Archaeology*, Offprint Series, Fall/Winter, 8.

NELSON, HARRY, & JURMAIN, ROBERT. 1988. *Introduction to Physical Anthropology*. St. Paul, MN: West.

NEWMAN, R.W., & MUNRO, E.H. 1955. "The relation of climate and body size in U.S. males." *American Journal of Physical Anthropology,* 13, 1–17.

NOEL HUME, IVOR. 1983. *Historical Archaeology: A Comprehensive Guide.* New York: Knopf.

OAKLEY, KENNETH P. 1964. *Man the Tool Maker.* Chicago: University of Chicago Press.

OFFICER, C.B. 1990. "Extinctions, iridium, and shocked minerals associated with the Cretaceous/Tertiary Transition." *Journal of Geological Education,* 38, 402–425.

OMOHUNDRO, JOHN. 1998. *Careers in Anthropology.* Mountain View, CA: Mayfield.

PAIGEN, B., GOLDMAN, L.R., MAGNANT, J.H., HIGHLAND, J.H., & STEEGMAN, A.T. 1987. "Growth and children living near the hazardous waste site, Love Canal." *Human Biology,* 59, 489–508.

PASSINGHAM, R.E. 1982. *The Human Primate.* San Francisco: W.H. Freeman.

PEARSALL, D.M. 1992. "The origins of plant cultivation in South America." In C. Wesley Cowan & Patty Jo Watson (Eds.), *The Origins of Agriculture: An International Perspective* (pp. 173–205). Washington, DC: Smithsonian Institution Press.

PERKINS, DEXTER. 1964. "The prehistoric fauna from Shanidar, Iraq." *Science,* 144, 1565–1566.

PHILLIPSON, DAVID W. 1993. *African Archaeology.* New York: Cambridge University Press.

PILBEAM, DAVID. 1972. *The Ascent of Man.* New York: Macmillan.

POIRER, FRANK E., STINI, WILLIAM A., & WREDEN, KATHY B. 1990. *In Search of Ourselves: An Introduction to Physical Anthropology* (4th ed.). Englewood Cliffs, NJ: Prentice Hall.

POLEDNAK, ANTHONY P. 1974: "Connective tissue responses in Negroes in relation to disease." *American Journal of Physical Anthropology,* 41, 49–57.

POSNANSKY, MERRICK. 1984. "Early agricultural societies in Ghana." In J. Clark & Steven A. Brandt (Eds.), *From Hunter to Farmer* (pp. 147–151). Berkeley: University of California Press.

POST, P.W., DANIELS, F., & BINFORD, R.T. 1975. "Cold injury and the evolution of white skin." *Human Biology,* 47, 65–80.

POTTS, RICHARD. 1988. *Early Hominid Activities at Olduvai.* New York: Aldine de Gruyter.

PRAG, JOHN, & NEAVE, RICHARD. 1997. *Making Faces.* College Station: Texas A&M University Press.

PRICE, DOUGLAS T., & BROWN, JAMES A. 1985. *Prehistoric Hunter-Gatherers: The Emergence of Cultural Complexity.* New York: Academic Press.

PROTHERO, DONALD R. 1989. *Interpreting the Stratigraphic Record.* New York: W.H. Freeman.

RADINSKY, LEONARD. 1967. "The oldest primate Endocast." *American Journal of Physical Anthropology,* 27, 358–388.

RASMUSSEN, TAB D., & SIMONS, ELWYN L. 1988. "New species of *Oligopithecus savagei,* early Oligocene primate from the Fayum, Egypt." *Folia Primatol,* 51, 182–208.

RATHBUN, T. A., & BUIKSTRA, J. E. 1984. *Human Identification: Case Studies in forensic Anthropology.* Springfield, IL: Charles C. Thomas.

RATHJE, WILLIAM. 1971. "The origin and development of lowland classic Maya civilization." *American Antiquity,* 36, 275–285.

————. 1984. "The garbage decade." *American Behavioral Scientist,* 28(1), 9-29.

————. 1992. "Five major myths about garbage, and why they are wrong." *Smithsonian* 23(4):113-122.

RATHJE, WILLIAM L., & RITENBAUGH, CHERYL K. 1984. "Household refuse analysis: Theory, method, and applications in social science." *American Behavioral Scientist,* 28(1), 5–153.

RAUP, DAVID M. 1986. *The Nemesis Affair: a Story of the Death of the Dinosaurs and the Ways of Science.* New York: W.W. Norton.

RELETHFORD, JOHN. 1997. *The Human Species: An Introduction to Biological Anthropology* (3rd ed.). Mountain View, CA: Mayfield.

RENFREW, COLIN. 1987. *Archaeology and Language.* London: Jonathan Cape.

RENFREW, COLIN, & BAHN, PAUL. 1996. *Archaeology: Theories, Methods and Practice.* New York: Thames & Hudson.

RENFREW, C., & COOKE, K.L. 1979. *Transformations: Mathematical Approaches to Culture Change.* New York: Academic Press.

RICE, PRUDENCE M. 1987. *Pottery Analysis: A Sourcebook.* Chicago: University of Chicago Press.

RICHARD, ALISON F. 1985. *Primates in Nature.* New York: W.H. Freeman.

RIESENFELD, ALPHONSE. 1973. "The effect of extreme temperatures and starvation on the body proportions of the rat." *American Journal of Physical Anthropology,* 39, 427–459.

RINDOS, DAVID. 1984. *The Origins of Agriculture: An Evolutionary Perspective.* New York: Academic Press.

ROGERS, J. DANIEL, & WILSON, SAMUEL M. 1988. *Ethnohistory and Archaeology: Approaches to Postcontact Change in the Americas.* New York: Plenum Press.

ROOSEVELT, ANNA CURTENIUS. 1984. "Population, health, and the evolution of subsistence: Conclusions from the conference." In Mark Nathan Cohen & George J. Armelagos (Eds.), *Paleopathology at the Origins of Agriculture* (pp. 559–584). New York: Academic Press.

ROSENBERG, MICHAEL. 1990. "The mother of invention: Evolutionary theory, territoriality, and the origins of agriculture." *American Anthropologist,* 92(2), 399–415.

ROWLEY-CONWY, PETER. 1993. "Was there a Neanderthal religion?" In G. Burenhult (Ed.), *The First Humans: Human Origins and History to 10,000 BC* (p. 70). New York: HarperCollins.

RUKANG, WU, & LIN, S. 1983. "Peking man." *Scientific American,* 248(6), 86–94.

SABLOFF, JEREMY, & LAMBERG-KARLOVSKY, C.C. (Eds.). 1975. *Ancient Civilization and Trade.* Albuquerque: University of New Mexico Press.

SACKETT, JAMES R. 1982. "Approaches to style in Lithic archaeology." *Journal of Anthropological Archaeology,* 1, 59–112.

SAHLINS, MARSHALL. 1976. *The Use and Abuse of Biology: An Anthropological Critique of Sociobiology.* Ann Arbor: University of Michigan Press.

SANDERS, WILLIAM T., & PRICE, BARBARA J. 1968. *Mesoamerica: The Evolution of a Civilization.* New York: Random House.

SARDESAI, D. R. 1989. *Southeast Asia: Past and Present.* Boulder, CO: Westview Press.

SARICH, V.M., & WILSON, A.C. 1967. "Rates of albumen evolution in primates." *Proceedings of the National Academy of Sciences*, 58, 142–148.

SAUER, CARL O. 1952. *Agricultural Origins and Dispersals*. New York: American Geographical Society.

SCARR, S. & WEINBERG, A. 1978. "Attitudes, interests, and IQ." *Human Nature* 1(4), 29-36.

SCHALLER, GEORGE. 1976. *The Mountain Gorilla-Ecology and Behavior*. Chicago: University of Chicago Press.

SCHIEFFELIN, BAMBI B. 1990. *The Give and Take of Everyday Life: Language and Socialization of Kaluli Children*. New York: Cambridge University Press.

SCHIEFFELIN, BAMBI B., & OCHS, ELINOR (Eds.). 1987. "Language Socialization Across Cultures." *Studies in the Social and Cultural Foundations of Language No. 3*. Cambridge: Cambridge University Press.

SCHIFFER, MICHAEL B. 1987. *Formation Processes of the Archaeological Record*. Albuquerque: University of New Mexico Press.

SCUPIN, RAYMOND. 1989. "Language, hierarchy and hegemony: Thai Muslim discourse strategies." *Language Sciences*, 10(2), 331–351.

SERVICE, ELMAN. 1971. (original 1962). *Primitive Social Organization: An Evolutionary Perspective*. New York: Random House.

———. 1975. *Origins of the State and Civilization: The Process of Cultural Evolution*. New York: W.W. Norton.

———. 1978. "Classical and modern theories of the origin of government." In Ronald Cohen & E.R. Service (Eds.), Origins of the State: The Anthropology of Political Evolution. Philadelphia: ISHI.

SHAW, THURSTAN. 1986. "Whose heritage?" *Museum*, 149:46-48.

SHEA, B.T., & GOMEZ, A.M. 1988. "Tooth scaling and evolutionary dwarfism: An investigation of allometry in human Pygmies." *American Journal of Physical Anthropology*, 77, 117–132.

SHIPMAN, PAT. 1984. "Scavenger hunt." *Natural History*, 4(84), 20–27.

———. 1986a. "Scavenging or hunting in early hominids: Theoretical frameworks and tests." *American Anthropologist*, 88, 27–43.

———. 1986b. "Baffling limb on the family tree." *Discover*, 7(9), 86–93.

SIGNOR, PHILIP W., & LIPPS, JERE H. 1982. "Sampling bias, gradual extinction patterns and catastrophes in the fossil record." *Geological Society of America Special Paper 190*.

SILBY, CHARLES G., & ALQUIST, JON E. 1984. "The phylogeny of the hominoid primates as indicated by DNA-DNA hybridization." *Journal of Molecular Evolution*, 20, 2–15.

SIMONS, ELWYN L. 1972. *Primate Evolution: An Introduction to Man's Place in Nature*. New York: Macmillan.

———. 1984. "Ancestor: Dawn ape of the Fayum." *Natural History*, 93(5), 18–20.

———. 1989a. "Description of two genera and species of Late Eocene Anthropoidea from Egypt." *Proceeding of the National Academy of Science*, 86, 9956–9960.

———. 1989b. "Human origins." *Science*, 245, 1343–1350.

———. 1990. "Discovery of the oldest known anthropoidean skull from the Paleogene of Egypt." *Science*, 247, 1567–1569.

SIMONS, ELWYN L., & RASMUSSEN, D. 1990. "Vertebrate paleontology of the Fayum: History of research, faunal review, and future prospects." In R. Said (Ed.), *The Geology of Egypt* (pp. 627–638). Rotterdam: Balkema Press.

SINGLETON, THERESA A. (Ed.). 1985. *The Archaeology of Slavery and Plantation Life*. New York: Academic Press.

———. 1999. "I, too, am America." *Archaeological Studies in African American Life*. Charlottesville: University of Virginia Press.

SMITH, ANDREW B. 1984. "Origins of the Neolithic in the Sahara." In Desmond J. Clark & Steven A. Brandt (Eds.), *From Hunter to Farmer* (pp. 84–92). Berkeley: University of California Press.

SMITH, BRUCE D. 1989. "Origins of agriculture in eastern North America." *Science*, 246, 1566–1571.

SMITH, F.H. 1984. "Fossil hominids from the Upper Pleistocene of central Europe and the origin of modern Europeans." In F.H. Smith & F. Spencer (Eds.), *The Origins of Modern Humans: A World Survey of Fossil Evidence* (pp. 137–210). New York: Alan R. Liss.

SMITH, STUART T. 1990. "Administration at the Egyptian Middle Kingdom frontier: Sealings from Uronarti and Askut." *Aegaeum*, 5, 197–219.

SMUTS, BARBARA. 1987. "What are friends for?" *Natural History*, 96(2).

SNOW, C. C., STOVER, E., & HANNIBAL, K. 1989. "Scientists as detectives investigating human rights." *Technology Review* 92, 2.

SNOW, C. C., & BIHURRIET, M. J. 1992. "An epidemiology of homicide: Ning'n nombre burials in the province of Buenos Aires from 1970 to 1984. In T. B. Jabine & R. P. Claude (Eds.), *Human Rights and Statistics: Getting the Record Straight*. Philadelphia: University of Pennsylvania Press.

SOLECKI, RALPH S. 1971. *Shanidar: The First Flower People*. New York: Knopf.

SOLHEIM, WILLIAM. 1971. "An earlier agricultural revolution." *Scientific American*, 133(11), 34–51.

SOWELL, THOMAS. 1994. *Race and Culture*. New York: Basic Books.

———. 1995. "Ethnicity and IQ." In Steven Fraser (Ed.), The Bell Curve Wars: Race, Intelligence and the Future of America (pp. 70–79). New York: Basic Books.

STEWARD, JULIAN H. 1955. *Theory of Culture Change: The Methodology of Multilinear Evolution*. Urbana: University of Illinois Press.

STEWART, T. D. (Ed.). 1970. Personal Identification in Mass Disasters. Washington, DC: Smithsonian Institution Press.

———. 1979. *Essentials of Forensic Anthropology*. Springfield, IL: Charles C. Thomas.

STIEBING, WILLIAM H., JR. 1984. *Ancient Astronauts: Cosmic Allusions and Popular Theories About Man's Past*. Buffalo, NY: Prometheus Books.

STINI, WILLIAM A. 1975. *Ecology and Human Adaptation*. Dubuque, IA: William C. Brown.

STONEKING, M., BHATIA, K. & WILSON, A.C. 1987. "Rate of sequence divergence estimated from restricted maps of mitochondrial DNAs from Papua, New Guinea." *Cold Sprin Harbor Symposia on Quantitative Biology*, 51, 433-439.

STOVER, E. 1981. "Scientists aid search for Argentina's ëdesaparacidos'." *Science* 211 (4486), 6.

———. 1992. "Unquiet graves: The search for the disappeared in Iraqi Kurdistan." A report published by Middle East Watch and Physicians for Human Rights.

STRASSER, ELIZABETH, & DAGOSTO, MARIAN (Eds.). 1988. *The Primate Postcranial Skeleton: Studies in Adaptation and Evolution.* New York: Academic Press.

STRINGER, C.B. 1985. "Middle Pleistocene hominid variability and the origin of late Pleistocene humans." In E. Delson (Ed.), *Ancestors: The Hard Evidence* (pp. 289–295). New York: Alan R. Liss.

STRINGER, C.B., & ANDREWS, P. 1988. "Genetic and fossil evidence for the origin of modern humans." *Science, 239,* 1263–1268.

STRUEVER, STUART (Ed.). 1970. *Prehistoric Agriculture.* Garden City, NY: Natural History Press.

STRUM, SHIRLEY C., & MITCHELL, WILLIAM. 1987. "Baboon models and muddles." In Warren G. Kinsey (Ed.), *The Evolution of Human Behavior* (pp. 87–114). Albany: State University of New York Press.

SUSSMAN, R.W. 1993. "A current controversy in human evolution: Overview." *American Anthropologist, 95,* 9–13.

SWISHER, C.C., CURTIS, G.H., JACOB, T., GETTY, A.G., & WIDIASMORO, A. SUPRIJO. 1994. "Age of the earliest known hominids in Java, Indonesia." *Science, 263,* 1118–1121.

SYMONS, DONALD. 1979. *The Evolution of Human Sexuality.* Oxford: Oxford University Press.

SZABO, G. 1967. "The regional anatomy of the human integument with special reference to the distribution of hair follicles, sweat glands and melanocytes." *Philosophical Transactions of the Royal Society of London, 252B,* 447–485.

TAINTER, JOSEPH A. 1990. *The Collapse of Complex Societies.* Cambridge: Cambridge University Press.

TANNER, NANCY M. 1981. *On Becoming Human.* London: Cambridge University Press.

———. 1987. "Gathering by females: The chimpanzee model revisited and the gathering hypothesis." In Warren G. Kinsey (Ed.), *The Evolution of Human Behavior* (pp. 3–27). Albany: State University of New York Press.

TARLING, D.H. 1985. *Continental Drift and Biological Evolution.* Burlington, NC: Carolina Biological Supply Co.

TATTERSALL, IAN. 1986. "Species recognition in human paleontology." *Journal of Human Evolution, 15,* 165–175.

TAYLOR, R.E. 1995. "Radiocarbon dating: The continuing revolution." *Evolutionary Anthropology, 4*(5), 169–181.

TEMPLETON, ALAN R. 1993. "The 'Eve' hypothesis: A genetic critique and reanalysis." *American Anthropologist, 95*(1), 51-72.

THORNE, A., & WOLPOFF, M.H. 1992. "The multiregional evolution of humans." *Scientific American, 266,* 76–83.

THROCKMORTON, PETER. 1962. "Oldest known shipwreck yields Bronze Age cargo." *National Geographic 121*(5), 697-711.

THROCKMORTON, PETER (Ed.). 1987. *The Sea Remembers: Shipwrecks and Archaeology from Homer's Greece to the Rediscovery of the Titanic.* New York: Weidenfeld & Nicholson.

TRIGGER, BRUCE. 1989. *A History of Archaeological Thought.* New York: Cambridge University Press.

———. 1993. *Early Civilizations: Ancient Egypt in Context.* Cairo: American University in Cairo Press.

TRINKAUS, ERIK, & SHIPMAN, PAT. 1994. *The Neandertals.* New York: Random House.

UCKO, PETER J., & DIMBLEBY, G.W. 1969. *The Domestication and Exploitation of Plants and Animals.* Chicago: Aldine.

VIANNA, N.J., & POLAN, A.K. 1984. "Incidence of low birth weight among Love Canal residents." *Science, 226,* 1217–1219.

VIGILANT. L., STONEKING, M., HARPENDING, H., HAWKES, K., & WILSON, A. C. 1991. "African populations and the evolution of human mitochondrial DNA." *Science 233,* 1303-1307.

VILLA, PAOLA. 1983. *Terra Amata and the Middle Pleistocene Archaeological Record of Southern France.* Berkeley: University of California Press.

VON DANIKEN, ERICH. 1970. *Chariots of the Gods.* New York: Bantam.

WALKER, A., LEAKEY, R.E., HARRIS, J.M., & BROWN, F.H. 1986. "2.5 MYR Australopithecus boisei from west of Lake Turkana, Kenya." *Nature, 322,* 517–522.

WALKER, A.C., & PICKFORD, M. 1983. "New post-cranial fossils of *Proconsul africanus* and *Proconsul nyanzae.*" In R.L. Ciochon & R. Corruccini (Eds.), *New Interpretations of Ape and Human Ancestry* (pp. 325–352). New York: Plenum Press.

WASHBURN, SHERWOOD. 1960. "Tools and human evolution." *Scientific American, 203*(3), 67–75.

WASSERMAN, H.P. 1965. "Human pigmentation and environmental adaptation." *Archives of Environmental Health, 11,* 691–694.

WATSON, PATTY JO, LEBLANC, STEVEN A., & REDMAN, CHARLES L. 1971. *Explanation in Archaeology: An Explicitly Scientific Approach.* New York: Columbia University Press.

———. 1984. *Archaeological Explanation: The Scientific Method in Archaeology.* New York: Columbia University Press.

WEINER, J. S. 1955. *The Piltdown Forgery.* London, Oxford University Press.

WENDORF, FRED, & SCHILD, ROMUALD. 1981. "The earliest food producers." *Archaeology, 34*(5), 30–36.

———. 1984. "The emergence of food production in the Egyptian Sahara." In Desmond J. Clark & Steven A. Brandt (Eds.), *From Hunter to Farmer* (pp. 93–101). Berkeley: University of California Press.

WHITE, JOHN PETER. 1993. "The settlement of ancient Australia." In G. Burenhult (Ed.), *The First Humans: Human Origins and History to 10,000 BC* (pp. 147–170). New York: HarperCollins.

WHITE, JOHN PETER, & O'CONNELL, JAMES F. 1982. *A Prehistory of Australia, New Guinea and Sahul.* New York: Academic Press.

WHITE, RANDALL. 1982. "Rethinking the Middle/Upper Paleolithic transition." *Current Anthropololgy, 23,* 169–192.

WHITE, TIM D., SUWA, GEN, & ASFAW, BERHANE. 1995. "Australopithecus ramidus, a new species of hominid from Aramis, Ethiopia." *Nature, 375,* 88.

WHITLEY, DAVID S., & DORN, RONALD I. 1993. "New perspectives on the Clovis vs. pre-Clovis controversy." *American Antiquity, 58*(4), 626–647.

WILLIAMS-BLANGERO, S., & BLANGERO, J. 1992. "Quantitative genetic analysis of skin reflectance: A multivariate approach." *Human Biology, 64,* 35–49.

WILSON, A.C., & CANN, R.L. 1992. "The recent African genesis of humans." *Scientific American, 266*(4), 68–73.

WILSON, B., GRIGSON, C., & PAYNE, S. (Eds.). 1982. *Aging and Sexing Animal Bones from Archaeological Sites*. Oxford: British Archaeological Reports, International Series 109.

WILSON, E.O. 1975. *Sociobiology: The New Synthesis*. Cambridge, MA: Harvard University Press.

———. 1978. *On Human Nature*. Cambridge, MA: Harvard University Press.

WITTFOGEL, KARL W. 1957. *Oriental Despotism: A Comparative Study of Total Power*. New Haven, CT: Yale University Press.

WOLF, ERIC R. 1982. *Europe and the People Without History*. Berkeley: University of California Press.

WOLPOFF, MILFORD H. 1983. "Ramapithecus and human origins: An anthropologist's perspective of changing interpretations." In R.L. Ciochon & R. Corruccini (Eds.), *New Interpretations of Ape and Human Ancestry* (pp. 651–676). New York: Plenum Press.

WORKMAN, P.L., BLUMBERG, B.S., & COOPER, A.J. 1963. "Selection, gene migration and polymorphic stability in a U.S. White and Negro population." *American Journal of Human Genetics*, 15, 71–84.

WRIGHT, HENRY T. 1977. "Recent researches of the origin of the state." *Annual Review of Anthropology*, 6, 355–370.

WRIGHT, HENRY T., & JOHNSON, G. 1975. "Population, exchange, and early state formation in southwestern Iran." *American Anthropologist*, 77, 267–289.

YOFFE, NORMAN. 1979. "The decline and rise of Mesopotamian civilization." *American Antiquity*, 44, 5–35.

ZEDER, MELINDA A. 1997. *The American Archaeologist: A Profile*. Walnut Creek, CA: Altamira Press.

Photo Credits

Index